MATHEMATICS STANDARD LEVEL

for the IB Diploma

Robert Smedley
Garry Wiseman

Course consultants:
Colin Jeavons
Sheila Messer

OXFORD
UNIVERSITY PRESS

OXFORD
UNIVERSITY PRESS

Great Clarendon Street, Oxford OX2 6DP

Oxford University Press is a department of the University of Oxford.
It furthers the University's objective of excellence in research, scholarship,
and education by publishing worldwide in

Oxford New York
Auckland Cape Town Dar es Salaam Hong Kong Karachi
Kuala Lumpur Madrid Melbourne Mexico City Nairobi
New Delhi Shanghai Taipei Toronto

With offices in
Argentina Austria Brazil Chile Czech Republic France Greece
Guatemala Hungary Italy Japan Poland Portugal Singapore
South Korea Switzerland Thailand Turkey Ukraine Vietnam

Oxford is a registered trade mark of Oxford University Press
in the UK and in certain other countries

British Library Cataloguing in Publication Data

Data available

ISBN-13: 978 0 19 914979 7

7 9 10 8

Typeset by Tech-Set Ltd.
Printed and bound in Great Britain by Bell and Bain Ltd.

Acknowledgements
The Publisher would like to thank the International Baccalaureate Organisation (IBO) for
permission to reproduce questions for Mathematical Methods Standard Level taken from
examination papers, 1996 to 2003, and from specimen examination papers 2003.

This book has been developed independently of the International Baccalaureate Organisation
(IBO). The text is in no way connected with, nor endorsed by, the IBO.

The photograph on the cover is reproduced courtesy of Alamy Images.

1 Algebra

Most of the work in this chapter is presumed knowledge.
It can be used as a revision chapter as necessary.

Objectives from the syllabus that are covered are:

2.6 The solution of

$$ax^2 + bx + c = 0, a \neq 0$$

The quadratic formula.
Use of the discriminant

$$\Delta = b^2 - 4ac.$$

1.1 Linear equations

Simple linear equations

A linear equation can be expressed in the form $ax + b = 0$.
Each of these equations is linear:

$$x + 1 = 9$$

$$3(x + 2) - 7 = x + 1$$

$$\frac{x}{3} + 7 = 4(x - 2)$$

> You can rearrange each of these equations into the form $ax + b = 0$.

You solve an equation by finding the value of any unknowns.

Whatever you do to the left-hand side (LHS) of an equation, you must also do to the right-hand side (RHS). For example, the equation $x - 4 = 7$ can be manipulated so that x (the unknown) is the only term on the LHS.

$$x - 4 = 7$$

Add 4 to both sides to obtain

$$x - 4 + 4 = 7 + 4$$

$$x + 0 = 11$$

$$\therefore \ x = 11$$

The solution is $x = 11$.

2   Algebra

Example 1

Solve the equation $3x - 7 = x + 3$.

....

$$3x - 7 = x + 3$$

Add 7 to both sides to obtain

$$3x - 7 + 7 = x + 3 + 7$$
$$\therefore\ 3x = x + 10$$

To get all the x terms on the LHS, subtract x from both sides.

$$3x - x = x + 10 - x$$
$$\therefore\ 2x = 10$$

Dividing both sides by 2 gives

$$x = 5$$

The solution is $x = 5$.

You can check that your solution works:
$$3x - 7 = x + 3$$
$$3(5) - 7 = 8$$
and $5 + 3 = 8$
So $x = 5$ works.

Example 2

Solve the equation $4(x + 1) - 3(x - 5) = 17$.

....

$$4(x + 1) - 3(x - 5) = 17$$
$$4x + 4 - 3x + 15 = 17$$
$$x + 19 = 17$$
$$x + 19 - 19 = 17 - 19$$
$$\therefore\ x = -2$$

The solution is $x = -2$.

Expand the brackets and simplify.

Subtract 19 from both sides.

Linear equations may contain fractions. Remove any fractions before proceeding.

Example 3

Solve the equation $\dfrac{x + 5}{2} = \dfrac{3x + 11}{5}$.

....

$$\frac{x + 5}{2} = \frac{3x + 11}{5}$$
$$\therefore\ 10\left(\frac{x + 5}{2}\right) = 10\left(\frac{3x + 11}{5}\right)$$

Simplify the fractions:

$$5(x + 5) = 2(3x + 11)$$

Expand the brackets:

$$5x + 25 = 6x + 22$$

The lowest common multiple of 2 and 5 is 10, so multiply throughout by 10.

Rearrange to obtain all the x terms on the LHS:

$$5x - 6x = 22 - 25$$
$$-x = -3$$
$$\therefore\ x = 3$$

The solution is $x = 3$.

Alternatively, you could use the technique of **cross-multiplication**.

$$\frac{x + 5}{2} = \frac{3x + 11}{5}$$

Cross-multiplying gives

$$5(x + 5) = 2(3x + 11)$$

Expanding and simplifying gives

$$5x + 25 = 6x + 22$$
$$5x - 6x = 22 - 25$$
$$\therefore\ x = 3$$

The solution is $x = 3$, as before.

Remember:

To cross-multiply

$$\frac{a}{b} = \frac{c}{d}$$

multiply:

$$\frac{a}{b} \times bd = \frac{c}{d} \times bd$$

$$\therefore\ ad = bc$$

Exercise 1A

In each of these questions, solve the given equations for x.

1 a) $3x + 2 = 20$
    b) $5x - 3 = 32$
    c) $16 + 7x = 2$
    d) $4 + 3x = 19$
    e) $6 - x - 4$
    f) $2x - 3 = 8$

2 a) $3x + 2 = x + 8$
    b) $2x - 3 = 6x + 5$
    c) $3x + 5 = 7x - 8$
    d) $6x + 9 = 8 - 4x$
    e) $2 - 5x = 8 - 3x$
    f) $2x + 7 = 3 - 10x$

3 a) $2(x - 3) + 5(x - 1) = 3$
    b) $3(5 - x) - 4(3x - 2) = 27$
    c) $2(4x - 1) - 3(x - 2) = 14$
    d) $3(x - 8) + 2(4x - 1) = 3$
    e) $6(x + 4) + 5(2x - 1) = 7$
    f) $3(2x + 5) - 4(x - 3) = 0$

4 a) $\dfrac{x + 2}{3} = \dfrac{2x + 1}{5}$
    b) $\dfrac{5x - 3}{4} = \dfrac{4x - 3}{3}$
    c) $\dfrac{3x + 1}{4} = \dfrac{2 - x}{3}$
    d) $\dfrac{2x + 3}{5} = \dfrac{4 + 3x}{3}$

Harder linear equations

Sometimes the unknown x may appear in the denominator.
If you cross-multiply you will remove the fractions.

Example 4

Solve the equation $\dfrac{2}{x+1} = \dfrac{3}{5}$.

....

$$\frac{2}{x+1} = \frac{3}{5}$$

Cross-multiply and expand the bracket to obtain

$$10 = 3(x+1)$$
$$\therefore \ 10 = 3x + 3$$

Rearranging gives

$$3x = 7$$
$$\therefore \ x = \tfrac{7}{3}$$

The solution is $x = \tfrac{7}{3}$.

The unknown may be in the denominator on both sides of the equation.

Example 5

Solve the equation $\dfrac{2}{x+1} = \dfrac{7}{5x-4}$.

....

Cross-multiply and expand the brackets to obtain

$$2(5x-4) = 7(x+1)$$
$$\therefore \ 10x - 8 = 7x + 7$$

Rearranging gives

$$10x - 7x = 7 + 8$$
$$3x = 15$$
$$\therefore \ x = 5$$

The solution is $x = 5$.

You can use the lowest common multiple (LCM) to get rid of numerical fractions before you solve the equation.

Example 6

Solve the equation $\frac{1}{2}(x + 3) - \frac{1}{3}(2x - 5) = 1$.

....

Since the lowest common multiple of 2 and 3 is 6, multiply throughout by 6:

$$6 \times \tfrac{1}{2}(x + 3) - 6 \times \tfrac{1}{3}(2x - 5) = 6 \times 1$$

Simplifying gives

$$3(x + 3) - 2(2x - 5) = 6$$
$$3x + 9 - 4x + 10 = 6$$
$$-x + 19 = 6$$
$$\therefore\ x = 13$$

The solution is $x = 13$.

Exercise 1B

....

Solve the given equations for x.

1 a) $\dfrac{2}{x + 3} = \dfrac{4}{5}$ b) $\dfrac{5}{x - 1} = \dfrac{2}{3}$

   c) $\dfrac{6}{2x - 3} = \dfrac{1}{3}$ d) $\dfrac{5}{2x - 3} = \dfrac{3}{8}$

2 a) $\dfrac{3}{x + 1} = \dfrac{4}{x}$ b) $\dfrac{2}{3x - 5} = \dfrac{5}{2x + 3}$

   c) $\dfrac{6}{x + 8} = \dfrac{5}{3x + 4}$ d) $\dfrac{7}{x - 1} = \dfrac{3}{x + 2}$

3 a) $\frac{1}{2}(2x - 1) + \frac{1}{4}(x - 2) = 4$ b) $\frac{1}{3}(x - 1) - \frac{1}{4}(2x - 3) = 1$

   c) $\frac{1}{5}(2x - 1) - \frac{1}{4}(3x - 4) = 0$ d) $\frac{2}{3}(x - 1) - \frac{1}{5}(x - 3) = x + 1$

   e) $\frac{2}{5}(2 - x) - \frac{1}{4}(3 - 5x) = x - 4$ f) $\frac{1}{3}(x - 1) - \frac{1}{6}(3x - 5) = 2x + 3$

4 $3(x - 2) - 4(2x - 3) - 2(3x - 1) = x + 4$

5 $3x - 1 + \dfrac{2x + 3}{5} = \dfrac{x - 4}{2}$

6 $5x + 7 = 2 - 3x$

7 $\dfrac{x + 5}{3} = \dfrac{2 - x}{4}$

8 $\frac{2}{5}(2x - 3) - \frac{1}{10}(x - 4) = 1$

9 $(x - 3)(x - 4) - 6 = x(x + 5)$

10 $\dfrac{3}{7 - 2x} = \dfrac{4}{x}$

11 $5(2 - x) = 3(6 - x) + 2$

12 $8x - 9 = 5 + 4x$

1.2 Linear inequalities

An inequality is a comparison between two mathematical expressions.

less than	is written as $<$
less than or equal to	is written as $\leqslant$
greater than	is written as $>$
greater than or equal to	is written as $\geqslant$

A **linear inequality** is a comparison between two linear expressions.

For example,

$$3x + 7 > x - 8 \qquad a < 3b + 2 \qquad 5x - 7y \leqslant 4$$

Inequalities can be solved using methods similar to those used for solving equations.

Example 1

Simplify the inequality $3x + 7 \geqslant x + 2$.

$$3x + 7 \geqslant x + 2$$

Subtract 7 from both sides:

$$3x + 7 - 7 \geqslant x + 2 - 7$$
$$\therefore \ 3x \geqslant x - 5$$

Subtract x from both sides:

$$3x - x \geqslant x - 5 - x$$
$$2x \geqslant -5$$
$$\therefore \ x \geqslant -\tfrac{5}{2}$$

This result tells you that provided $x \geqslant -\frac{5}{2}$, the original inequality will be satisfied.

You can illustrate inequalities graphically. Just sketch the graphs of $y = 3x + 7$ and $y = x + 2$ and find the x-coordinate where they meet.

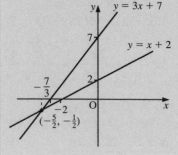

Inequalities can contain brackets.

Example 2

Simplify the inequality $4(3x + 1) - 3(x + 2) < 3x + 1$.

···

$$4(3x + 1) - 3(x + 2) < 3x + 1$$

Expand the brackets:

$$12x + 4 - 3x - 6 < 3x + 1$$

Simplifying and rearranging gives

$$9x - 2 < 3x + 1$$
$$6x < 3$$
$$x < \frac{3}{6}$$
$$\therefore \; x < \frac{1}{2}$$

You can write this inequality as $x < \frac{1}{2}$.

You need to take care when you multiply or divide by a negative number.

$7 > 5$ is true.

Multiply both sides by -1:

$-7 > -5$ is false.

However, now reverse the inequality:

$-7 < -5$ is true.

When you multiply or divide an inequality by a negative number you must reverse the inequality sign.

Example 3

Simplify the inequality $3x + 7 \geqslant 5x - 3$.

··

The x terms can be rearranged so that they are all on the RHS and therefore positive.

$$3x + 7 \geqslant 5x - 3$$
$$7 + 3 \geqslant 5x - 3x$$
$$\therefore \; 10 \geqslant 2x$$

Divide both sides by 2:

$$5 \geqslant x$$

Reading from right to left gives

$$x \leqslant 5$$

You can illustrate this solution graphically:

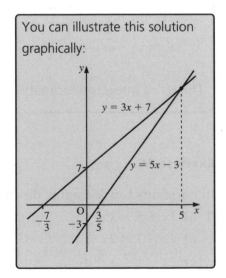

Alternatively, you can rearrange the x terms so that they are all on the LHS:

$$-2x \geqslant -10$$

Multiply both sides by -1, but remember to reverse the inequality sign:

$$2x \leqslant 10$$

Dividing both sides by 2 gives

$$x \leqslant 5$$

> Remember to reverse the inequality:
>
> $\geqslant$ becomes $\leqslant$.

You may be asked to find values of a variable that satisfy two different inequalities at the same time (simultaneously).

Example 4

Find the set of integers which satisfy simultaneously both of the following inequalities:

$$2x - 1 \geqslant 4 \qquad\qquad\qquad [1]$$
and $\quad \frac{1}{3}x + 1 < 3 \qquad\qquad\qquad [2]$

First, simplify [1]: $\qquad 2x - 1 \geqslant 4$

$$2x \geqslant 5$$

$$\therefore\ x \geqslant \tfrac{5}{2}$$

Therefore the integers which satisfy [1] are 3, 4, 5, 6, …

Next, simplify [2]: $\qquad \frac{1}{3}x + 1 < 3$

$$\tfrac{1}{3}x < 2$$

$$\therefore\ x < 6$$

Therefore the integers which satisfy [2] are 5, 4, 3, 2, …

Here are both results shown on the number line.

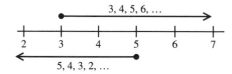

The set of integers which satisfy both inequalities is {3, 4, 5}.

Exercise 1C

In questions **1** to **3** simplify the inequalities.

1 a) $3x + 5 > x + 13$ b) $2x - 3 \leqslant 5x + 9$

    c) $4x - 7 \geqslant 2x + 4$ d) $5x - 8 > x + 7$

    e) $2x - 1 < x + 4$ f) $7x - 3 \geqslant 2x - 1$

2 a) $2(x + 3) - 3(x - 2) > 8$ b) $6(2x - 1) + 5(x + 1) < 33$

   c) $5(x - 3) < 6(x - 4)$ d) $3(x + 4) \geqslant 6(x + 2)$

   e) $3(x - 2) - 2(4 - 3x) > 5$ f) $7(1 - x) + 3(4 - 5x) \leqslant 41$

3 a) $\frac{1}{2}x + 2 < 7$ b) $\frac{1}{6}(x - 1) \geqslant \frac{1}{3}(x - 4)$

   c) $\frac{1}{2}(x + 3) \leqslant \frac{1}{3}(x - 5)$ d) $\frac{1}{7}(2x + 5) > \frac{1}{8}(x + 3)$

   e) $\dfrac{x - 2}{4} < \dfrac{2x - 3}{3}$ f) $\dfrac{4 - x}{2} \geqslant \dfrac{2 - x}{3}$

   g) $\frac{1}{3}(6 - x) \leqslant \frac{1}{5}(2 - 3x)$ h) $\frac{1}{9}(2x - 1) > \frac{1}{3}(3 - x)$

4 Find the integers which simultaneously satisfy each of the following pairs of inequalities.

   a) $4x + 3 \geqslant 2x + 5$ $x + 4 \leqslant 7$

   b) $5x + 3 > 3 - x$ $3x + 5 < 2x + 7$

   c) $5 - 2x \leqslant 3 - x$ $1 - 2x \leqslant 11 - 4x$

   d) $3x + 2 \geqslant 2x - 1$ $7x + 3 < 5x + 2$

   e) $5x - 4 \geqslant 4x - 3$ $\frac{1}{3}x < 1$

   f) $\frac{1}{2}(x + 1) > 1$ $5x + 1 < 4(x + 2)$

5 Show that there are no real numbers which simultaneously satisfy the two inequalities
$2x + 1 \geqslant x + 1$ and $\frac{1}{2}(x + 5) \leqslant 2$.

6 Show that there is just one number which simultaneously satisfies these three inequalities and find it.

   $\frac{1}{2}(x - 1) > 1$ $2 \quad 3x < 7 \quad 4x$ $\frac{1}{3}x \leqslant 1$

....

1.3   Straight-line graphs

Consider the straight line with gradient m, passing through the point A$(0, c)$ on the y-axis. Let point P(x, y) be a general point on the line.

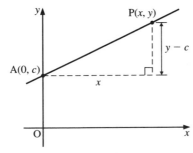

$$\text{Gradient} = m = \dfrac{y - c}{x - 0}$$

$$\therefore \ y = mx + c$$

> The general Cartesian form of the equation of a straight line, where m is the gradient of the line and c is the y-intercept, is
> $y = mx + c$

An alternative form for the equation of a straight line is

$$ax + by + c = 0$$

which is generally used when the gradient is a fraction.

To sketch the graph of a straight line you need only two points through which the line passes.

The y-intercept is normally one of these points, and you can find the second by substituting an x value.

Example 1

Write down the gradient and y-intercept for each of these straight lines:
a) $y = 5x - 2$
b) $2y = 1 - 6x$
c) $3x + 2y - 8 = 0$

....

a) Comparing $y = 5x - 2$ with $y = mx + c$ gives

$m = 5$ and $c = -2$.

Therefore the gradient is 5 and the y-intercept is -2.

b) Rearranging $2y = 1 - 6x$ into the form $y = mx + c$ gives

$$2y = 1 - 6x$$
$$2y = -6x + 1$$
$$\therefore y = -3x + \tfrac{1}{2}$$

Comparing with $y = mx + c$ gives

$m = -3$ and $c = \tfrac{1}{2}$.

Therefore the gradient is -3 and the y-intercept is $\tfrac{1}{2}$.

c) Rearranging $3x + 2y - 8 = 0$ into the form $y = mx + c$ gives

$$2y = -3x + 8$$
$$\therefore y = -\tfrac{3}{2}x + 4$$

Comparing with $y = mx + c$ gives

$m = -\tfrac{3}{2}$ and $c = 4$.

Therefore the gradient is $-\tfrac{3}{2}$ and the y-intercept is 4.

Example 2

Sketch the graphs of these equations:

a) $y = 2x - 1$

b) $3y + 2x - 3 = 0$

c) $2y - x + 1 = 0$

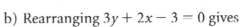

a) Comparing $y = 2x - 1$ with $y = mx + c$ gives the gradient of the line as 2 and the y-intercept as -1.
Therefore one point that the line passes through is $(0, -1)$.
To find a second point, let $x = 1$ which gives

$$y = 2(1) - 1$$

$$\therefore\ y = 1$$

Therefore a second point is $(1, 1)$.
The graph of $y = 2x - 1$ is shown.

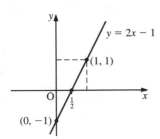

b) Rearranging $3y + 2x - 3 = 0$ gives

$$3y = -2x + 3$$

$$\therefore\ y = -\tfrac{2}{3}x + 1$$

Comparing $y = -\tfrac{2}{3}x + 1$ with $y = mx + c$ gives the gradient of the line as $-\tfrac{2}{3}$ and the y-intercept as 1.
Therefore one point that the line passes through is $(0, 1)$.
To find a second point, let $x = 3$ which gives

$$y = -\tfrac{2}{3}(3) + 1$$

$$\therefore\ y = -1$$

Therefore a second point is $(3, -1)$.
The graph of $3y + 2x - 3 = 0$ is shown.

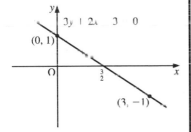

c) Rearranging $2y - x + 1 = 0$ gives

$$2y = x - 1$$

$$\therefore\ y = \tfrac{1}{2}x - \tfrac{1}{2}$$

Comparing $y = \tfrac{1}{2}x - \tfrac{1}{2}$ with $y = mx + c$ gives the gradient of the line as $\tfrac{1}{2}$ and the y-intercept as $-\tfrac{1}{2}$.
Therefore one point that the line passes through is $(0, -\tfrac{1}{2})$.
To find a second point, let $x = 2$ which gives

$$y = \tfrac{1}{2}(2) - \tfrac{1}{2}$$

$$\therefore\ y = \tfrac{1}{2}$$

Therefore a second point is $(2, \tfrac{1}{2})$.
The graph of $2y - x + 1 = 0$ is shown.

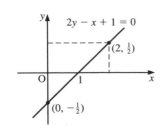

Exercise 1D

1 Write down the gradient and y-intercept for the graph of each equation.

a) $y = 3x - 2$ b) $y = 7x + 3$ c) $3y = 5x - 2$

d) $6y = 3x + 1$ e) $4y = 2 - x$ f) $3y = 4x + 5$

g) $2y = 5 - 3x$ h) $2x + 3y - 4 = 0$ i) $3x - 5y + 6 = 0$

j) $4x - y = 7$ k) $6x + 3y = 4$ l) $8x - 5y = 12$

2 Sketch the graphs of each of these equations.

a) $y = 2x - 3$ b) $y = 3x + 4$ c) $2y = 4x - 5$

d) $3x + 4y - 24 = 0$ e) $6x - 5y = 30$ f) $x + 2y = 10$

3 Write down the equations of each of these straight lines, expressing your answers in the form $ax + by + c = 0$ where a, b and c are integers.

a)
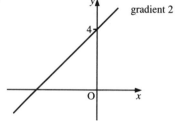

b)
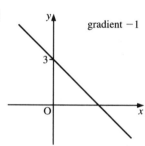

c)
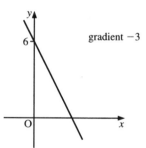

d)
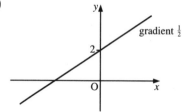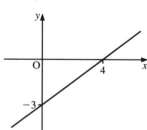

e)
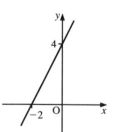

f)

g)
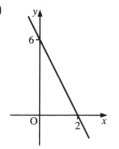

h)
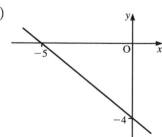

4 Write down the equation of the straight line which passes through the points $A(2, 5)$ and $B(6, 17)$.

....

1.4 Simultaneous linear equations

Consider the equation $x + y = 9$.

The diagram shows its graph.

Each point on the line is a solution of the equation.

Examples are: $x = 0, y = 9$

$x = 4, y = 5$

$x = 0.25, y = 8.75$

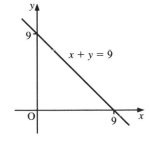

You cannot solve the equation uniquely unless you are given further information, or another equation.

Suppose the second equation is

$2x + y = 13$

The diagram shows both graphs on the same axes.

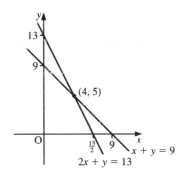

$x + y = 9$ and $2x + y = 13$ are **simultaneous linear equations**.

The point $(4, 5)$ lies on both graphs. So $x = 4, y = 5$ is the solution of the simultaneous linear equations because it satisfies both equations.

> You can check by substitution:
> $4 + 5 = 9$
> $2 \times 4 + 5 = 8 + 5 = 13$

> You can solve a pair of simultaneous linear equations graphically by:
> ❖ drawing their graphs on the same axes
> ❖ finding their point of intersection.

There are two common algebraic methods for solving simultaneous linear equations.

Elimination

You can eliminate either the x terms or the y terms by adding or subtracting the two equations.

Example 1

Solve the simultaneous equations
$$2x + y = 7$$
$$2x - y = 5$$

....

Number the equations:
$$2x + y = 7 \qquad\qquad [1]$$
$$2x - y = 5 \qquad\qquad [2]$$

Subtract [2] from [1]:
$$2y = 2 \qquad\qquad [1] - [2]$$
$$\therefore\; y = 1$$

Substitute $y = 1$ into [1] to obtain x:
$$2x + 1 = 7$$
$$2x = 6$$
$$\therefore\; x = 3$$

The solution is $x = 3$, $y = 1$.

> Alternatively you could add the equations:
> $4x = 12$.

> Alternatively you could substitute $y = 1$ into [2].

In some cases, the coefficients of the x terms or the y terms may not be the same. You need to multiply the equations by a suitable constant to make the coefficients of either the x or y terms the same. This process is called **balancing the coefficients**.

Example 2

Solve the simultaneous equations
$$5x - 7y = 27 \qquad\qquad [1]$$
$$2x + 3y = 5 \qquad\qquad [2]$$

....

To balance the coefficients of the y terms, multiply equation [1] by 3 and multiply equation [2] by 7. This gives
$$15x - 21y = 81 \qquad\qquad [3]$$
$$14x + 21y = 35 \qquad\qquad [4]$$

Adding [3] and [4] gives
$$29x = 116$$
$$\therefore\; x = 4$$

> Note that the y terms match up.

> To solve a pair of simultaneous equations you need to find x **and** y.

Substituting $x = 4$ into [2] to find the corresponding y value gives

$$2(4) + 3y = 5$$
$$8 + 3y = 5$$
$$3y = -3$$
$$\therefore\; y = \frac{-3}{3} = -1$$

The solution is $x = 4$, $y = -1$.

Alternatively, you can balance the coefficients of the x terms by multiplying equation [1] by 2 and multiplying equation [2] by 5. This gives

$$10x - 14y = 54 \qquad\qquad [5]$$
$$10x + 15y = 25 \qquad\qquad [6]$$

Subtracting [6] from [5] gives

$$-29y = 29$$
$$\therefore\; y = -1$$

Substituting $y = -1$ into [2] to find the corresponding x value gives

$$2x + 3(-1) = 5$$
$$\therefore\; x = 4$$

The solution is $x = 4$, $y = -1$, as before.

You can check the solution by substituting into [1]:
$$5(4) - 7(-1) = 20 + 7$$
$$= 27$$

Note that x has been eliminated.

Substitution

This method involves rearranging one of the equations so that you have either

❖ y in terms of x, in which case you then substitute this expression for y into the second equation,

or

❖ x in terms of y, in which case you substitute this expression for x into the second equation.

The choice usually depends on which gives the simpler expression after rearranging.

Example 3

Solve the simultaneous equations

$$3x + 4y - 27 = 0 \qquad\qquad [1]$$
$$5x + y - 11 = 0 \qquad\qquad [2]$$

Rearranging [2] for y gives

$$y = 11 - 5x \qquad\qquad [3]$$

Substituting [3] into [1] gives

$$3x + 4(11 - 5x) - 27 = 0$$
$$3x + 44 - 20x - 27 = 0$$
$$-17x = -17$$
$$\therefore\; x = 1$$

Substituting $x = 1$ into [3] to find the corresponding y value gives

$$y = 11 - 5(1)$$
$$\therefore \ y = 6$$

The solution is $x = 1$, $y = 6$.

Graphical solution

This method involves drawing graphs of the two linear equations and finding the coordinates of their point of intersection. The coordinates of the intersection point give the solution to the pair of equations.

> To draw the graphs of the linear equations, use the methods described in section 1.3.

Example 4

Solve the simultaneous equations

$$x + y = 9$$
$$2x + y = 13$$

....

Rearranging $x + y = 9$ gives $y = -x + 9$.
One point the line passes through is $(0, 9)$, the y-intercept.
Letting $x = 1$ gives a second point as $(1, 8)$.

Rearranging $2x + y = 13$ gives $y = -2x + 13$.
One point the lines passes through is $(0, 13)$, the y-intercept.
Letting $x = 1$ gives a second point as $(1, 11)$.

If you plot these sets of points you get the two straight lines shown.

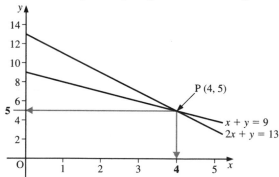

> You can use the trace function on your GDC to check the intersection point.
> Some GDCs enable you to solve simultaneous equations by directly typing in the equations.

The coordinates of P, where the two lines intersect, satisfy both equations and are therefore the solution to *both* equations.

The point P has coordinates $(4, 5)$, therefore the solution is $x = 4$, $y = 5$.

Exercise 1E

....

1 Solve each pair of simultaneous equations. Use the elimination method.

a) $2x - 3y = 7$ b) $3x - y = 1$ c) $2x - 7y = 1$ d) $3x - 4y = 5$
    $2x + 3y = 1$ $5x + y = 7$ $2x + 3y = 11$ $6x - 4y = 2$

e) $5x + 2y = 7$ f) $x - 2y = 5$ g) $2x + 3y = 1$ h) $3x - 2y = 5$
    $2x + y = 2$ $3x + y = 8$ $3x + y = 5$ $2x + y = 8$

i) $3x + 5y = 1$ j) $5x + 4y = 1$ k) $6x - 5y = 12$ l) $3x + 2y = 3$
    $2x + 3y = 0$ $7x + 5y = 2$ $5x - 4y = 11$ $x - 6y = 11$

2 Solve each pair of simultaneous equations. Use either the elimination or the substitution method, but show your working clearly.

a) $2x + y - 10 = 0$
   $3x - 2y - 8 = 0$

b) $5x - 4y + 1 = 0$
   $3x + y - 13 = 0$

c) $4x - 3y + 1 = 0$
   $3x - 4y + 6 = 0$

d) $5x - 7y + 3 = 0$
   $3x + 2y + 8 = 0$

e) $7x - 3y - 8 = 0$
   $5x + 7y + 8 = 0$

f) $2x - 5y + 4 = 0$
   $3x - 5y + 1 = 0$

g) $x - y - 1 = 0$
   $5x - 2y - 3 = 0$

h) $3x + 2y - 2 = 0$
   $x - 6y + 1 = 0$

3 Solve each pair of simultaneous equations. Use the substitution method, and in each case illustrate your solution graphically.

a) $y = x + 3$
   $y = 2x + 1$

b) $y = x - 4$
   $y = 3x - 16$

c) $y = 5 - 2x$
   $y = x + 8$

d) $y = 4 - x$
   $y = 2x + 10$

e) $y = x - 3$
   $y = 6x + 2$

f) $y = 2x - 10$
   $y = \frac{1}{3}x$

g) $2x - 5y = 4$
   $3x + 2y - = -13$

h) $7x + 3y = 6$
   $5x + 4y = 8$

1.5 Factorisation

The product of x and $x + 3$ is $x^2 + 3x$.

Conversely, the factors of $x^2 + 3x$ are x and $x + 3$.

To **factorise** an expression, you must find quantities which, when multiplied together, give the original expression.

Example 1

Factorise each of these expressions.

a) $3x^2 + x$ b) $4xy + 2y$ c) $2x^2 - xy$

a) One obvious factor is x:
$$3x^2 + x = x(3x + 1)$$

b) Two obvious factors are 2 and y:
$$4xy + 2y = 2y(2x + 1)$$

c) There is an obvious factor of x:
$$2x^2 - xy = x(2x - y)$$

You can check that you have factorised correctly by expanding the brackets. For example,

$x(3x + 1) = 3x^2 + x$

Trinomials

A **trinomial** is a three-termed expression such as $y^4 + 2xy - 3$.
A special case of trinomials are quadratic expressions $ax^2 + bx + c$, where a, b and c are non-zero constants. You may be able to factorise such expressions by trial and error.

> An example of a quadratic trinomial is $x^2 + 2x + 6$.

Example 2

Factorise $x^2 + 5x + 6$.

....

If this expression factorises then
$$x^2 + 5x + 6 = (x + ?)(x + ?)$$
Therefore look for two numbers whose product is 6 and whose sum is 5. These are 2 and 3 and the factors are
$$(x + 2) \text{ and } (x + 3)$$

If the constant c is negative, there must be a minus sign in one of the factors.

Example 3

Factorise $x^2 + 4x - 21$.

....

In this case look for two numbers whose product is -21 and whose sum is $+4$. These numbers are $+7$ and -3.
The factors are
$$(x + 7) \text{ and } (x - 3)$$

The coefficient of x^2 may not be equal to 1.

Example 4

Factorise $2x^2 - x - 6$.

....

The coefficient of x^2 is 2, which can be obtained by multiplying x and $2x$. If this expression factorises then
$$2x^2 - x - 6 = (2x + ?)(x + ?)$$
Therefore look for two numbers whose product is -6 and which fit this arrangement. If you try 3 and -2 and expand, you get the required expression. The factors are
$$(2x + 3) \text{ and } (x - 2)$$
The other combinations don't give the required result; for example:
$$(2x - 2)(x + 3) = 2x^2 + 4x - 6$$

In Example 4 the coefficient of the x^2 term is not equal to 1.
To factorise $6x^2 - x - 2$, first look at the quadratic term:

$$6x^2 - x - 2 = (6x + ?)(x + ?) \text{ or } (2x + ?)(3x + ?)$$

Now look at the constant term. The constant -2 must have been
generated from -2 and 1 or 2 and -1. Trying the different
combinations shows that the factors are:

$$(2x + 1) \text{ and } (3x - 2)$$

Difference of two squares

Expanding and simplifying $(a + b)(a - b)$ gives

$$(a + b)(a - b) = a^2 - ab + ba - b^2$$
$$= a^2 - b^2$$

The identity $a^2 - b^2 = (a + b)(a - b)$ is called the **difference of
two squares**.

Example 5

Factorise $x^2 - 25$.

....

Notice that the expression is the difference of the two squares
x^2 and 5^2. Using the identity $a^2 - b^2 = (a + b)(a - b)$ gives

$$x^2 - 25 = (x + 5)(x - 5).$$

Example 6

Factorise $9x^2 - y^2$.

....

Notice that the expression is the difference of the two squares
$(3x)^2$ and y^2. Using the identity $a^2 - b^2 = (a + b)(a - b)$ gives

$$9x^2 - y^2 = (3x + y)(3x - y).$$

Example 7

Factorise $3x^2 - 12y^2$.

....

Notice that neither 3 nor 12 are square numbers. However,
there is an obvious factor of 3. Factorising gives

$$3(x^2 - 4y^2)$$

Notice that the expression $x^2 - 4y^2$ is the difference of the two
squares x^2 and $(2y)^2$.
Using the identity $a^2 - b^2 = (a + b)(a - b)$ gives

$$x^2 - 4y^2 = (x + 2y)(x - 2y).$$

Therefore the full factorisation is

$$3x^2 - 12y^2 = 3(x + 2y)(x - 2y).$$

Exercise 1F

In questions **1** to **6** factorise each of the given expressions.

1 a) $2x + 4y$ b) $6a + 9b$ c) $5p - 10q$

   d) $8y - 12x$ e) $2m + 6n$ f) $3a - 9b$

   g) $2a + 4b - 6c$ h) $9x - 3y + 6z$

2 a) $5x^2 + x$ b) $2a^2 + 3a$ c) $6p^2 - 5p$

   d) $4y^2 - 7y$ e) $4x^2 + 6xy$ f) $3a^2 - 6ab$

   g) $2bc - 8ba$ h) $4a^2 - 6ab + 8ac$

3 a) $ax + ay + bx + by$ b) $ax - ay + bx - by$

   c) $px + py + 2qx + 2qy$ d) $cx + 2cy - 2dx - 4dy$

   e) $a^2 + ac + ba + bc$ f) $c^2 + cd - bc - bd$

   g) $p^2 + 5py - 2px - 10xy$ h) $b^2 - 2by - 3bx + 6xy$

4 a) $x^2 - 3x + 2$ b) $x^2 + 4x + 3$ c) $x^2 + 5x - 6$

   d) $x^2 + 6x + 9$ e) $x^2 - x - 12$ f) $x^2 - 4x - 12$

   g) $x^2 + 9x + 20$ h) $x^2 + x - 6$ i) $x^2 - 8x + 7$

   j) $x^2 - 3x - 10$ k) $x^2 + 4x - 5$ l) $x^2 + 7x + 10$

5 a) $2x^2 + x - 6$ b) $2x^2 + 3x + 1$ c) $3x^2 - 5x - 2$

   d) $3x^2 - 10x + 3$ e) $5x^2 - 11x + 2$ f) $5x^2 + 23x + 12$

   g) $2x^2 - 17x + 30$ h) $3x^2 + 26x + 35$ i) $7x^2 - 19x - 6$

   j) $4x^2 + 4x - 3$ k) $4x^2 - 5x - 6$ l) $6x^2 + 13x + 5$

6 a) $x^2 - 16$ b) $x^2 - 9$ c) $x^2 - 81$

   d) $x^2 - 144$ e) $4x^2 - 9$ f) $25x^2 - 4$

   g) $16 - 9x^2$ h) $49 - 64x^2$ i) $4x^2 - 81y^2$

   j) $25x^2 - 16y^2$ k) $x^2 - 169y^2$ l) $4x^2 - 36y^2$

1.6   Quadratic equations

An expression of the form $ax^2 + bx + c$, where a, b and c are constants with $a \neq 0$, is called a **quadratic expression**.

> $\neq$ means 'not equal to'.

An equation of the form $ax^2 + bx + c = 0$, where a, b and c are constants with $a \neq 0$, is called a **quadratic equation**.
A quadratic equation will have at most two real solutions.
The solutions are generally called the **roots** of the quadratic.
There are three types of quadratic equation:

❖ $b = 0$, $a \neq 0$, $c \neq 0$ that is $ax^2 + c = 0$

❖ $c = 0$, $a \neq 0$, $b \neq 0$ that is $ax^2 + bx = 0$

❖ $a \neq 0$, $b \neq 0$, $c \neq 0$ that is $ax^2 + bx + c = 0$

$ax^2 + c = 0$

Example 1

Solve the equation $x^2 - 16 = 0$ using

a) algebraic methods

b) a GDC to draw the graph of $y = x^2 - 16$ and find its roots.

Check your solutions by using the Equation Solving Function of the GDC.

> **Remember:**
> GDC = graphic display calculator

....

a) Rearrange the equation to obtain

$$x^2 = 16$$
$$x = \pm\sqrt{16}$$
$$\therefore \ x = \pm 4$$

The solutions are $x = \pm 4$.

> You must include -4 as a solution, since $(-4)^2 = 16$.

This equation can also be solved by factorisation using 'the difference of two squares'. That is:

$$x^2 - y^2 = (x - y)(x + y) \tag{1}$$

The equation $x^2 - 16 = 0$ can be written as the difference of two squares:

$$x^2 \quad 4^2 = 0$$

Using [1] gives

$$x^2 - 4^2 = (x - 4)(x + 4) = 0$$

Using the result 'if $ab = 0$ then $a = 0$ or $b = 0$' gives

$$x - 4 = 0 \quad \text{or} \quad x + 4 = 0$$
$$\therefore \ x = 4 \quad \text{or} \quad x = -4$$

The solutions are $x = \pm 4$, as before.

b) Using a GDC gives the graph of $y = x^2 - 16$ as shown.

The solutions of the equation $x^2 - 16 = 0$ are the x values of the intersection points of the graph and the x-axis.

From the GDC these are $x = 4$ and $x = -4$, which agree with the solutions from a).

Finally, the GDC Equation Solving Function gives the solutions as $x = 4$ and $x = -4$.

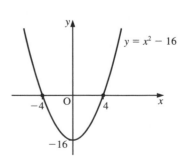

You may need to leave the answer in exact form.

Example 2

a) Solve the equation $4x^2 - 24 = 0$.

b) Sketch the graph of $y = 4x^2 - 24$, showing clearly where the curve intersects the axes.

....

a) Rearranging gives

$$4x^2 = 24$$
$$x^2 = 6$$
$$\therefore\ x = \pm\sqrt{6}$$

The solutions are $x = \pm\sqrt{6}$.

b)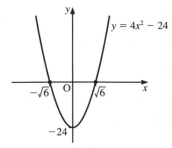

$ax^2 + bx = 0$

This type of quadratic equation can be solved by simple factorisation, where x is a factor.

Example 3

a) Solve the equation $x^2 - 7x = 0$.

b) Sketch the graph of $y = x^2 - 7x$, showing clearly where the curve intersects the axes.

....

a) Factorising the LHS gives

$$x(x - 7) = 0$$
$$x = 0 \quad \text{or} \quad x - 7 = 0$$
$$\therefore\ x = 0 \quad \text{or} \quad x = 7$$

The solutions are $x = 0$ and $x = 7$.

b)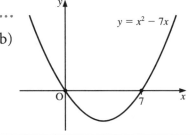

$ax^2 + bx + c = 0$

Many quadratic equations of this form can be solved by factorisation methods. However, not all quadratics of this form will factorise. When a quadratic expression will not factorise, the quadratic is said to be **irreducible**. Other methods are required for solving irreducible quadratics, namely 'completing the square' and the formula method (see pages 26–36).

Example 4

a) Solve the equation $x^2 + 5x + 6 = 0$.

b) Sketch the graph of $y = x^2 + 5x + 6$, showing clearly where the curve intersects the x-axis.

....

a) Factorising the LHS gives

$$(x + 2)(x + 3) = 0$$
$$x + 2 = 0 \quad \text{or} \quad x + 3 = 0$$
$$\therefore\ x = -2 \quad \text{or} \quad x = -3$$

The solutions are $x = -2$ and $x = -3$.

b)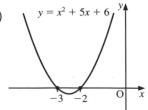

Example 5

Solve the equation $2x^2 - 13x - 24 = 0$.

Factorising the LHS gives

$$(2x + 3)(x - 8) = 0$$

$$2x + 3 = 0 \quad \text{or} \quad x - 8 = 0$$

$$\therefore \ x = -\tfrac{3}{2} \quad \text{or} \quad x = 8$$

The solutions are $x = -\tfrac{3}{2}$ and $x = 8$.

You may need to simplify and rearrange an equation so that there is a quadratic expression on the LHS and zero on the RHS.

Example 6

Solve the equation $x^2 - x - 10 = x + 5$.

Rearranging and factorising the LHS gives

$$x^2 - 2x - 15 = 0$$

$$(x + 3)(x - 5) = 0$$

$$x + 3 = 0 \quad \text{or} \quad x - 5 = 0$$

$$\therefore \ x = -3 \quad \text{or} \quad x = 5$$

The solutions are $x = -3$ and $x = 5$.

> Use your GDC to draw the graphs of $y = x^2 - x - 10$ and $y = x + 5$ and confirm that they cross at $x = -3$ and $x = -5$.

Example 7

Solve the equation $(3x + 1)(2x - 1) - (x + 2)^2 = 5$.

Expanding the LHS and simplifying gives

$$6x^2 - x - 1 - (x^2 + 4x + 4) = 5$$

$$5x^2 - 5x - 5 = 5$$

$$5x^2 - 5x - 10 = 0$$

$$5(x^2 - x - 2) = 0$$

$$5(x + 1)(x - 2) = 0$$

$$x + 1 = 0 \quad \text{or} \quad x - 2 = 0$$

$$\therefore \ x = -1 \quad \text{or} \quad x = 2$$

The solutions are $x = -1$ and $x = 2$.

Solving problems with quadratic equations

You can often apply quadratic equations to real-life situations.

Example 8

A piece of wire of length 1 metre is cut into two parts and each part is bent to form a square. If the total area of the two squares formed is 325 cm², find the perimeter of each square.

....

Let one of the pieces of wire be of length x cm.
Then the other piece is of length $(100 - x)$ cm.

A _____ B _____ C
|← x cm →|← $(100 - x)$ cm →|

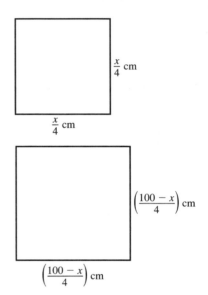

The square formed from the piece AB has sides of length $\dfrac{x}{4}$ cm.

The area, A_1, of this square is given by

$$A_1 = \left(\frac{x}{4}\right)\left(\frac{x}{4}\right) = \frac{x^2}{16} \text{ cm}^2$$

The square formed from the piece BC has sides of length

$$\left(\frac{100 - x}{4}\right) \text{ cm.}$$

The area, A_2, of this square is given by

$$A_2 = \left(\frac{100 - x}{4}\right)\left(\frac{100 - x}{4}\right) = \frac{(100 - x)^2}{16} \text{ cm}^2$$

Since the total area of the two squares is 325 cm²,

$$A_1 + A_2 = 325$$

So $\dfrac{x^2}{16} + \dfrac{(100 - x)^2}{16} = 325$

Multiplying throughout by 16 gives

$$x^2 + (100 - x)^2 = 5200$$
$$x^2 + 10\,000 - 200x + x^2 = 5200$$
$$2x^2 - 200x + 4800 = 0$$
$$2(x^2 - 100x + 2400) = 0$$
$$2(x - 40)(x - 60) = 0$$
$$x - 40 = 0 \quad \text{or} \quad x - 60 = 0$$
$$\therefore \; x = 40 \quad \text{or} \quad x = 60$$

If $x = 40$ cm, the square formed from the piece of wire AB has perimeter 40 cm, and the square formed from the piece of wire BC has perimeter 60 cm.

If $x = 60$ cm, the square formed from the piece of wire AB has perimeter 60 cm, and the square formed from the piece of wire BC has perimeter 40 cm.

The perimeters of the squares are 40 cm and 60 cm.

> In practical problems, remember to check that the answers make sense.

Exercise 1G
....

In questions **1** to **4**, solve each of the given quadratic equations for x.

1   a) $x^2 - 5x + 6 = 0$ b) $x^2 - 3x - 4 = 0$

    c) $x^2 - 7x + 10 = 0$ d) $x^2 + 5x + 6 = 0$

    e) $x^2 - 6x + 8 = 0$ f) $x^2 - 5x - 6 = 0$

    g) $x^2 = 9$ h) $x^2 + 2x = 8$

    i) $x^2 = x + 12$ j) $x^2 + 20 = 9x$

    k) $x^2 = 4x$ l) $x^2 - 8 = 7x$

2   a) $2x^2 + 5x + 2 = 0$ b) $3x^2 - 7x + 2 = 0$

    c) $2x^2 - 3x - 5 = 0$ d) $5x^2 + 14x - 3 = 0$

    e) $4x^2 + 5x + 1 = 0$ f) $6x^2 - 5x + 1 = 0$

    g) $3x^2 = 10x + 8$ h) $2x^2 + x = 15$

    i) $16x^2 = 9$ j) $3x^2 - x = 10$

    k) $5x^2 + 13x = 6$ l) $8x^2 + 3 = 14x$

3   a) $(x + 1)(x + 3) = 8$ b) $(x + 2)^2 = 2x + 12$

    c) $(2x + 3)(x - 1) = 2(5x + 1)$

    d) $(x - 3)(x - 4) + 7 = (2x + 5)(x - 1)$

4   a) $\dfrac{2}{x + 1} = \dfrac{x}{3 - 2x}$ b) $\dfrac{x + 2}{4} = \dfrac{x}{4 - x}$

    c) $\dfrac{x^2}{x + 3} = \dfrac{2 - 3x}{2}$ d) $\dfrac{2x - 1}{8 - x} = \dfrac{5}{x + 2}$

> Hint: Cross-multiply to remove the fractions.

5   The perimeter of a rectangle is 34 cm. Given that the diagonal is of length 13 cm, and that the width is x cm, derive the equation $x^2 - 17x + 60 = 0$. Hence find the dimensions of the rectangle.

6   A garden is in the shape of a rectangle, 20 metres by 8 metres. Around the outside is a border of uniform width, and in the middle is a square pond. The width of the border is the same as the width of the pond. The size of the area which is not occupied by either border or pond is 124 m². Letting the width of the border be x m, derive the equation $3x^2 - 56x + 36 = 0$. Solve this equation to find the value of x.

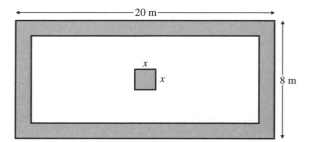

7 A metal sleeve of length 20 cm has rectangular cross-section 10 cm by 8 cm. The metal has uniform thickness, x cm, along the sleeve, and the total volume of metal in the sleeve is 495 cm³.

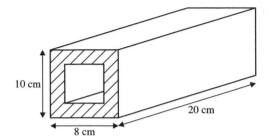

Derive the equation $16x^2 - 144x + 99 = 0$, and solve it to find the value of x.

8 A strand of wire of length 32 cm is cut into two pieces. One piece is bent to form a rectangle of width x cm and length $(x + 2)$ cm, and the other piece is bent to form a square.

a) Show that the square has sides of length $(7 - x)$ cm.

b) Given that the total of the areas enclosed by both the rectangle and the square is 31 cm², form an equation for x and solve it to find the value of x.

···

1.7 Completing the square

Some examples of perfect squares are

$$25 = 5^2 \quad x^4 = (x^2)^2$$
$$x^2 + 12x + 36 = (x + 6)^2$$
$$(x + y)^6 = [(x + y)^3]^2$$

You can make the quadratic expression $x^2 + 8x$ into a perfect square by adding half the coefficient of the x term squared. That is:

$$x^2 + 8x + \left(\tfrac{8}{2}\right)^2 = x^2 + 8x + (4)^2$$
$$= x^2 + 8x + 16$$

Now

$$x^2 + 8x + 16 = (x + 4)(x + 4)$$
$$= (x + 4)^2$$

which is a perfect square.

This process of adding half the coefficient of the x term squared is called **completing the square**.

In general, if you want to make $x^2 + bx$ into a perfect square, add $\left(\dfrac{b}{2}\right)^2$:

$$x^2 + bx + \left(\frac{b}{2}\right)^2 = x^2 + bx + \frac{b^2}{4}$$
$$= \left(x + \frac{b}{2}\right)\left(x + \frac{b}{2}\right)$$
$$= \left(x + \frac{b}{2}\right)^2$$

which is a perfect square.

The process of completing the square is used to express a quadratic expression $ax^2 + bx + c$ in the form

$$a(x + p)^2 + q$$

where p and q are constants.

First, we will look at those quadratic expressions in which $a = 1$. In general,

$$x^2 + bx + c = \left(x + \frac{b}{2}\right)^2 - \left(\frac{b}{2}\right)^2 + c$$
$$= \left(x + \frac{b}{2}\right)^2 - \frac{b^2}{4} + c$$

In the examples which follow, you will see why it is useful to express a quadratic expression in this form.

> Because you have added $\left(\dfrac{b}{2}\right)^2$ to make $x^2 + bx$ into a perfect square, you must also subtract it so that LHS = RHS.

Example 1

Express $x^2 + 6x - 1$ in the form $a(x + p)^2 + q$
Hence solve the equation $x^2 + 6x - 1 = 0$, leaving your answers in exact form.

....

Completing the square gives
$$x^2 + 6x - 1 = \left(x + \tfrac{6}{2}\right)^2 - \left(\tfrac{6}{2}\right)^2 - 1$$
$$= (x + 3)^2 - 9 - 1$$
$$= (x + 3)^2 - 10$$
Rewriting the equation $x^2 + 6x - 1 = 0$ gives
$$(x + 3)^2 - 10 = 0$$
$$(x + 3)^2 = 10$$
$$x + 3 = \pm\sqrt{10}$$
$$\therefore\ x = -3 \pm \sqrt{10}$$
The two solutions are $x = -3 + \sqrt{10}$ and $x = -3 - \sqrt{10}$.

Example 2

Use the method of completing the square to solve the equation $x^2 - 3x + 1 = 0$, leaving your answers in exact form.

....

Completing the square gives

$$x^2 - 3x + 1 = \left(x - \frac{3}{2}\right)^2 - \left(-\frac{3}{2}\right)^2 + 1$$

$$= \left(x - \frac{3}{2}\right)^2 - \frac{9}{4} + 1$$

$$= \left(x - \frac{3}{2}\right)^2 - \frac{5}{4}$$

Rewriting the equation $x^2 - 3x + 1 = 0$ gives

$$\left(x - \frac{3}{2}\right)^2 - \frac{5}{4} = 0$$

$$\left(x - \frac{3}{2}\right)^2 = \frac{5}{4}$$

$$x - \frac{3}{2} = \pm\sqrt{\frac{5}{4}} = \pm\frac{\sqrt{5}}{2}$$

$$\therefore \ x = \frac{3}{2} \pm \frac{\sqrt{5}}{2}$$

The solutions are $x = \frac{3}{2} + \frac{\sqrt{5}}{2}$ and $x = \frac{3}{2} - \frac{\sqrt{5}}{2}$.

If the equation is changed slightly to $x^2 - 3x + 3 = 0$, something interesting happens. Completing the square gives

$$x^2 - 3x + 3 = \left(x - \frac{3}{2}\right)^2 + \frac{3}{4}$$

So $\left(x - \frac{3}{2}\right)^2 + \frac{3}{4} = 0$,

$$\left(x - \frac{3}{2}\right)^2 = \frac{-3}{4}$$

$$x - \frac{3}{2} = \pm\sqrt{\frac{-3}{4}}$$

The square root of a negative number is not a real number. The equation $x^2 - 3x + 3 = 0$ has no real roots.

To use this technique when the coefficient of the x^2 term is not 1 you must factorise the expression, as in the next two examples.

Example 3

Express $2x^2 + 8x + 5$ in the form $a(x + p)^2 + q$ and state the values of a, p and q.

....

$$2x^2 + 8x + 5 = 2\left(x^2 + 4x + \tfrac{5}{2}\right)$$

Now proceed as before with the expression $x^2 + 4x + \frac{5}{2}$.

$$2\left(x^2 + 4x + \tfrac{5}{2}\right) = 2\left[(x + 2)^2 - (2)^2 + \tfrac{5}{2}\right]$$

$$= 2\left[(x + 2)^2 - 4 + \tfrac{5}{2}\right]$$

$$= 2\left[(x + 2)^2 - \tfrac{3}{2}\right]$$

$$= 2(x + 2)^2 - 3$$

Therefore, $a = 2$, $p = 2$ and $q = -3$.

Example 4

Express $3x^2 + 15x - 20$ in the form $a(x + p)^2 + q$.
Hence solve the equation $3x^2 + 15x - 20 = 0$, giving your
answers to one decimal place.

Taking out a factor of 3 gives
$$3x^2 + 15x - 20 = 3\left(x^2 + 5x - \tfrac{20}{3}\right)$$
Completing the square gives
$$3\left(x^2 + 5x - \tfrac{20}{3}\right) = 3\left[\left(x + \tfrac{5}{2}\right)^2 - \left(\tfrac{5}{2}\right)^2 - \tfrac{20}{3}\right]$$
$$= 3\left[\left(x + \tfrac{5}{2}\right)^2 - \tfrac{25}{4} - \tfrac{20}{3}\right]$$
$$= 3\left[\left(x + \tfrac{5}{2}\right)^2 - \tfrac{155}{12}\right]$$
$$= 3\left(x + \tfrac{5}{2}\right)^2 - \tfrac{155}{4}$$
Rewriting the equation $3x^2 + 15x - 20 = 0$ gives
$$3\left(x + \tfrac{5}{2}\right)^2 - \tfrac{155}{4} = 0$$
$$3\left(x + \tfrac{5}{2}\right)^2 = \tfrac{155}{4}$$
$$\left(x + \tfrac{5}{2}\right)^2 = \tfrac{155}{12}$$
$$x + \tfrac{5}{2} = \pm\sqrt{\tfrac{155}{12}}$$
$$\therefore\ x = -\tfrac{5}{2} + \sqrt{\tfrac{155}{12}} \quad \text{or} \quad x = -\tfrac{5}{2} - \sqrt{\tfrac{155}{12}}$$
$$x = 1.1 \qquad\qquad \text{or} \quad x = -6.1$$

The solutions are $x = 1.1$ and $x = -6.1$ to 1 d.p.

Exercise 1H

1 Express each of these in the form $(x + p)^2 + q$.

a) $x^2 + 4x + 6$ b) $x^2 - 6x + 13$

c) $x^2 - 10x + 40$ d) $x^2 - x - 5$

e) $x^2 - 5x + 9$ f) $x^2 - 20x + 3$

2 Use the method of completing the square to express the
solutions to each of these quadratic equations in the form
$a \pm b\sqrt{n}$, where a and b are rational, and n is an integer.

a) $x^2 - 4x - 1 = 0$ b) $x^2 + 6x + 2 = 0$

c) $x^2 - 2x - 1 = 0$ d) $x^2 - 8x - 3 = 0$

e) $x^2 + x - 1 = 0$ f) $x^2 + 3x + 1 = 0$

g) $x^2 - 5x - 2 = 0$ h) $x^2 - x - 3 = 0$

i) $x^2 + 5x + 1 = 0$ j) $x^2 + 12x + 5 = 0$

k) $x^2 - 9x + 10 = 0$ l) $x^2 - \tfrac{1}{2}x - \tfrac{1}{4} = 0$

3 Complete the square for each of these quadratic expressions.

a) $2x^2 + 8x - 13$ b) $3x^2 - 6x + 2$

c) $5 - 4x - x^2$ d) $5 + 4x - 2x^2$

e) $2x^2 - 6x + 5$ f) $23 - 10x - 5x^2$

4 Use the method of completing the square to solve each of these quadratic equations, expressing your solutions in the form $a \pm b\sqrt{n}$, where a, b and n are rational.

a) $2x^2 - 3x - 3 = 0$ b) $3x^2 - 6x + 1 = 0$

c) $4x^2 + 4x - 5 = 0$ d) $3x^2 + 5x - 1 = 0$

e) $5x^2 + x - 3 = 0$ f) $2x^2 - 3x - 1 = 0$

g) $2x^2 - x - 2 = 0$ h) $4x^2 + 3x - 2 = 0$

i) $7x^2 - 14x + 5 = 0$ j) $6x^2 + 4x - 3 = 0$

k) $5x^2 - 20x + 17 = 0$ l) $2x^2 + 18x + 21 = 0$

5 Express $x^2 + 4x + 7$ in the form $(x + p)^2 + q$.
Hence show that the equation $x^2 + 4x + 7 = 0$ has no real root.

6 Express $5x^2 - 30x + 47$ in the form $a(x + p)^2 + q$.
Hence show that the equation $5x^2 - 30x + 47 = 0$ has no real root.

7 Show that the equation $\dfrac{x + 2}{x - 3} = \dfrac{x + 4}{2x + 3}$ has no real root.

8 Given that the equation $x^2 + ax = b$, where a and b are real numbers, has a unique solution, prove that $a^2 + 4b = 0$.

....

1.8   Quadratic formula

The general quadratic equation has the form $ax^2 + bx + c = 0$, where a, b and c are constants with $a \neq 0$. Solving this general equation in terms of a, b and c will give a formula for the roots of any quadratic equation $ax^2 + bx + c = 0$:

> A root x is a solution of the equation $f(x) = 0$.

If $ax^2 + bx + c = 0$, where a, b and c are constants with $a \neq 0$, then

$$x = \frac{-b \pm \sqrt{b^2 - 4ac}}{2a}$$

> This is called the **quadratic formula**.

You can prove the formula like this:
Factorising out a in $ax^2 + bx + c = 0$ gives

$$a\left(x^2 + \frac{b}{a}x + \frac{c}{a}\right) = 0$$

You will not need to learn this proof for your examination.

Completing the square gives

$$a\left[\left(x + \frac{b}{2a}\right)^2 - \frac{b^2}{4a^2} + \frac{c}{a}\right] = 0$$

$$a\left[\left(x + \frac{b}{2a}\right)^2 + \left(\frac{-b^2 + 4ac}{4a^2}\right)\right] = 0$$

$$a\left(x + \frac{b}{2a}\right)^2 + \frac{4ac - b^2}{4a} = 0$$

$$a\left(x + \frac{b}{2a}\right)^2 = \frac{b^2 - 4ac}{4a}$$

$$\left(x + \frac{b}{2a}\right)^2 = \frac{b^2 - 4ac}{4a^2}$$

$$x + \frac{b}{2a} = \pm\sqrt{\frac{b^2 - 4ac}{4a^2}}$$

$$\therefore x = -\frac{b}{2a} \pm \frac{\sqrt{b^2 - 4ac}}{2a}$$

$$\therefore x = \frac{-b \pm \sqrt{b^2 - 4ac}}{2a}$$

as required.

You can use the quadratic formula to solve any quadratic equation.

Example 1

Solve the equation $2x^2 + 2x - 1 = 0$, giving your solutions in an exact form.

....

In this case $a = 2$, $b = 2$, $c = -1$, which gives

$$x = \frac{-2 \pm \sqrt{2^2 - 4(2)(-1)}}{2(2)}$$

$$= \frac{-2 \pm \sqrt{12}}{4}$$

$$= \frac{-2 \pm 2\sqrt{3}}{4}$$

$$= \frac{-1 \pm \sqrt{3}}{2}$$

$$\therefore x = \frac{-1 + \sqrt{3}}{2} \quad \text{or} \quad x = \frac{-1 - \sqrt{3}}{2}$$

These are the exact solutions and are left in this form.

You could get the same solutions by completing the square. Using the formula is generally quicker.

Example 2

Solve the equation $x^2 - 8x + 4 = 0$, giving your answers correct to three significant figures.

Using the quadratic formula with $a = 1$, $b = -8$ and $c = 4$ gives

$$x = \frac{-(-8) \pm \sqrt{(-8)^2 - 4(1)(4)}}{2(1)} = \frac{8 \pm \sqrt{48}}{2}$$

$\therefore\ x = 7.46$ or 0.536 (3 s.f.)

Use your GDC to draw the graph of $y = x^2 - 8x + 4$ and confirm that the x-intercepts are those found by using the formula.

Not all quadratic equations *have* real roots.

Example 3

Using the quadratic formula, show that the equation $5x^2 + 4x + 10 = 0$ has no real roots.

Using the quadratic formula with $a = 5$, $b = 4$ and $c = 10$ gives

$$x = \frac{-4 \pm \sqrt{(4)^2 - 4(5)(10)}}{2(5)}$$

$$\therefore\ x = \frac{-4 \pm \sqrt{-184}}{10}$$

Since $\sqrt{-184}$ is not real, the equation $5x^2 + 4x + 10 = 0$ has no real roots.

The square root of a negative number is not a real number.

Some quadratic equations have two identical roots. These are called **repeated roots**.

Example 4

Solve the equation $x^2 - 10x + 25 = 0$, and show that it has a repeated root.

Using the quadratic formula with $a = 1$, $b = -10$ and $c = 25$ gives

$$x = \frac{-(-10) \pm \sqrt{(-10)^2 - 4(1)(25)}}{2(1)}$$

$$= \frac{10 \pm \sqrt{0}}{2}$$

$\therefore\ x = 5$, a repeated root.

Discriminant of a quadratic equation

The quantity
$$\Delta = b^2 - 4ac$$
is called the **discriminant** of the quadratic equation $ax^2 + bx + c = 0$. The type of roots that a quadratic equation has depends on the value of the discriminant.

❖ In Example 2, $\Delta = 48$ and the associated equation has two real roots.
❖ In Example 3, $\Delta = -184$ and the associated equation has no real roots.
❖ In Example 4, $\Delta = 0$ and the associated equation has a repeated root.

> Consider the general quadratic equation
> $$ax^2 + bx + c = 0$$
> ❖ When $b^2 - 4ac > 0$, the equation has two distinct real roots.
> ❖ When $b^2 - 4ac < 0$, the equation has no real roots.
> ❖ When $b^2 - 4ac = 0$, the equation has one repeated root.

Δ is the Greek capital letter 'delta'.

The discriminant of a quadratic equation tells you whether the associated graph of the quadratic expression cuts the x-axis at two different points, does not cut the x-axis at all or touches the x-axis at one point. Each case is illustrated, for $a > 0$.

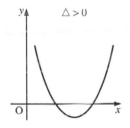
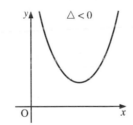
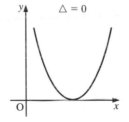

The equation has two roots.　　The equation has no roots.　　The equation has one repeated root.

Example 5

Use the discriminant to determine which of these quadratic equations has two distinct real roots, equal roots or no real roots.

a) $3x^2 - x + 2 = 0$　　　　　b) $x^2 - 3x - 28 = 0$

c) $4x^2 - 4x + 1 = 0$

....

a) In this case $a = 3$, $b = -1$, $c = 2$. The discriminant is given by
$$\Delta = b^2 - 4ac$$
$$= (-1)^2 - 4(3)(2)$$
$$= -23$$
Since $\Delta < 0$ the equation has no real roots.

b) In this case $a = 1$, $b = -3$, $c = -28$. The discriminant is given by
$$\Delta = (-3)^2 - 4(1)(-28)$$
$$= 9 + 112$$
$$= 121$$
Since $\Delta > 0$ the equation has two distinct real roots.

c) In this case $a = 4$, $b = -4$, $c = 1$. The discriminant is given by
$$\Delta = (-4)^2 - 4(4)(1)$$
$$= 0$$
Since $\Delta = 0$ the equation has equal roots.

> Δ is a perfect square ($\sqrt{121} = 11$), so the quadratic function will factorise.

> Use your GDC to draw the graphs for a), b) and c), illustrating the cases of 0, 1 and 2 roots.

If the graph of $y = ax^2 + bx + c$ has no real roots, this means that y is always positive or always negative. The curve never crosses the x-axis.

Example 6

Calculate the discriminant of the quadratic expression $2x^2 + 7x + 7$. Hence show that $2x^2 + 7x + 7$ is always positive.

. . . .

Calculating the discriminant with $a = 2$, $b = 7$ and $c = 7$ gives
$$\Delta = b^2 - 4ac$$
$$= (7)^2 - 4(2)(7)$$
$$= -7$$
The discriminant is -7.

Since $\Delta = -7 < 0$, the equation $2x^2 + 7x + 7 = 0$ has no real roots. Therefore, $y = 2x^2 + 7x + 7$ is never zero and the graph of $y = 2x^2 + 7x + 7$ does not cut the x-axis.

Since the coefficient of the x^2 term is positive, you know that the curve is U-shaped. Therefore, the entire curve lies above the x-axis. Therefore $2x^2 + 7x + 7$ is always positive.

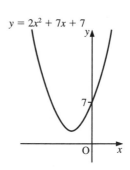

Example 7

Find the values of the constant k given that the equation $(5k + 1)x^2 - 8kx + 3k = 0$ has a repeated root.

. . . .

The equation $(5k + 1)x^2 - 8kx + 3k = 0$ has a repeated root if the discriminant of the equation is zero.

Calculating the discriminant of the equation with $a = 5k + 1$, $b = -8k$ and $c = 3k$ gives

$$\Delta = b^2 - 4ac$$
$$= (-8k)^2 - 4(5k + 1)(3k)$$
$$= 64k^2 - 12k(5k + 1)$$
$$\therefore \Delta = 4k^2 - 12k$$

Putting $\Delta = 0$ and factorising gives

$4k^2 - 12k = 0$

$4k(k - 3) = 0$

$\therefore k = 0$ or $k - 3 = 0$

$\therefore k = 0$ or $k = 3$

The required values of k are 0 and 3.

Exercise 1i

1 Use the quadratic formula to solve each of these equations, giving your answers correct to three significant figures.

a) $x^2 + 2x - 1 = 0$ b) $x^2 + 4x + 2 = 0$

c) $x^2 - 3x - 5 = 0$ d) $x^2 - 7x + 4 = 0$

e) $x^2 + 3x - 5 = 0$ f) $x^2 + 8x - 10 = 0$

2 Use your GDC to solve each of these equations, giving your answers correct to three significant figures.

a) $x^2 + x - 1 = 0$ b) $x^2 - 6x - 10 = 0$

c) $x^2 + 5x + 3 = 0$ d) $x^2 - 6x + 6 = 0$

e) $x^2 - 10x + 15 = 0$ f) $x^2 + 12x - 20 = 0$

3 Use the quadratic formula to solve each of these equations, giving your answers correct to three significant figures.

a) $2x^2 + 3x - 4 = 0$ b) $3x^2 + x - 3 = 0$

c) $4x^2 + 5x - 7 = 0$ d) $2x^2 + 7x + 4 = 0$

e) $5x^2 + 2x - 1 = 0$ f) $6x^2 + 5x - 3 = 0$

4 Use your GDC to solve each of these equations, giving your answers correct to three significant figures.

a) $2x^2 + x - 8 = 0$ b) $6x^2 + 3x - 1 = 0$

c) $3x^2 + 7x + 3 = 0$ d) $2x^2 - 3x - 8 = 0$

e) $6x^2 + 9x + 2 = 0$ f) $5x^2 + 4x - 3 = 0$

5 Solve each of these equations, giving your answers correct to three significant figures.

a) $(x + 2)(x - 1) - 3x = 4$ b) $(x + 2)^2 + 5x = 6$

c) $\dfrac{2x + 5}{x - 2} = \dfrac{x + 3}{x + 6}$ d) $\dfrac{2 - x}{3 - x} = \dfrac{4 - 3x}{5 + x}$

6 Use the discriminant to determine the number of real roots of each of these quadratic equations.

a) $5x^2 - 3x + 7 = 0$ b) $6x^2 - 5x - 3 = 0$

c) $9x^2 - 12x + 4 = 0$ d) $2x^2 + 6x + 3 = 0$

e) $4x^2 + 25 = 20x$ f) $3x^2 = 1 - 2x$

7 Calculate the discriminant of the quadratic $3x^2 + 5x + 8$. Hence show that $3x^2 + 5x + 8 > 0$, for all values of x.

8 Calculate the discriminant of the quadratic $5x^2 + 2x + 1$. Hence show that $5x^2 + 2x + 1 > 0$, for all values of x.

9 a) By completing the square, show that $x^2 + 3x + 5 > 0$ for all values of x.

b) Illustrate your answer on your GDC.

10 a) By completing the square, show that $2x^2 - 4x + 5 > 0$ for all values of x.

b) Illustrate your answer on your GDC.

11 Prove that the inequality $3x^2 + 13 < 12x$ has no real solution.

12 Find the possible values of the constant a given that the equation $ax^2 + (8 - a)x + 1 = 0$ has a repeated root.

13 Given that the equation $x^2 - 3bx + (4b + 1) = 0$ has a repeated root, find the possible values of the constant b.

14 Show that there is no real value of the constant c for which the equation

$$cx^2 + (4c + 1)x + (c + 2) = 0$$

has a repeated root.

15 Given that the roots of the equation $x^2 + ax + (a + 2) = 0$ differ by 2, find the possible values of the constant a. Hence state the possible values of the roots of the equation.

1.9 Disguised quadratic equations

Some equations do not appear to be quadratic, but in fact are.

Example 1

Solve the equation $x^4 + 5x^2 - 14 = 0$.

This equation does not appear to be a quadratic, but writing it in the form
$$(x^2)^2 + 5(x^2) - 14 = 0$$
and letting $y = x^2$ gives
$$y^2 + 5y - 14 = 0$$
which is a quadratic equation in y.
To solve the equation in y, factorise in the usual way giving
$$(y + 7)(y - 2) = 0$$
Solving gives $y = -7$ or $y = 2$.
Now replacing y with x^2 gives $x^2 = -7$ or $x^2 = 2$.
You can see that $x^2 = -7$ gives no real solutions.
However, $x^2 = 2$ gives $x = \pm\sqrt{2}$.

You can say that the original equation is a 'quadratic equation in x'.

A negative number has no real square root.

An equation with no x^2 in it may still be quadratic.

Example 2

Solve the equation $x - 9\sqrt{x} + 20 = 0$.

Rewrite the equation as
$$(\sqrt{x})^2 - 9(\sqrt{x}) + 20 = 0$$
The original equation is a quadratic in $\sqrt{x}$. Letting $y = \sqrt{x}$ gives
$$y^2 - 9y + 20 = 0$$
Factorising and solving gives
$$(y - 4)(y - 5) = 0$$
$$\therefore y = 4 \quad \text{or} \quad y = 5$$
Replacing y with $\sqrt{x}$ gives $\sqrt{x} = 4$ or $\sqrt{x} = 5$.
Solving gives $x = 16$ or $x = 25$.

Exercise 1J

1 Solve each of these equations for x.
a) $x^4 - 13x^2 + 36 = 0$
b) $x^4 - 2x^2 - 3 = 0$
c) $x^6 - 28x^3 + 27 = 0$
d) $x^6 + 5x^3 - 24 = 0$
e) $x - 5\sqrt{x} + 6 = 0$
f) $x - 6\sqrt{x} + 5 = 0$

2 Solve each of these equations for x.

a) $x^2 + 1 = \dfrac{6}{x^2}$

b) $x^3 + 7 = \dfrac{8}{x^3}$

c) $x = 12\sqrt{x} - 35$

d) $x^3 - 6x + \dfrac{8}{x} = 0$

e) $\sqrt{x} + \dfrac{10}{\sqrt{x}} = 7$

f) $x^2 + 3 = \dfrac{18}{x^2}$

3 Solve $(x + 3)^2 - 5(x + 3) + 4 = 0$.

4 Solve $(3x - 1)^2 + 6(3x - 1) - 7 = 0$.

5 a) Solve $y^2 - 7y + 10 = 0$.

b) Hence find the solutions to $(x^2 + 1)^2 - 7(x^2 + 1) + 10 = 0$.

6 a) Solve $y^2 - 5y - 14 = 0$.

b) Hence find the solutions to $(x^3 - 1)^2 - 5(x^3 - 1) - 14 = 0$.

7 a) By using the substitution $p = x + \dfrac{1}{x}$, show that the equation

$$2x^4 + x^3 - 6x^2 + x + 2 = 0$$

reduces to $2p^2 + p - 10 = 0$.

b) Hence solve $2x^4 + x^3 - 6x^2 + x + 2 = 0$.

....

1.10 Sketching the graph of a quadratic function

The simplest quadratic function is $y = x^2$.
If y is plotted against x then the result is the graph shown here.

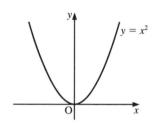

The shape of the graph of a quadratic function is called a parabola.

Important features of the graph to note are:

❖ It is ∪ shaped

❖ It has a line of symmetry, which is the y-axis

❖ It has a vertex at $(0, 0)$.

You can produce variations in the graph of this simplest quadratic function by translations of the graph of $y = x^2$.

If you plot the graph of $y = (x - 1)(x - 3)$ you still get a ∪-shaped graph but the line of symmetry and vertex are different. It is easy to see that when $x = 1$ and $x = 3$, then $y = 0$ and hence the graph intersects the x-axis at $x = 1$ and $x = 3$. The line of symmetry will be $x = 2$: half-way between the x intercepts, because the graph is symmetrical.

To find the lowest point, the y-coordinate of the vertex, find y when $x = 2$, which gives $y = (2 - 1)(2 - 3) = -1$.
This gives the curve shown.

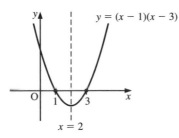

You can check this using your GDC and plotting $y = (x - 1)(x - 3)$ directly.

Example 1

Given that $y = (x + 2)(x - 6)$ write down:

a) the equation of the axis of symmetry

b) the coordinates of the vertex

c) the minimum value of y.

Hence sketch the curve.

....

a) You can see that when $x = -2$ and $x = 6$ the curve intersects the x-axis. The equation of the axis of symmetry is therefore $x = 2$ (half way between -2 and 6).

b) The y-coordinate of the vertex is found by substituting $x = 2$ into the equation of the curve, giving

$$y = (2 + 2)(2 - 6) = -16$$

Therefore the vertex has coordinates $(2, -16)$.

c) The minimum value of y occurs at the vertex and is therefore -16.

The sketch of the curve is shown.

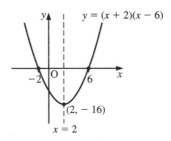

You will usually meet the quadratic function in the more general form $y = ax^2 + bx + c$, and be asked to sketch the graph of the function. It is useful to remember these facts:

❖ If $a > 0$ then the graph is ∪ shaped and the vertex is where the function reaches a minimum value.

❖ If $a < 0$ then the graph is ∩ shaped and the vertex is where the function reaches a maximum value.

❖ If $y = 0$ has roots then these will be the x-coordinates of the points where the graph intersects the x-axis.

If you express the function in the form $a(x + p)^2 + q$, you can deduce details about the graph of $y = ax^2 + bx + c$.

Example 2

Given the quadratic function $y = x^2 - 2x - 15$,
a) use the GDC to plot the graph of $y = x^2 - 2x - 15$
b) express y in the form $(x + p)^2 + q$ and write down:
   i) the equation of the axis of symmetry
   ii) the coordinates of the vertex
   iii) the minimum value of y
c) find the coordinates of the x-intercepts.

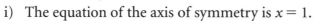

a) Using the GDC gives the graph shown.
b) Completing the square gives
$$x^2 - 2x - 15 = (x - 1)^2 - (-1)^2 - 15$$
$$= (x - 1)^2 - 16$$
   Therefore, $y = (x - 1)^2 - 16$.
   i) The equation of the axis of symmetry is $x = 1$.
   ii) The coordinates of the vertex are $(1, -16)$.
   iii) The graph is ∪ shaped and hence y attains a *minimum* value at the vertex of -16.
c) The x-intercepts occur when $y = 0$.
   So: $x^2 - 2x - 15 = 0$
   Factorising and solving gives
   $$(x + 3)(x - 5) = 0$$
   $$\therefore \ x = -3 \quad \text{or} \quad x = 5$$
Therefore the x-intercepts are $(-3, 0)$ and $(5, 0)$.

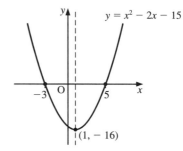

Here, $p = -1$, $q = -16$
Axis of symmetry is
$x = -p = 1$
Vertex is $(-p, q) = (1, -16)$

Example 3

Express $y = x^2 - 2x - 8$ in the form $(x + p)^2 + q$. Hence state:
a) the coordinates of the vertex
b) whether y attains a maximum or minimum value, and state its value
c) the equation of the axis of symmetry.
By solving the equation $x^2 - 2x - 8 = 0$ write down the x-intercepts.
Hence sketch the graph of $y = x^2 - 2x - 8$.

Completing the square gives
$$x^2 - 2x - 8 = (x-1)^2 - (-1)^2 - 8$$
$$= (x-1)^2 - 9$$

Therefore, $y = (x-1)^2 - 9$.

a) The coordinates of the vertex of the graph of $y = x^2 - 2x - 8$ are $(1, -9)$.

b) Since the coefficient of x^2 is positive the graph of y is $\cup$ shaped and hence y attains a minimum value. The minimum value occurs at the vertex and from a) you can see its value is -9.

c) The equation of the axis of symmetry is $x = 1$.

Using factorisation to solve the equation $x^2 - 2x - 8 = 0$ gives
$$x^2 - 2x - 8 = 0$$
$$(x+2)(x-4) = 0$$
$$\therefore \quad x = -2 \text{ or } x = 4$$

Therefore the x-intercepts are $(-2, 0)$ and $(4, 0)$.

Using this information gives the graph of $y = x^2 - 2x - 8$.

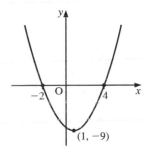

Example 4

Express $y = -x^2 + 10x - 21$ in the form $a(x+p)^2 + q$ and hence:

a) state the coordinates of the vertex

b) determine the maximum or minimum value of y.

Find the x-intercepts on the graph of $y = -x^2 + 10x - 21$ and hence sketch the graph.

....

Completing the square gives
$$-x^2 + 10x - 21 = -(x^2 - 10x + 21)$$
$$= -[(x-5)^2 - (-5)^2 + 21]$$
$$= -[(x-5)^2 - 4]$$
$$= -(x-5)^2 + 4$$

Therefore, $y = -(x-5)^2 + 4$.

a) The coordinates of the vertex are $(5, 4)$.

b) The coefficient of the x^2 term is negative, therefore the graph of y is $\cap$ shaped and y attains a maximum value. This maximum value occurs at the vertex and its value is 4.

The x-intercepts are found by solving the equation $-x^2 + 10x - 21 = 0$. Factorising and solving gives

$$-x^2 + 10x - 21 = 0$$
$$\therefore \quad x^2 - 10x + 21 = 0$$
$$(x - 3)(x - 7) = 0$$
$$\therefore \quad x = 3 \text{ or } x = 7$$

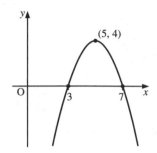

The x-intercepts are $(3, 0)$ and $(7, 0)$.
The graph of $y = -x^2 + 10x - 21$ is shown.

Maxima and minima

Many practical situations give rise to quadratic expressions. Sometimes these involve finding the maximum or minimum value of a quantity. The next example shows how to do this by the method of completing the square.

Example 5

Find the maximum area of a rectangle which has a perimeter of 28 units.

....

Since the perimeter is 28 units, the sum of the length and the width is 14 units.
Let the rectangle have length x.
Then its width is $14 - x$.

The area, A, of the rectangle is given by
$$A = x(14 - x)$$
$$= 14x - x^2$$

Express A in the form $a(x + p)^2 + q$ to determine the maximum value of A:
$$A = -x^2 + 14x = -(x^2 - 14x)$$
$$= -[(x - 7)^2 - 49]$$
$$= -(x - 7)^2 + 49$$

Therefore, $A = -(x - 7)^2 + 49$.

The maximum value of A is 49 units^2, which occurs when $x = 7$ units.

Notice that when $x = 7$, $y = 14 - 7 = 7$. In other words, the rectangle is a square when it attains its maximum area.

Use your GDC to sketch the graph of $y = 14x - x^2$ and confirm that a maximum of 49 takes place when $x = 7$.

Exercise 1K

1 Sketch the graph of each of the following quadratic functions, and label all axis crossings.

a) $y = (x - 3)(x + 3)$ b) $y = (x - 5)(x - 9)$

c) $y = (2x - 3)(x + 2)$ d) $y = x^2 - 4$

e) $y = 25 - x^2$ f) $y = x^2 - 3x - 4$

g) $y = x^2 + 7x + 12$ h) $y = 2x^2 - 3x - 2$

i) $y = -2x^2 - 5x + 7$

2 Use the method of completing the square to find the minimum value of y and the value of x at which it occurs.

a) $y = x^2 + 4x + 6$ b) $y = x^2 - 6x + 13$

c) $y = x^2 - 10x + 40$ d) $y = x^2 + 2x - 5$

c) $y = x^2 + 3x + 8$ f) $y = x^2 - 7x + 15$

3 Use the method of completing the square to sketch the graphs of these quadratics, marking the coordinates of the vertex and the x- and y-intercepts on your diagrams.

a) $y = x^2 + 2x - 5$ b) $y = x^2 + 8x + 3$

c) $y = x^2 + 3x + 5$ d) $y = x^2 + 10x + 23$

e) $y = x^2 - 3x - 6$ f) $y = x^2 - 5x + 4$

4 Use the method of completing the square to find the minimum value of y and the value of x at which it occurs.

a) $y = 2x^2 + 10x - 5$ b) $y = 3x^2 + 6x + 14$

c) $y = 4x^2 + x - 7$ d) $y = 3x^2 - 2x + 9$

e) $y = 6x^2 + x + 5$ f) $y = 5x^2 - 2x + 8$

5 Use the method of completing the square to find the maximum value of y and the value of x at which it occurs.

a) $y = 3 - 2x - x^2$ b) $y = 5 + 4x - x^2$

c) $y = 8 + 2x - x^2$ d) $y = 3 + 2x - 2x^2$

e) $y = 7 - 3x - 4x^2$ f) $y = 6 - 5x - 2x^2$

6 Use the method of completing the square to sketch the graphs of these quadratics.

a) $y = 2x^2 - 4x - 4$ b) $y = 3x^2 + 6x - 15$

c) $y = 8 + 2x - x^2$ d) $y = 4 + 8x - 2x^2$

e) $y = 2x^2 - 4x + 5$ f) $y = 6 - 6x - 3x^2$

7 A farmer has 40 m of fencing with which to enclose a rectangular pen. Given the pen is x m wide,

a) show that its area is $(20x - x^2)$ m²

b) deduce the maximum area that he can enclose.

8 Another farmer also has 40 m of fencing, and he also wishes to enclose a rectangular pen of maximum area, but one side of his pen will consist of part of a wall which is already in place.

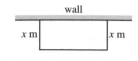

Given that the two sides of his pen touching the wall each have length x m, find an expression, in terms of x, for the area that he can enclose. Deduce that the maximum area is 200 m².

9 A third farmer also has 40 m of fencing but he decides to use a right-angled corner of a building, as in the diagram.

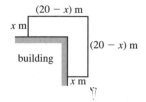

Show that the area which he can enclose is given by the expression $(40x - 3x^2)$ m², and deduce the maximum value of this area.

10 When a stone is projected vertically into the air with an initial speed of 30 m s⁻¹ its height, h metres, above the point of projection, at a time t seconds after the instant of projection, can be approximated by the formula $h = 30t - 5t^2$.

Find the maximum height reached by the stone, and the time at which this occurs.

11 A strip of wire of length 28 cm is cut into two pieces. One piece is bent to form a square of side x cm, and the other piece is bent to form a rectangle of width 3 cm.

a) Show that the lengths of the other two sides of the rectangle are given by $(11 - 2x)$ cm.

b) Deduce that the total combined area of the square and the rectangle is $(x^2 - 6x + 33)$ cm².

c) Prove that the minimum total area which can be enclosed in this way is 24 cm².

12 It is required to fit a rectangle of maximum area inside a triangle, PQR, in which PR = 1 metre, RQ = 2 metres, and $P\hat{R}Q = 90°$. The diagram shows an arbitrary rectangle, RSTU, in which TU = x metres and ST = y metres.

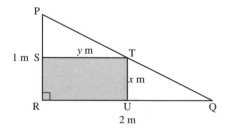

a) Show that $y = 2 - 2x$.

b) Find an expression, in terms of x, for the area of the rectangle, and deduce that the rectangle of maximum area which fits inside triangle PQR has area $\frac{1}{2}$ m².

13 Show that, in general, for any rectangle drawn inside any right-angled triangle, the area of the rectangle cannot exceed half the area of the triangle.

1.11 Algebraic fractions

You can add and subtract algebraic fractions in the same way as you add and subtract numerical fractions. To find the sum or difference of two numerical fractions, each fraction must be expressed in terms of the same denominator, called the **lowest common denominator**. For example, in calculating

$$\tfrac{1}{3} + \tfrac{1}{6}$$

the lowest common denominator is 6, since 3 is a factor of 6. Therefore,

$$\tfrac{1}{3} + \tfrac{1}{6} = \tfrac{2}{6} + \tfrac{1}{6} = \tfrac{3}{6} = \tfrac{1}{2}$$

When the only common factor of the denominators is 1, you can find their product and use the product as the common denominator. For example, in calculating

$$\tfrac{1}{3} + \tfrac{1}{7}$$

the lowest common denominator is $3 \times 7 = 21$, since 3 and 7 have no common factor other than 1. Therefore,

$$\tfrac{1}{3} + \tfrac{1}{7} = \tfrac{7}{21} + \tfrac{3}{21} = \tfrac{10}{21}$$

A numerical fraction whose numerator is greater than or equal to the denominator is called an **improper fraction**.
For example, $\tfrac{7}{5}$ is an improper fraction which can be written as $1\tfrac{2}{5}$.

An algebraic fraction is called improper if the degree of the numerator is greater than or equal to the degree of the denominator. For example,

$$\frac{x^2 + 2x \quad 8}{3x + 7}$$

is an improper fraction because the degree of the numerator is 2 and the degree of the denominator is 1.

> The degree of an algebraic expression is the highest power of x that appears in it.

Example 1

Express $\dfrac{4}{x + 6} - \dfrac{2}{x + 7}$ as a single fraction.

··

The lowest common denominator is $(x + 6)(x + 7)$. Therefore:

$$\frac{4}{x + 6} - \frac{2}{x + 7} \equiv \frac{4(x + 7) - 2(x + 6)}{(x + 6)(x + 7)}$$

$$\equiv \frac{4x + 28 - 2x - 12}{(x + 6)(x + 7)}$$

$$\equiv \frac{2x + 16}{(x + 6)(x + 7)}$$

$$\therefore \frac{4}{x + 6} - \frac{2}{x + 7} \equiv \frac{2(x + 8)}{(x + 6)(x + 7)}$$

> $\equiv$ means 'is identical to'.
> For example, $x + x \equiv 2x$

Example 2

Express $\dfrac{5}{x+3} + \dfrac{2}{x-1}$ as a single fraction.

Hence solve the equation

$$\frac{5}{x+3} + \frac{2}{x-1} = 3$$

....

Expressing the LHS as one single algebraic fraction gives

$$\frac{5(x-1) + 2(x+3)}{(x+3)(x-1)} \equiv \frac{7x+1}{(x+3)(x-1)}$$

$$\therefore \quad \frac{7x+1}{(x+3)(x-1)} = 3$$

Multiply throughout by $(x+3)(x-1)$ to give

$$7x + 1 = 3(x+3)(x-1)$$

$$7x + 1 = 3(x^2 + 2x - 3)$$

$$7x + 1 = 3x^2 + 6x - 9$$

$$\therefore \ 3x^2 - x - 10 = 0$$

Factorising and solving gives

$$3x^2 - x - 10 = 0$$

$$(3x + 5)(x - 2) = 0$$

$$\therefore \ x = -\tfrac{5}{3} \quad \text{or} \quad x = 2$$

Exercise 1L

....

1 Express each of these as a single fraction.

a) $\dfrac{3}{x-1} + \dfrac{2}{x+3}$

b) $\dfrac{2}{x+4} - \dfrac{1}{x-3}$

c) $\dfrac{4}{x-5} - \dfrac{2}{x+4}$

d) $\dfrac{x}{x-3} - \dfrac{2}{x+2}$

e) $\dfrac{2x}{x+2} - \dfrac{1}{x-2}$

f) $\dfrac{5}{2x-3} - \dfrac{3}{2x+5}$

g) $\dfrac{2x+1}{x-2} + \dfrac{3x}{x+4}$

h) $\dfrac{5x}{x^2+3} + \dfrac{6}{2x-5}$

2 Show that

$$\frac{3}{x-2} + \frac{4}{x-1} \equiv \frac{7x-11}{(x-2)(x-1)}$$

Hence solve the equation

$$\frac{3}{x-2} + \frac{4}{x-1} = \frac{7}{x}$$

3 Express $\dfrac{2}{x-3} + \dfrac{1}{2x+1}$ as a single fraction.

Hence solve the equation

$$\frac{2}{x-3} + \frac{1}{2x+1} = \frac{5}{2x-3}$$

4 Given

$$\frac{a}{6x-1} - \frac{1}{3x+1} \equiv \frac{b}{(6x-1)(3x+1)}$$

where a and b are both constants, find the values of a and b.

5 Find the values of the constants A and B for which

$$\frac{A}{x-5} + \frac{B}{x+5} \equiv \frac{2x}{x^2-25}$$

Summary

You should know how to ...

► Solve a linear equation

► Solve a linear inequality

► Solve a pair of simultaneous equations

► Factorise a quadratic expression

► Solve a quadratic equation:
  ▷ by using a GDC
  ▷ by factorising
  ▷ by completing the square
  ▷ by using the formula $x = \dfrac{-b \pm \sqrt{b^2 - 4ac}}{2a}$

► Use the discriminant $\Delta = b^2 - 4ac$

► Manipulate algebraic fractions

Revision exercise 1

1 The quadratic equation $4x^2 + 4kx + 9 = 0$, $k > 0$, has exactly one solution for x. Find the value of k.

© *IBO*[2000]

2 Show that the equation $(k-1)x^2 + 2x - (k-3) = 0$ has real roots for all values of k.

3 Consider the function $f(x) = 2x^2 - 8x + 5$,

a) Express $f(x)$ in the form $a(x-p)^2 + q$, where $a, p, q \in \mathbb{Z}$.

b) Find the minimum value of $f(x)$.

> $\in$ means 'belongs to'
> $\mathbb{Z}$ = the set of integers
> So $x \in \mathbb{Z}$ means 'x is an integer'.

© *IBO*[2002]

4 The diagram shows part of the curve $y = a(x-h)^2 + k$, where $a, h, k \in \mathbb{Z}$.

a) The vertex is at the point $(3, 1)$. Write down the value of h and of k.

b) The point P$(5, 9)$ is on the graph. Show that $a = 2$.

c) Hence show that the equation of the curve can be written as $y = 2x^2 - 12x + 19$.

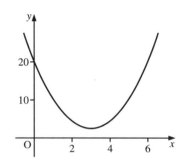

© *IBO*[2003]

5 The diagram shows part of the graph with equation $y = x^2 + px + q$. The graph cuts the x-axis at -2 and 3.

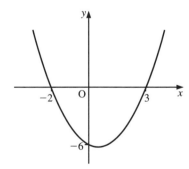

Find the value of

a) p b) q

© *IBO*[2001]

6 The diagram shows the graph with equation $y = \frac{1}{2}x(6 - x)$.

a) Write down

  i) the coordinates of Q

  ii) the equation of the line of symmetry of the graph.

b) Find the coordinates of V, the maximum point.

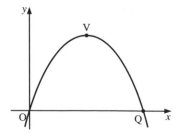

© *IBO*[1998]

7 The curve shown in the diagram has equation $y = 2 + x - x^2$.

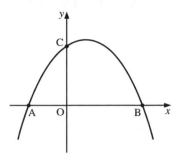

Find the coordinates of the points

a) A   b) B   c) C   © *IBO* [*1997*]

8 The diagram represents the graph of the function
$f: x \mapsto (x - p)(x - q)$.

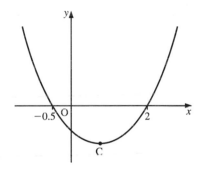

a) Write down the values of p and q.

The function has a minimum value at the point C.

b) Find the x-coordinate of C.   © *IBO* [*1999*]

9 The diagram shows part of the graph of the function
$y = ax^2 + bx + c$.

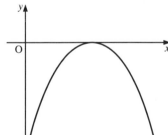

State whether each of the following expressions is positive, negative or zero.

a) a b) c c) $b^2 - 4ac$ d) b © *IBO* [*2000*]

2 Circular functions and trigonometry

3.1 The circle: radian measure of angles; length of an arc; area of a sector.

3.4 The circular functions sin x, cos x and tan x.

3.6 Solution of triangles.

The cosine rule: $c^2 = a^2 + b^2 - 2ab \cos C$.

The sine rule: $\dfrac{a}{\sin A} = \dfrac{b}{\sin B} = \dfrac{c}{\sin C}$.

Area of a triangle as $\dfrac{1}{2} ab \sin C$.

A plane shape bounded by straight lines is called a polygon.

When the shape is bounded by three straight lines, the polygon is called a triangle.

When the shape is bounded by four straight lines, the polygon is called a quadrilateral.

2.1 Triangles

You should know **Pythagoras' theorem**:

> In any right-angled triangle ABC, the square of the hypotenuse is equal to the sum of the squares of each of the other two sides.
>
> $$AB^2 = BC^2 + AC^2$$

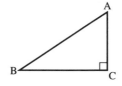

Trigonometric ratios

In the right-angled triangle ABC shown below:

❖ side AB is called the **hypotenuse**

❖ side BC is called the **adjacent**, since it is adjacent to (next to) the angle $A\hat{B}C$ (θ)

❖ side AC is called the **opposite** (relative to the angle $A\hat{B}C$), since it is opposite the angle $A\hat{B}C$.

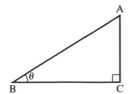

The trigonometric ratios are

$$\sin \theta = \frac{\text{opposite}}{\text{hypotenuse}} \qquad \cos \theta = \frac{\text{adjacent}}{\text{hypotenuse}} \qquad \tan \theta = \frac{\text{opposite}}{\text{adjacent}}$$

'sin' is short for sine;
'cos' is short for cosine;
'tan' is short for tangent.

Example 1

In the triangle ABC, BC = 6 cm, angle $A\hat{B}C = 60°$, and angle $A\hat{C}B = 90°$.
Calculate a) the length AB,
       b) the length AC (to one decimal place).

··

a) The side AB is the hypotenuse and BC is adjacent to the 60° angle, so use the cosine ratio:

$$\cos\theta = \frac{\text{adjacent}}{\text{hypotenuse}}$$

Therefore, $\cos 60° = \dfrac{6}{AB}$

$\therefore AB = \dfrac{6}{0.5} = 12$

The length AB is 12 cm.

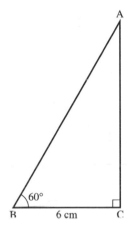

b) To find the length AC, you could use trigonometry again but using Pythagoras' theorem is easier:

$$AB^2 = BC^2 + AC^2$$
$$12^2 = 6^2 + AC^2$$
$$AC^2 = 144 - 36$$
$$\therefore AC = \sqrt{108} = 10.4$$

The length AC is 10.4 cm (to 1 d.p.)

Trigonometric ratios of 45°

Consider the right-angled isosceles triangle ABC.

Using Pythagoras' theorem gives
$$AB^2 = BC^2 + AC^2$$
$$= 1^2 + 1^2$$
$$\therefore AB = \sqrt{2}$$

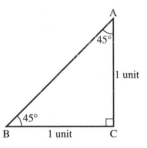

The trigonometric ratios are:

$$\sin 45° = \frac{AC}{AB} = \frac{1}{\sqrt{2}}$$

$$\cos 45° = \frac{BC}{AB} = \frac{1}{\sqrt{2}}$$

$$\tan 45° = \frac{AC}{BC} = 1$$

$$\sin 45° = \frac{1}{\sqrt{2}} \qquad \cos 45° = \frac{1}{\sqrt{2}} \qquad \tan 45° = 1$$

Trigonometric ratios of 30° and 60°

Consider the equilateral triangle ABC with sides of length 2 units.

Let D be the point where the perpendicular from A meets the base BC.

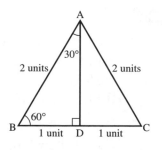

From Pythagoras' theorem:

$$AD^2 = AB^2 - BD^2$$
$$= 2^2 - 1^2$$
$$\therefore AD = \sqrt{3}$$

The trigonometric ratios are:

$$\sin 30° = \frac{BD}{AB} \quad \text{and} \quad \sin 60° = \frac{AD}{AB}$$
$$= \frac{1}{2} \qquad\qquad\qquad = \frac{\sqrt{3}}{2}$$

$$\sin 30° = \frac{1}{2} \text{ and } \sin 60° = \frac{\sqrt{3}}{2}$$

$$\cos 30° = \frac{AD}{AB} \quad \text{and} \quad \cos 60° = \frac{BD}{AB}$$
$$= \frac{\sqrt{3}}{2} \qquad\qquad\qquad = \frac{1}{2}$$

$$\cos 30° = \frac{\sqrt{3}}{2} \text{ and } \cos 60° = \frac{1}{2}$$

$$\tan 30° = \frac{BD}{AD} \quad \text{and} \quad \tan 60° = \frac{AD}{BD}$$
$$= \frac{1}{\sqrt{3}} \qquad\qquad\qquad = \sqrt{3}$$

$$\tan 30° = \frac{1}{\sqrt{3}} \text{ and } \tan 60° = \sqrt{3}$$

The trigonometric ratios of 0°, 30°, 45°, 60° and 90° are summarised in the table.

Ratio	0°	30°	45°	60°	90°
$\sin \theta$	0	$\frac{1}{2}$	$\frac{1}{\sqrt{2}}$	$\frac{\sqrt{3}}{2}$	1
$\cos \theta$	1	$\frac{\sqrt{3}}{2}$	$\frac{1}{\sqrt{2}}$	$\frac{1}{2}$	0
$\tan \theta$	0	$\frac{1}{\sqrt{3}}$	1	$\sqrt{3}$	∞

You should memorise these ratios, as many questions involve these angles.

Example 2

Find the perimeter of triangle ABC shown, expressing your answer in the form $a + b\sqrt{c}$, where a, b and c are integers.

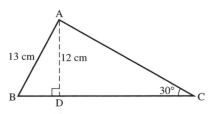

....

In the right-angled triangle ABD:

$$AB^2 = AD^2 + BD^2$$
$$13^2 = 12^2 + BD^2$$
$$BD^2 = 25$$
$$\therefore\ BD = 5$$

You need to find the lengths BC and AC.

BC = BD + CD

In triangle ACD:

$$\tan A\hat{C}D = \frac{AD}{CD}$$
$$\tan 30° = \frac{12}{CD}$$
$$\therefore\ CD = \frac{12}{\left(\frac{1}{\sqrt{3}}\right)} = 12\sqrt{3}$$

Therefore, BC = BD + CD = $(5 + 12\sqrt{3})$ cm.

In triangle ACD:

$$\sin A\hat{C}D = \frac{AD}{AC}$$
$$\sin 30° = \frac{12}{AC}$$
$$\therefore\ AC = \frac{12}{\left(\frac{1}{2}\right)} = 24$$

The length AC is 24 cm.

The perimeter, P, of triangle ABC is given by

$$P = AB + BC + CA$$
$$= 13 + (5 + 12\sqrt{3}) + 24$$
$$= (42 + 12\sqrt{3})$$

The perimeter of the triangle is $(42 + 12\sqrt{3})$ cm.

Area of a triangle

The area of triangle ABC is given by

$$\text{Area} = \tfrac{1}{2} \times b \times h$$

where b is the length of the base and h is the perpendicular height. That is,

$$\text{Area} = \tfrac{1}{2}bh$$

in the usual notation.

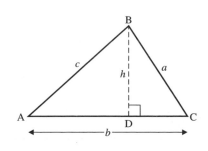

In triangle BDC,

$$\sin C = \frac{BD}{BC}$$

$$\therefore \ \sin C = \frac{h}{a}$$

$$\therefore \ h = a \sin C$$

Substituting $h = a \sin C$ into Area $= \frac{1}{2}bh$ gives

$$\text{Area} = \frac{1}{2}b(a \sin C)$$

$$\therefore \ \text{Area} = \frac{1}{2}ab \sin C$$

where C is the **included angle** between sides a and b.

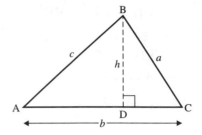

> a, b and c are used to denote the sides of a triangle opposite the angles A, B and C.

The area of a triangle ABC is given by

$$\text{Area} = \frac{1}{2}ab \sin C$$

where C is the **included angle** between sides a and b.

Example 3

Find the area of the triangle ABC, where C = 30°, AC = 10 cm and BC = 14 cm.

....

First sketch triangle ABC, as shown.

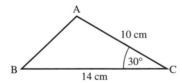

In this case the angle you know is C, and $a = 14$ and $b = 10$.

Using Area $= \frac{1}{2}ab \sin C$ gives

$$\text{Area} = \frac{1}{2} \times 14 \times 10 \times \sin 30°$$

$$= \frac{1}{2} \times 140 \times \frac{1}{2}$$

$$= 35 \text{ cm}^2$$

The area of the triangle is 35 cm².

Example 4

The angle of elevation of the top of a building of height 50 m from a point X on the same level as the foot of the building is 30°.

Calculate the angle of elevation to the nearest degree of the top of the building from another point Y, 15 m nearer the foot of the building.

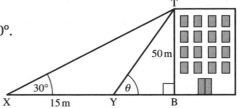

In $\triangle$ XBT, $\tan 30° = \dfrac{50}{XB}$

$\therefore$ $XB = \dfrac{50}{\left(\dfrac{1}{\sqrt{3}}\right)} = 50\sqrt{3}$

$XB = 86.6$

The angle of elevation from Y is 35°.

In $\triangle$ YBT, $\tan \theta = \dfrac{50}{YB}$

$\therefore$ $\tan \theta = \dfrac{50}{(86.6 - 15)}$

$\tan \theta = 35°$, to nearest degree

Example 5

A van moves from a point X on a bearing of 025° for 4 km. The van then changes direction and travels on a bearing of 115° for 7 km. Calculate:

a) the bearing of the van from X,
b) the direct distance of the van from X, to the nearest km.

Sketching the course of the van gives the diagram shown.

Notice that $X\hat{Y}P = 25°$ by standard geometry and $P\hat{Y}Z = 180° - 115° = 65°$.
Therefore $X\hat{Y}Z = 25° + 65° = 90°$, and $\triangle XYZ$ is right-angled.

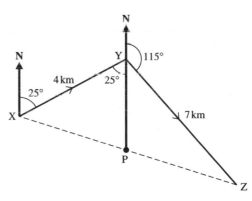

a) The bearing of the van from X is given by $25° + Y\hat{X}Z$.

Now $\tan Y\hat{X}Z = \dfrac{7}{4}$

$\therefore$ $Y\hat{X}Z = 60.3°$

Therefore the bearing of the van from X is 085.3° or 085° to the nearest degree.

b) The direct distance is given by XZ:

$XZ^2 = 7^2 + 4^2$

$XZ^2 = 49 + 16$

$XZ = \sqrt{65} = 8.062$

The direct distance is 8 km, to the nearest km.

Exercise 2A

State your answers **correct to three significant figures** where appropriate.

1 Find the perimeter of triangle ABC, expressing your answer in the form $a + b\sqrt{2} + c\sqrt{3}$, where a, b and c are integers.

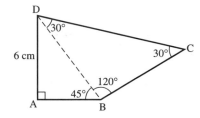

2 Find the perimeter of the quadrilateral ABCD, expressing your answer in the form $a + b\sqrt{2} + c\sqrt{6}$, where a, b and c are integers.

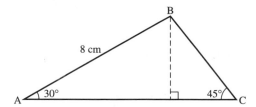

3 A surveyor is attempting to calculate the height of a point, P, on a building, by taking measurements on horizontal, level ground. From a point A he records the angle of elevation of P as 30°. He then advances 20 m to a point B, from which he records the angle of elevation of P as 45°.

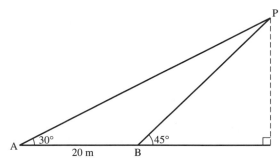

Calculate the height of P above the ground.

4 A man is attempting to calculate the height of a kite, K, which is flying above horizontal ground. From a point A he records the angle of elevation of K as 23°. He then advances 80 m to a point B from which he records the angle of elevation of K as 34°.

Calculate the height of the kite above the ground.

5 A ship sails 100 km from a port X on a bearing of 055° and then 150 km on a bearing of 145°. It reaches its destination at port Y. Find:

a) the bearing of port Y from port X

b) the distance between port X and port Y.

6 Find the area of each of the following triangles.

a)

6 cm, 4 cm

b)

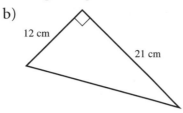
12 cm, 21 cm

c)

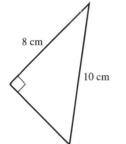
8 cm, 10 cm

d)

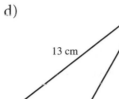
13 cm, 5 cm

e)

7 cm, 40°, 9 cm

f)

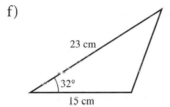
23 cm, 32°, 15 cm

g)

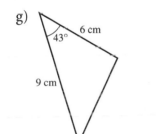
6 cm, 43°, 9 cm

h)

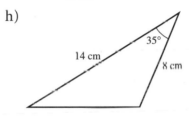
14 cm, 35°, 8 cm

i)

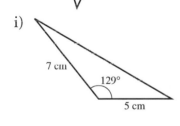
7 cm, 129°, 5 cm

j)

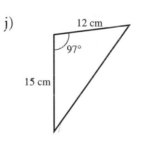
12 cm, 97°, 15 cm

7 In the triangle ABC, AB = 9.2 cm, BC = 7.8 cm and
A$\hat{\text{B}}$C = 48°. Calculate the area of the triangle.

8 In the triangle PQR, PQ = 5.9 cm, QR = 7.6 cm and
P$\hat{\text{Q}}$R = 142°. Calculate the area of the triangle.

9 In the triangle ABC, AB = 6.2 cm, BC = 8.7 cm and
C$\hat{\text{A}}$B = 93°.

a) Calculate the size of the angle A$\hat{\text{B}}$C.

b) Hence calculate the area of the triangle ABC.

10 A pennant in the shape of the letter C is formed by cutting a rhombus from a semicircular plate, as in the diagram. Calculate the area of the pennant.

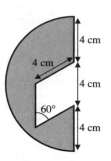

11 The diagram shows the component for a model, which is to be made from a square of metal of side *b* units, by removing a circular hole of radius *a* units. Given that the area of the circle removed is to be half the area of the square, show that

$$\frac{b}{a} = \sqrt{2\pi}$$

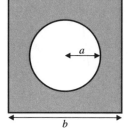

12 The diagram shows a regular pentagon ABCDE with centre O and sides of length 6 cm.

a) Explain why $A\hat{O}B = 72°$.

b) Calculate the area of the triangle AOB.

c) Hence find the area of the pentagon ABCDE.

13 Calculate the area of a regular hexagon with sides of length 8 cm.

14 Calculate the area of a regular octagon with sides of length 9 cm.

15 In the triangle ABC, AB = 4.8 cm, BC = 3.8 cm and $A\hat{B}C = 42°$. P is the foot of the perpendicular from A to BC.

a) Calculate the area of triangle ABC.

b) Deduce the length of AP.

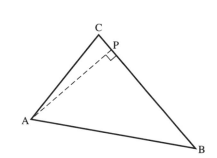

16 PQR is a triangle in which PQ = 5.9 cm, QR = 6.4 cm and $P\hat{Q}R = 63°$. Calculate the area of triangle PQR and the length of the perpendicular from P to QR.

17 In triangle KLM, KL = 9.3 cm, LM = 7.2 cm and $K\hat{L}M = 82°$. Calculate the area of triangle KLM and length of the perpendicular from K to ML.

2.2   Radian measure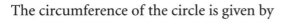

Consider an arc of length 1 unit of a circle of radius 1 unit.

> The angle θ subtended at the centre of the circle by the arc of length 1 unit is called 1 **radian**, written as 1 rad.

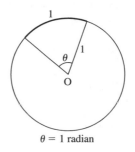

$\theta = 1$ radian

The circumference of the circle is given by

$$C = 2\pi r$$

Substituting $r = 1$ gives

$$C = 2\pi(1)$$
$$\therefore\ C = 2\pi$$

An arc of length 1 subtends an angle of 1 radian, so the circumference of length 2π subtends an angle of 2π radians. That is, there are 2π radians at the centre O of the circle. In other words, 2π radians are equivalent to 360°.

Since 2π radians = 360°,

$$\pi \text{ radians} = 180° \quad \text{and} \quad \frac{\pi}{2} \text{ radians} = 90°$$

> 1 radian $\approx$ 57.3°

❖ To convert degrees to radians, multiply by $\dfrac{\pi}{180}$.

❖ To convert radians to degrees, multiply by $\dfrac{180}{\pi}$.

Example 1

Express each of these angles in radians.

a) 45°   b) 60°   c) 270°

....

a) $45° = 45 \times \dfrac{\pi}{180} = \dfrac{\pi}{4} \text{ rad}$ b) $60° = 60 \times \dfrac{\pi}{180} = \dfrac{\pi}{3} \text{ rad}$

c) $270° = 270 \times \dfrac{\pi}{180} = \dfrac{3\pi}{2} \text{ rad}$

Example 2

Express each of these angles in degrees.

a) $\dfrac{\pi}{6}$ radians   b) $\dfrac{5\pi}{6}$ radians   c) $\dfrac{4\pi}{3}$ radians

....

a) $\dfrac{\pi}{6} \text{ rad} = \dfrac{\pi}{6} \times \dfrac{180}{\pi} = 30°$ b) $\dfrac{5\pi}{6} \text{ rad} = \dfrac{5\pi}{6} \times \dfrac{180}{\pi} = 150°$

c) $\dfrac{4\pi}{3} \text{ rad} = \dfrac{4\pi}{3} \times \dfrac{180}{\pi} = 240°$

Sectors and segments

Consider the sector of a circle, of radius r, which subtends an angle of θ at the centre.

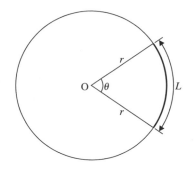

The length, L, of the arc is given by

$$L = \frac{\theta}{360} \times 2\pi r = \frac{\theta \pi r}{180}$$

where θ is measured in degrees.

L is also given by

$$L = \frac{\theta}{2\pi} \times 2\pi r = \theta r$$

where θ is measured in radians.

> The length, L, of an arc that subtends an angle θ at the centre of a circle is given by
>
> $$L = \frac{\theta \pi r}{180} \ (\theta \text{ in degrees}) \quad \text{or} \quad r\theta \ (\theta \text{ in radians})$$

The area, A, of the sector is given by

$$A = \frac{\theta}{360} \times \pi r^2 = \frac{\theta \pi r^2}{360}$$

where θ is measured in degrees.

A is also given by

$$A = \frac{\theta}{2\pi} \times \pi r^2 = \frac{\theta r^2}{2}$$

where θ is measured in radians.

> The area, A, of a sector of a circle with angle θ is given by
>
> $$A = \frac{\theta \pi r^2}{360} \ (\theta \text{ in degrees}) \quad \text{or} \quad \tfrac{1}{2}r^2\theta \ (\theta \text{ in radians})$$

Example 3

The sector of a circle of radius 3 cm subtends an angle of $\frac{5\pi}{18}$ radians at the centre. Find

a) the length of the arc of the sector

b) the area of the sector of the circle.

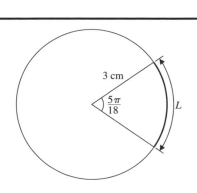

a) The length, L, of the arc is given by

$$L = \theta r$$
$$= \left(\frac{5\pi}{18}\right)(3) = \frac{5\pi}{6} = 2.6 \text{ (to 1 d.p.)}$$

The length of the arc is 2.6 cm (to 1 d.p.).

b) The area, A, of the sector is given by

$$A = \frac{\theta r^2}{2}$$
$$= \frac{\left(\frac{5\pi}{18}\right)(3)^2}{2} = \frac{5\pi}{4} = 3.9 \text{ (to 1 d.p.)}$$

The area of the sector is 3.9 cm² (to 1 d.p.).

Example 4

The shaded area in the diagram is a segment of a circle of radius r. Show that the area of the segment is given by

$$\frac{r^2(\pi - 3)}{12}$$

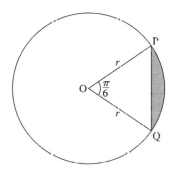

· ·

The area, A_s, of the sector of the circle, is given by

$$A_s = \frac{\theta r^2}{2} = \frac{\left(\frac{\pi}{6}\right)r^2}{2}$$

$$\therefore A_s = \frac{\pi r^2}{12}$$

The area of triangle OPQ, A_t, is given by

$$A_t = \frac{1}{2}ab \sin C = \frac{1}{2}(r)(r) \sin\left(\frac{\pi}{6}\right)$$

$$\therefore A_t = \frac{r^2}{4}$$

$$\boxed{\sin \frac{\pi}{6} = \sin 30° = \frac{1}{2}}$$

The area, A, of the shaded segment, is given by

$$A = A_s - A_t = \frac{\pi r^2}{12} - \frac{3r^2}{12}$$

$$\therefore A = \frac{r^2(\pi - 3)}{12}$$

as required.

Exercise 2B

1 Express each of these angles in radians, giving your answers in terms of π.

a) 30° b) 90° c) 120° d) 10°

e) 80° f) 300° g) 36° h) 240°

i) 72° j) 360° k) 342° l) 1°

2 Express each of these angles in degrees.

a) π rad b) $\dfrac{\pi}{4}$ rad c) 3π rad d) $\dfrac{\pi}{6}$ rad

e) $\dfrac{4\pi}{5}$ rad f) $\dfrac{\pi}{12}$ rad g) $\dfrac{5\pi}{3}$ rad h) π rad

i) $\dfrac{5\pi}{12}$ rad j) $\dfrac{\pi}{90}$ rad k) $\dfrac{3\pi}{2}$ rad l) $\dfrac{7\pi}{6}$ rad

3 Express each of these angles in degrees correct to 1 decimal place.

a) 4 rad b) 0.2 rad c) 4.3 rad d) 0.5 rad

e) 0.7 rad f) 3 rad g) 5.2 rad h) 2.1 rad

i) 5 rad j) 0.04 rad k) 16 rad l) 1 rad

4 A sector of a circle of radius 5 cm subtends an angle of $\dfrac{3\pi}{10}$ rad at the centre. Calculate

a) the length of the arc of the sector

b) the area of the sector.

5 A circle of radius 9 cm is divided into three equal sectors. Calculate

a) the length of the arc of each sector

b) the area of each sector.

6 A sector of angle $\dfrac{5\pi}{12}$ rad is cut from a circle of radius 6 cm. Calculate

a) the perimeter of the sector

b) the area of the sector.

7 OAB is a sector of a circle, centre O, and is such that OA = OB = 7 cm and $A\hat{O}B = \dfrac{5\pi}{14}$ rad. Calculate

a) the perimeter of the sector OAB

b) the area of the sector OAB.

8 The sector of a circle of radius 8 cm subtends an angle of 30° at the centre. Calculate

a) the length of the arc of the sector

b) the area of the sector.

9 OPQ is a sector of a circle, centre O, and is such that
OP = OQ = 12 cm and PÔQ = 45°. Calculate

a) the perimeter of the sector POQ

b) the area of the sector POQ.

10 Calculate the area of a segment of angle $\frac{\pi}{2}$ rad cut from a
circle of radius 5 cm.

11 Calculate the area of a segment of angle $\frac{\pi}{3}$ rad cut from a
circle of radius 10 cm.

12 OMN is a sector of a circle, centre O, and is such that
OM = ON = 12 cm, and MÔN = $\frac{\pi}{6}$ rad. S is the segment
bounded by the chord MN and the arc MN. Calculate

a) the area of S   b) the perimeter of S.

13 OAB is a sector of a circle, centre O, and is such that
OA = OB = 8 cm, and AÔB = $\frac{\pi}{5}$ rad. S is the segment
bounded by the chord AB and the arc AB. Calculate

a) the area of S   b) the perimeter of S.

14 The diagram shows a pennant ABC, which has a triangular
hole in the middle. The hole is an equilateral triangle ABC of
side 8 cm. AB, BC and CA are circular arcs with centres at C,
A and B respectively. Calculate

a) the area of the triangle ABC

b) the area of the sector ABC

c) the area of the shaded region.

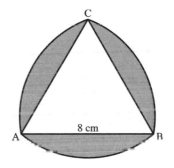

2.3 The functions sin θ, cos θ and tan θ

The sine function

Section 2.1 described the sine function for positive values of θ less
than 90°. However, it is defined for values of θ outside this range.
For example, the calculator will give a value for sin 270°, or in
radian mode, a value for sin $\frac{3\pi}{2}$.

If you plot values of sin θ against θ you get the graphs shown.

> There is no restriction on the
> size of angle θ but if you use
> −360° ≤ θ ≤ 360° you can see
> the main features of the graphs.

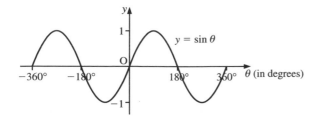

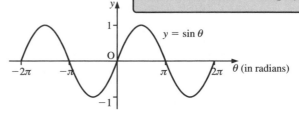

❖ The function $f(\theta) = \sin\theta$ is periodic, of period $360°$ or 2π rad. That is

$$\sin(\theta + 360°) = \sin\theta \quad \text{or} \quad \sin(\theta + 2\pi) = \sin\theta$$

❖ The function $f(\theta) = \sin\theta$ has rotational symmetry about the origin of order 2.

❖ The maximum value of $f(\theta)$ is 1 and its minimum value is -1. In other words $-1 \le f(\theta) \le 1$.

From the graph of $y = \sin\theta$ you can see that there is an infinite number of values of θ for any value of $\sin\theta$ between -1 and 1. For example, $\sin 150° = \sin 30°$.

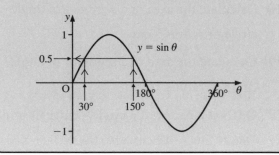

The cosine function

If you plot values of $\cos\theta$ against θ you get the graphs shown.

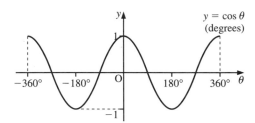 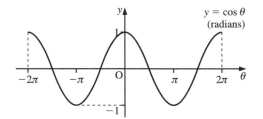

❖ The function $f(\theta) = \cos\theta$ is periodic, of period $360°$ or 2π rad. That is

$$\cos(\theta + 360°) = \cos\theta \quad \text{or} \quad \cos(\theta + 2\pi) = \cos\theta$$

❖ The graph of $f(\theta) = \cos\theta$ is symmetrical about the y-axis.

❖ The maximum value of $f(\theta)$ is 1 and its minimum value is -1. In other words $-1 \le \cos\theta \le 1$.

Try plotting accurate graphs of $\sin\theta$, $\cos\theta$ and $\tan\theta$ for $-360° \le \theta \le 360°$ using a GDC.

The tangent function

If you plot values of $\tan\theta$ against θ you get the graphs shown.

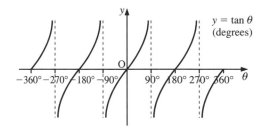 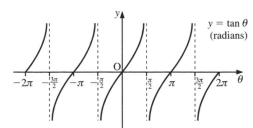

❖ The function $f(\theta) = \tan\theta$ is periodic, of period $180°$ or π rad. That is

$$\tan(\theta + 180°) = \tan\theta \quad \text{or} \quad \tan(\theta + \pi) = \tan\theta$$

❖ The graph of $f(\theta) = \tan\theta$ has rotational symmetry about the origin of order 2.

❖ The function $f(\theta) = \tan\theta$ is not defined when $\theta = \pm90°$, $\pm270°$, …, or in radians, $\theta = \pm\dfrac{\pi}{2}, \pm\dfrac{3\pi}{2}, \ldots$

$\tan\theta \to \infty$ at $90°$, $270°$, …
The lines $x = 90°$, $x = 270°$, … are **asymptotes** to the curve: the curve gets nearer and nearer to these lines but never actually reaches them. $y = \tan\theta$ has no maximum or minimum value.

2.4   Sine and cosine rules

The sine rule

In any triangle ABC,

$$\frac{a}{\sin A} = \frac{b}{\sin B} = \frac{c}{\sin C}$$

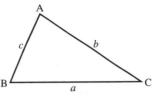

a is the side opposite angle A.

To prove this rule you need to consider two cases:

i) when triangle ABC is an acute-angled triangle

ii) when triangle ABC is not an acute-angled triangle.

i) Triangle ABC is acute-angled

Let the perpendicular from A meet BC at D. In triangle ABD:

$$\sin B = \frac{AD}{c}$$

$$\therefore\ AD = c \sin B \qquad [1]$$

In triangle ADC:

$$\sin C = \frac{AD}{b}$$

$$\therefore\ AD = b \sin C \qquad [2]$$

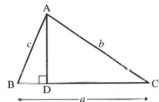

Eliminating AD from [1] and [2] gives

$$c \sin B = b \sin C$$

$$\therefore\ \frac{b}{\sin B} = \frac{c}{\sin C} \qquad [3]$$

Similarly, dropping a perpendicular from B to meet AC gives

$$\frac{c}{\sin C} = \frac{a}{\sin A} \qquad [4]$$

Therefore, from [3] and [4],

$$\frac{a}{\sin A} = \frac{b}{\sin B} = \frac{c}{\sin C}$$

ii) Triangle ABC is not acute-angled

Let the perpendicular from A meet CB extended at D.
In triangle ABD:

$$\sin A\hat{B}D = \frac{AD}{c}$$

$$\therefore\ AD = c \sin A\hat{B}D$$

Now $A\hat{B}D = (180° - B)$ and $\sin(180° - B) = \sin B$. So:

$$AD = c \sin B \qquad [5]$$

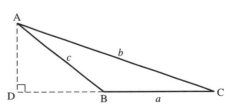

$A\hat{B}D + A\hat{B}C = 180°$ (angles on a straight line)

> Look back at the sine graph on page 63 to see that
> $\sin(180° - B) = \sin B$.

In triangle ACD:

$$\sin C = \frac{AD}{b}$$

$$\therefore \; AD = b \sin C \qquad\qquad [6]$$

Eliminating AD from [5] and [6] gives

$$c \sin B = b \sin C$$

$$\therefore \; \frac{c}{\sin C} = \frac{b}{\sin B} \qquad\qquad [7]$$

If you drop a perpendicular from B to meet AC at E and proceed as in case i), you get:

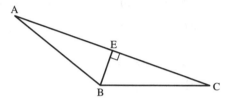

$$\frac{a}{\sin A} = \frac{c}{\sin C} \qquad\qquad [8]$$

Therefore, from [7] and [8],

$$\frac{a}{\sin A} = \frac{b}{\sin B} = \frac{c}{\sin C}$$

as required.

Example 1

In triangle ABC, A = 40°, B = 75° and AB = 6 cm.

Calculate   a) the length AC

                  b) the length BC.

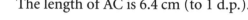

a) C= 180° − (40° + 75°) = 65°.

Therefore, applying the sine rule gives

$$\frac{AC}{\sin 75^\circ} = \frac{6}{\sin 65^\circ}$$

$$\therefore \; AC = \frac{6 \sin 75^\circ}{\sin 65^\circ} = 6.4$$

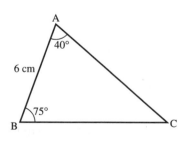

The length of AC is 6.4 cm (to 1 d.p.).

It is helpful to draw a diagram.

b) Applying the sine rule again gives

$$\frac{BC}{\sin 40^\circ} = \frac{6}{\sin 65^\circ}$$

$$\therefore \; BC = \frac{6 \sin 40^\circ}{\sin 65^\circ} = 4.3$$

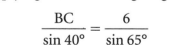

The length of BC is 4.3 cm (to 1 d.p.).

Example 2

In triangle ABC, angle $A\hat{B}C = \dfrac{\pi}{6}$, angle $A\hat{C}B = \dfrac{\pi}{4}$ and

BC = 10 cm. Calculate

a) angle $B\hat{A}C$ b) length AC   c) the area of triangle ABC.

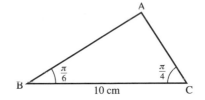

a) Now $B\hat{A}C = \pi - \left(\dfrac{\pi}{6} + \dfrac{\pi}{4}\right) = \dfrac{7\pi}{12}$

Angle $B\hat{A}C$ is $\dfrac{7\pi}{12}$ rad.

b) By the sine rule:

$$\frac{AC}{\sin B} = \frac{BC}{\sin A}$$

$$\frac{AC}{\sin\left(\dfrac{\pi}{6}\right)} = \frac{10}{\sin\left(\dfrac{7\pi}{12}\right)}$$

$$\therefore \ AC = \frac{10\sin\left(\dfrac{\pi}{6}\right)}{\sin\left(\dfrac{7\pi}{12}\right)} = 5.2$$

The length of AC is 5.2 cm (to 1 d.p.).

c) The area of triangle ABC is given by

$$\text{area} = \tfrac{1}{2}ab\sin C$$

$$= \tfrac{1}{2} \times 10 \times 5.2 \times \sin\left(\dfrac{\pi}{4}\right)$$

$$= 18.4 \text{ cm}^2$$

The area of the triangle ABC is 18.4 cm² (to 1 d.p.).

The ambiguous case

Suppose you are given triangle ABC such that AC = 7 cm, BC = 12 cm and B = 30°. Constructing this triangle with ruler and compasses gives this diagram.

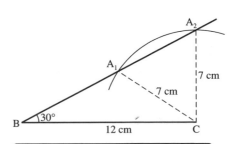

Since triangle A_1CA_2 is isosceles, $A_1\hat{A}_2C = A_2\hat{A}_1C$.
Angles $B\hat{A}_1C$ and $A_2\hat{A}_1C$ are supplementary (their sum is 180°).
In other words, there are two possible positions for vertex A, namely A_1 and A_2.

This is known as the **ambiguous case**. This situation arises when you are given two sides and a non-included angle.

> A non-included angle is one that is not the angle between the two given sides.

Example 3

In triangle ABC, AB = 8 cm, BC = 10 cm and angle
A$\hat{C}$B = 42°. Calculate the length of AC.

....

The sketch on the right shows the two possible triangles
which can be drawn from this information.

Applying the sine rule gives

$$\frac{10}{\sin A} = \frac{8}{\sin 42°}$$

$$\sin A = \frac{10 \sin 42°}{8}$$

$$\therefore \ A = 56.8°$$

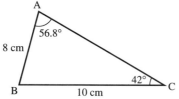

Therefore, B$\hat{A}'$C = 180° − 56.8° = 123.2°.

The two possible cases are:

i)

ii)

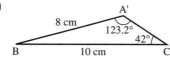

In case i), A$\hat{B}$C = 180° − (42° + 56.8°) = 81.2°

Applying the sine rule to triangle ABC gives

$$\frac{AC}{\sin 81.2°} = \frac{8}{\sin 42°}$$

$$\therefore \ AC = \frac{8 \sin 81.2°}{\sin 42°} = 11.8$$

In case ii), A'$\hat{B}$C = 180° − (42° + 123.2°) = 14.8°

Applying the sine rule to triangle A'BC gives

$$\frac{A'C}{\sin 14.8°} = \frac{8}{\sin 42°}$$

$$\therefore \ A'C = \frac{8 \sin 14.8°}{\sin 42°} = 3.1$$

The two possible lengths of AC are 11.8 cm and 3.1 cm.

Exercise 2C

....

1 Use the sine rule to find each of the unknown labelled sides or
angles. In any ambiguous cases, give both alternatives. State
your answers **correct to three significant figures**.

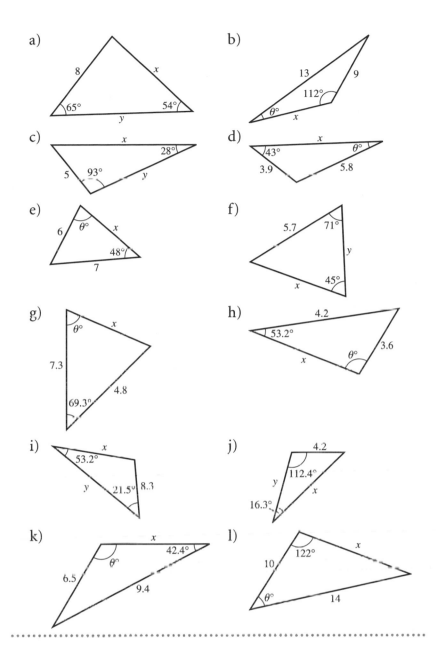

The cosine rule

In any triangle ABC,

$$c^2 = a^2 + b^2 - 2ab\cos C$$

and similarly

$$a^2 = b^2 + c^2 - 2bc\cos A$$

and $b^2 = a^2 + c^2 - 2ac\cos B$

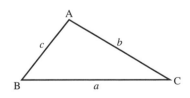

To prove this rule, you need to consider two cases:

i) when triangle ABC is an acute-angled triangle

ii) when triangle ABC is not an acute-angled triangle.

i) Triangle ABC is acute-angled

Let the perpendicular from A meet BC at D.

Let $DC = x$, then $BD = a - x$.

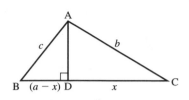

In triangle ABD,

$$AD^2 = c^2 - (a - x)^2$$
$$= c^2 - a^2 + 2ax - x^2 \qquad [1]$$

In triangle ADC,

$$AD^2 = b^2 - x^2 \qquad [2]$$

Eliminating AD^2 from [1] and [2] gives

$$b^2 - x^2 = c^2 - a^2 + 2ax - x^2$$
$$b^2 = c^2 - a^2 + 2ax$$
$$\therefore \ c^2 = a^2 + b^2 - 2ax \qquad [3]$$

In triangle ADC,

$$\cos C = \frac{x}{b}$$
$$\therefore \ x = b \cos C$$

Substituting $x = b \cos C$ into [3] gives

$$c^2 = a^2 + b^2 - 2a(b \cos C)$$
$$\therefore \ c^2 = a^2 + b^2 - 2ab \cos C$$

as required.

ii) Triangle ABC is not acute-angled

Let the perpendicular from A meet CB extended at D.
Let $DB = x$.

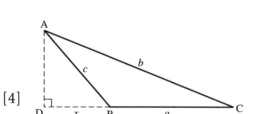

In triangle ADB,

$$AD^2 = c^2 - x^2 \qquad [4]$$

In triangle ADC,

$$AD^2 = b^2 - (a + x)^2$$
$$= b^2 - a^2 - 2ax - x^2 \qquad [5]$$

Eliminating AD^2 from [4] and [5] gives

$$c^2 - x^2 = b^2 - a^2 - 2ax - x^2$$
$$c^2 = b^2 - a^2 - 2ax$$
$$\therefore \ b^2 = a^2 + c^2 + 2ax \qquad [6]$$

In triangle ADB,

$$\cos A\hat{B}D = \frac{x}{c}$$
$$\therefore \ x = c \cos A\hat{B}D$$

Now $A\hat{B}D = 180° - B$ and since $\cos(180° - B) = -c\cos B$,
$x = -c\cos B$.
Substituting $x = -c\cos B$ into [6] gives

$$b^2 = a^2 + c^2 + 2a(-c\cos B)$$
$$\therefore b^2 = a^2 + c^2 - 2ac\cos B$$

as required.

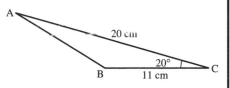

Look back at the cosine graph
on page 64 to see
$\cos(180° - B) = -\cos B$.

Example 4

In triangle ABC, AC = 20 cm, BC = 11 cm and
angle $A\hat{C}B = 20°$.
Calculate   a) the length AB   b) angle $A\hat{B}C$.

a) Apply the cosine rule to triangle ABC:
$$AB^2 = 11^2 + 20^2 - 2 \times 11 \times 20 \cos 20°$$
$$= 121 + 400 - 413.46 = 107.54$$
$$\therefore AB = \sqrt{107.54} = 10.37$$

The length AB is 10.37 cm.

b) To find angle $A\hat{B}C$, rearrange the cosine formula for cos B.
Now
$$b^2 = a^2 + c^2 - 2ac\cos B$$
$$\therefore \cos B = \frac{a^2 + c^2 - b^2}{2ac}$$

Applying this formula to triangle ABC gives
$$\cos B = \frac{11^2 + 10.37^2 - 20^2}{2(11)(10.37)} = -0.7516$$
$$\therefore B = 138.7°$$

Exercise 2D

1 Use the cosine rule to find each of the unknown labelled sides
or angles. Give your answers to 3 s.f.

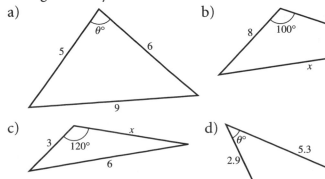

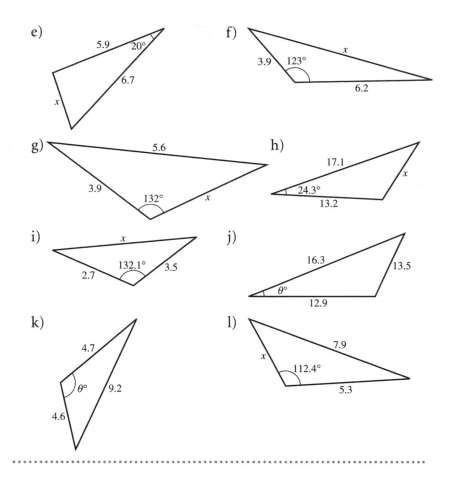

e)

5.9 20°

6.7

x

f)

x

3.9 123°

6.2

g)

5.6

3.9

132° x

h)

17.1

24.3° x

13.2

i)

x

132.1° 3.5

2.7

j)

16.3 13.5

θ°

12.9

k)

4.7

θ° 9.2

4.6

l)

7.9

x

112.4°

5.3

Bearings

You can use the sine and cosine rules to solve problems involving bearings.

Example 5

A ship sails 6 km from S to T on a bearing of 063° and then 9 km from T to U on a bearing of 148°. Calculate

a) the distance SU

b) the bearing of U from S.

....

a) The bearing of U from T is 148°, so the angle β between TU and the South line is $180° - 148° = 32°$. Also, by the properties of parallel lines, $\alpha = 63°$.

So, $S\hat{T}U = 63° + 32° = 95°$.

Apply the cosine rule to triangle STU:

$$SU^2 = 6^2 + 9^2 - 2 \times 6 \times 9 \cos 95°$$
$$= 36 + 81 + 9.41 = 126.41$$
$$\therefore \ SU = \sqrt{126.41} = 11.24$$

Hence, the distance SU is 11.24 km.

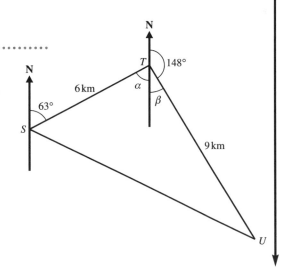

b) Applying the sine rule to triangle STU:

$$\frac{\sin T\hat{S}U}{9} = \frac{\sin 95°}{11.24}$$

$$\sin T\hat{S}U = \frac{9 \sin 95°}{11.24}$$

$$\therefore \ T\hat{S}U = 52.9°$$

Hence, the bearing of U from S is $63° + 52.9° = 115.9°$.

Exercise 2E

1 In this question the angles are given in **radians**. Use the sine rule or cosine rule to find the unknown marked sides or angles.

a)
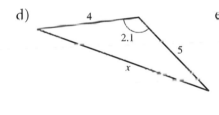

b)
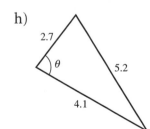

c)
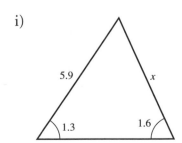

d)
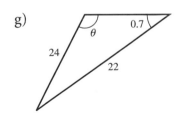

e)
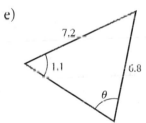

f)
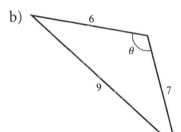

g)

h)

i)

j)
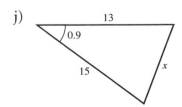

k)
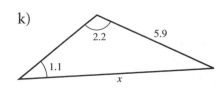

l)
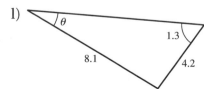

2 In triangle ABC, AB = 6 cm, BC = 8 cm and angle B = 0.93 rad.
   a) Find the length of AC.
   b) Find the size of angle C.

3 In triangle PQR, PQ = 12 cm, QR = 14 cm and PR = 11 cm.
   a) Find the size, in radians, of angle P.
   b) Find the size, in radians, of angle Q.

4 In triangle LMN, LM = 5 cm, MN = 8 cm and
   angle N = 20°. Given that angle L is obtuse,
   a) calculate the size of angle L
   b) calculate the length of LN.

5 Given that PQR is a triangle in which PQ = 5 cm, QR = 8 cm
   and angle R = 30°, calculate the two possible values of the
   length of PR.

6 A ship leaves a harbour, H, and sails for 32 km on a bearing of
   025° to a point X. At X it changes course and then sails for
   45 km on a bearing of 280° to a port P.
   a) Sketch a diagram showing H, X and P.
   b) Calculate the direct distance from H to P.
   c) Calculate the bearing of P from H.

7 A bird leaves a nest, N, and flies 800 m on a bearing of 132° to
   a tree, T. It then leaves T and flies 650 m on a bearing of 209° to
   a pylon P. Assuming that N, T and P are at the same height
   above the ground, calculate the distance and bearing on which
   the bird must fly in order to return directly from P to N.

8 An army cadet is involved in a compass exercise. He leaves a
   point O and walks 50 m due west to a point A. He then walks
   80 m due north to a point B, and finally 60 m, on a bearing of
   320°, to a point C.
   a) Illustrate this information on a sketch.
   b) Calculate the distance and bearing of B from O.
   c) Calculate the distance and bearing on which he must walk
      in order to return from C to O.

9 A ship travelling south-west with constant speed observes the
   flash of a lighthouse on a bearing of 240°. 8 km further on the
   ship observes the flash of the same lighthouse, due west.
   a) How far is the ship from the lighthouse at this time?
   b) How close to the lighthouse will it pass?

Summary

You should know how to ...

▶ Solve problems with right-angled triangles:

▷ using Pythagoras' theorem

$$a^2 + b^2 = c^2$$

▷ using trigonometry

$$\sin x = \frac{b}{c}, \cos x = \frac{a}{c}, \tan x = \frac{b}{a}$$

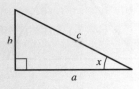

▶ Solve problems with triangles that do not contain a right angle:

▷ Using the formula

$$\text{Area} = \frac{1}{2} ab \sin C$$

▷ using the sine rule

$$\frac{a}{\sin A} = \frac{b}{\sin B} = \frac{c}{\sin C}$$

▷ using the cosine rule

$$c^2 = a^2 + b^2 - 2ab \cos C$$

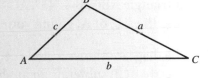

▶ Use radian measure.

π radians = 180°

Length of an arc, $l = r\theta$

Area of a sector, $A = \frac{1}{2} r^2 \theta$

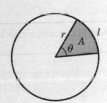

Revision exercise 2

1 Two boats A and B start moving from the same point P. Boat A moves in a straight line at 20 km h^{-1} and boat B moves in a straight line at 32 km h^{-1}. The angle between their paths is 70°.

Find the distance between the boats after 2.5 hours.

© IBO [2002]

2 The diagram shows a triangle with sides 5 cm, 7 cm and 8 cm.
Find

a) the size of the smallest angle, in degrees

b) the area of the triangle.

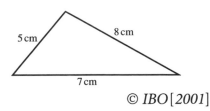

© IBO [2001]

3 The diagrams show two triangles both satisfying the conditions
AB = 20 cm, AC = 17 cm, angle $A\hat{B}C$ = 50°.

Triangle 1   Triangle 2

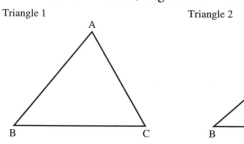

a) Calculate the size of angle $A\hat{C}B$ in triangle 2.

b) Calculate the area of triangle 1.   © *IBO* [*2001*]

4 In triangle ABC, $\hat{B}$ = 43°, AC = 6.8 cm and AB = 4.3 cm.
Find the size of $\hat{A}$, giving your answer to the nearest degree.

5 The diagram shows a circle centre O and radius 15 cm.
The arc ACB subtends an angle of 2 radians at the centre O.

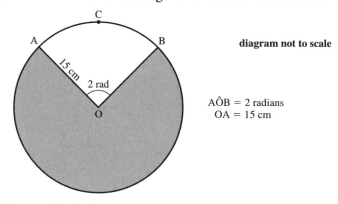

diagram not to scale

$A\hat{O}B$ = 2 radians
OA = 15 cm

Find   a) the length of the arc ACB

b) the area of the shaded region.   © *IBO* [*2002*]

6 The diagram shows two concentric circles with radii 1 cm and 4 cm.

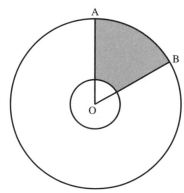

The angle $A\hat{O}B = \dfrac{\pi}{3}$.

Find   a) the area of the shaded region

b) the perimeter of the shaded region.   © *IBO* [*1996*]

7 In the diagram, O is the centre of the circle and AT is the tangent to the circle at T.

The radius of the circle is 6 cm and OA = 12 cm.

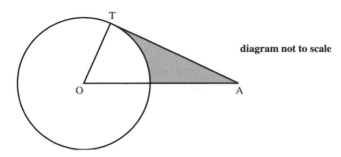

diagram not to scale

Find the area of the shaded region.

8 The first diagram shows a circular sector of radius 10 cm and an angle θ radians. The second diagram shows it formed into a cone of slant height 10 cm. The vertical height, h, of the cone is equal to the radius, r, of its base.

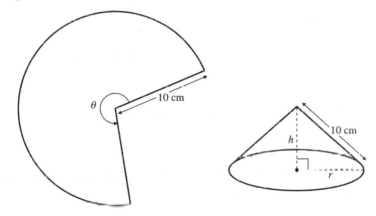

Find the size of the angle θ in radians.   © IBO[1999]

9 The diagram shows a triangle ABC in which

$AC = 7\dfrac{\sqrt{2}}{2}$, $BC = 6$, $A\hat{B}C = 45°$.

a) Use the fact that $\sin 45° = \dfrac{\sqrt{2}}{2}$ to show that $\sin B\hat{A}C = \frac{6}{7}$.

diagram not to scale

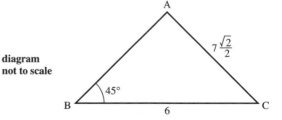

The point D is on AB, between A and B, such that $\sin B\hat{D}C = \frac{6}{7}$.

b) i) Write down the value of $B\hat{D}C + B\hat{A}C$.
   ii) Calculate the angle $B\hat{C}D$.
   iii) Find the length of BD.

c) Show that
$$\frac{\text{area of } \triangle BDC}{\text{area of } \triangle BAC} = \frac{BD}{BA}$$

Note: $7\dfrac{\sqrt{2}}{2}$ is a mixed number.

It means $7 + \dfrac{\sqrt{2}}{2}$.

© IBO[2001]

10 In the diagram, the points O(0, 0) and A(8, 6) are fixed. The angle OP̂A varies as the point P(x, 10) moves along the horizontal line $y = 10$.

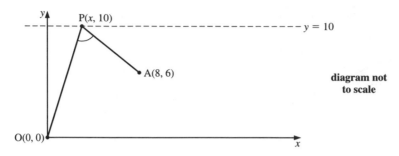

a) i) Show that $AP = \sqrt{x^2 - 16x + 80}$.

   ii) Write down a similar expression for OP in terms of x.

b) Hence, show that

$$\cos O\hat{P}A = \frac{x^2 - 8x + 40}{\sqrt{(x^2 - 16x + 80)(x^2 + 100)}}$$

c) Find, in degrees, the angle OP̂A when $x = 8$.

d) Find the positive value of x such that OP̂A $= 60°$.

Let the function f be defined by

$$f(x) = \cos O\hat{P}A = \frac{x^2 - 8x + 40}{\sqrt{(x^2 - 16x + 80)(x^2 + 100)}}, 0 \leqslant x \leqslant 15$$

e) Consider the equation $f(x) = 1$.

   i) Explain, in terms of the position of the points O, A and P, why this equation has a solution.

   ii) Find the **exact** solution to the equation.

© IBO [2002]

3 Functions and equations

2.1 Concept of function $f : x \mapsto f(x)$: domain, range; image (value).
Composite functions $f \circ g$; identity function.
Inverse function f^{-1}.

2.2 The graph of a function; its equation $y = f(x)$.
Function graphing skills:
use of a graphic display calculator to graph a variety of functions;
investigation of key features of graphs;
solution of equations graphically.

2.3 Transformations of graphs: translations; stretches; reflections in the axes.
The graph of $y = f^{-1}(x)$ as the reflection in the line $y = x$ of the graph of $y = f(x)$.

2.4 The reciprocal function $x \mapsto \dfrac{1}{x}$, $x \neq 0$: its graph; its self-inverse nature.

So far you have seen the equation of a curve written in the form

$y = $ some expression in x

Another way of writing this is to use **functional notation**. That is,

$y = f(x)$

> $f(x)$ means 'a function of x'

For example, you could write $y = x^2$ as $f(x) = x^2$.
To evaluate the function when $x = 3$ you would write

$f(3) = 3^2$
$\therefore f(3) = 9$

You can say that 9 is the image of 3 under the function f.
This is the same as saying that when $x = 3$, $y = 9$.

In general:

❖ $f(x)$ is called the **image** of x.

❖ The set of permitted x values is called the **domain** of the function.

❖ The set of all images is called the **range** of the function.

When a function is defined for all real values, we write the domain of f as

The set of x such that $x \in \mathbb{R} = \{x : x \in \mathbb{R}\}$

or simply $x \in \mathbb{R}$, where $\mathbb{R}$ is the set of all real numbers.

If a function f is defined for all real values except one particular value, say c, then we write the domain of f as $x \in \mathbb{R}$, $x \neq c$.

> **Remember:**
> $\in$ means 'belongs to'.
> $\mathbb{R}$ is the set of all real numbers.
> $x \in \mathbb{R}$ means x is a real number.

3.1   Transforming the graph of a function

Here are the graphs of some standard functions which you need to know:

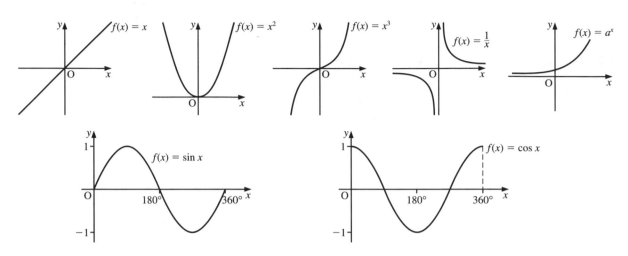

Try sketching each of these graphs using your GDC.

Translations parallel to the *y*-axis

Consider the function f defined by $f(x) = x^2$, $x \in \mathbb{R}$. Plotting the graph of f gives the curve shown on the first diagram below.

If you now simplify expressions for i) $f(x) + 1$ and ii) $f(x) - 1$, you have

i) $f(x) + 1 = x^2 + 1$ and   ii) $f(x) - 1 = x^2 - 1$

Using the GDC to plot graphs of both i) and ii) on the same set of axes gives the curves shown on the second diagram.

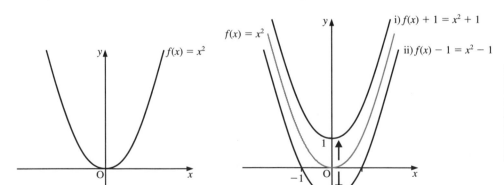

> In i) the graph moves up the *y*-axis.
> In ii) the graph moves down the *y*-axis.

In case i), the graph of f has been translated 1 unit parallel to the *y*-axis.

In case ii), the graph of f has been translated -1 unit parallel to the *y*-axis.

Try using your GDC to plot the graph of $f(x) = x^2$ together with the graphs of i) and ii) on page 80. In addition, plot the graphs of $f(x) + 5$ and $f(x) - 5$. The graphs are shown below.

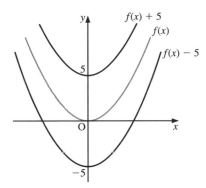

In this case the translations are 5 units and -5 units respectively, parallel to the y-axis.

> In general:
> ❖ The algebraic transformation $f(x) + a$, where a is a constant, causes a geometric transformation of the graph of f, namely a **translation of a units parallel to the y-axis**.
> ❖ The algebraic transformation $f(x) - a$, where a is a constant, causes a geometric transformation of the graph of f, namely a **translation of $-a$ units parallel to the y-axis**.

Example 1

The function f is defined by $f(x) = \dfrac{1}{x}, x \neq 0$.

Sketch the graph of f. Hence sketch the graph of $g(x) = \dfrac{1}{x} - 2$.

· ·

The graph of $f(x)$ is shown on the right.

Since $g(x) = f(x) - 2$, you obtain the graph of g by translating the graph of f by -2 units parallel to the y-axis.

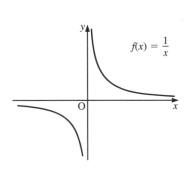

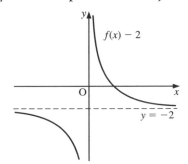

Translations parallel to the x-axis

Look again at the function defined by $f(x) = x^2$.

If you plot the graph of $f(x)$ you get the standard curve as shown.

If you now simplify expressions for i) $f(x + 2)$ and ii) $f(x - 2)$, you have

i) $f(x + 2) = (x + 2)^2$ and   ii) $f(x - 2) = (x - 2)^2$
$= x^2 + 4x + 4$ $= x^2 - 4x + 4$

Using the GDC to plot graphs of both i) and ii) on the same set of axes gives the graphs shown.

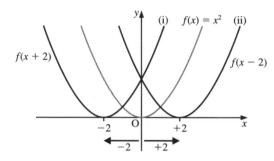

In i) the graph moves left.
In ii) the graph moves right.

In case i), the graph of f has been translated -2 units parallel to the x-axis.

In case ii), the graph of f has been translated 2 units parallel to the x-axis.

> In general:
> ❖ The algebraic transformation $f(x + a)$, where a is a constant, causes a geometric transformation of the graph of f, namely a **translation of $-a$ units parallel to the x-axis**.
> ❖ The algebraic transformation $f(x - a)$, where a is a constant, causes a geometric transformation of the graph of f, namely a **translation of a units parallel to the x-axis**.

Example 2

The function f is defined by $f(x) = x^2 + 1, x \in \mathbb{R}$.
a) Sketch the graph of f.
b) The function g is defined by $g(x) = f(x + 3)$. Find $g(x)$ in its simplest form and hence sketch the graph of g.

....

a) Sketching the graph of $f(x) = x^2 + 1$ gives the curve shown on the right.

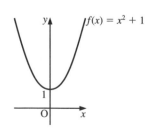

b) The function g is given by

$$g(x) = f(x + 3)$$
$$= (x + 3)^2 + 1$$
$$= x^2 + 6x + 9 + 1$$
$$\therefore g(x) = x^2 + 6x + 10$$

You can get the graph of g by translating the graph of f by -3 units parallel to the x-axis. The graph of $g(x)$ is sketched as shown.

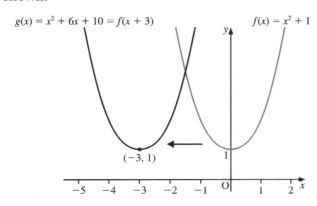

Try sketching the graph of $f(x - 3)$ using your GDC.
Predict the translation first.

Stretch transformations

Consider the function f defined by $f(x) = \sin x, 0 \leqslant x \leqslant 360°$.
If you plot the graph of f on your GDC you get the graph shown.

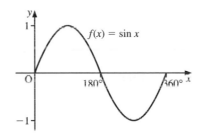

Now consider the functions i) $f(2x)$ and ii) $2f(x)$:

i) $f(2x) = \sin 2x$ and   ii) $2f(x) = 2 \sin x$

If you plot the graphs of i) and ii) on your GDC you get the diagrams shown.

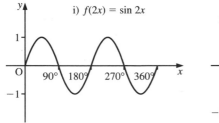
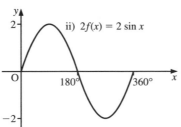

In case i), the graph of f has been stretched parallel to the x-axis by a scale factor of $\frac{1}{2}$.

In case ii), the graph of f has been stretched parallel to the y-axis by a scale factor of 2.

Remember:
A scale factor of 2 means all lengths are doubled.
A scale factor of $\frac{1}{2}$ means all lengths are halved.

In general:

❖ The algebraic transformation $f(ax)$, where a is a constant, causes a geometric transformation of the graph of f, namely a **stretch parallel to the x-axis by a scale factor of $\dfrac{1}{a}$.**

❖ The algebraic transformation $af(x)$, where a is a constant, causes a geometric transformation of the graph of f, namely a **stretch parallel to the y-axis by a scale factor of a.**

Example 3

The function f is defined by $f(x) = \cos x, 0 \leqslant x \leqslant 360°$.

a) Sketch the graph of f using the GDC.

b) Describe the geometric transformation which when applied to the graph of f will give the graph of $g(x) = \cos 3x$.

c) Sketch the graph g stating clearly the x intercepts.

....

a) Sketching the graph of $f(x) = \cos x, 0 \leqslant x \leqslant 360°$ gives the graph shown on the right.

b) The geometric transformation applied to the graph of f is a stretch parallel to the x-axis by a scale factor of $\frac{1}{3}$.

c) Sketching the graph of $g(x) = \cos 3x$ gives this graph:

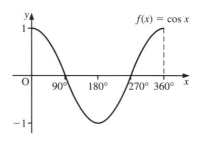

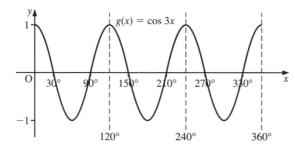

Reflection transformations

Consider the function f defined by $f(x) = x + 1, x \in \mathbb{R}$.
If you plot the graph of f you get the line shown.

Now consider the functions

 i) $-f(x) = -(x + 1) = -x - 1$

 ii) $f(-x) = (-x) + 1 = -x + 1$

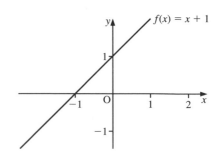

If you plot graphs of both i) and ii) on the same set of axes, you get the lines shown here:

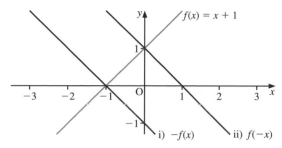

In case i), the graph of f has been reflected in the x-axis.

In case ii), the graph of f has been reflected in the y-axis.

> In general:
> * The algebraic transformation $-f(x)$ causes a geometric transformation of the graph of f, namely a **reflection in the x-axis**.
> * The algebraic transformation $f(-x)$ causes a geometric transformation of the graph of f, namely a **reflection in the y-axis**.

Example 4

The function g is defined for all real values of x and given by $g(x) = 2^{-x}$.

a) Sketch the graph of g.

b) Describe the geometric transformation which when applied to the graph of $f(x) = 2^x$ gives the graph of g.

> See page 300 for graphs of the form $y = a^x$.

. . . .

a) Plotting the graph of g using the GDC gives the graph shown:

b) The graph of $f(x) = 2^x$ looks like this:

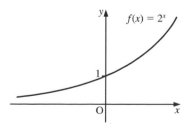

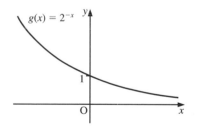

The graph of $g(x) = 2^{-x}$ is obtained by reflecting the graph of 2^x in the y-axis. The two graphs are shown together here.

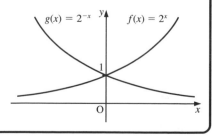

Exercise 3A

In each part of questions **1** to **5** sketch the curves of the given functions on the same set of axes. In each case, comment on the transformation.

1 a) $y = x^2$; $y = x^2 + 3$ b) $y = x^2$; $y = (x - 1)^2$
   c) $y = x^2$; $y = 3x^2$ d) $y = x^2$; $y = 2x^2 + 1$

2 a) $y = x^3$; $y = x^3 + 8$ b) $y = x^3$; $y = (x - 3)^3$
   c) $y = x^3$; $y = -x^3$ d) $y = x^3$; $y = 1 - x^3$

3 a) $y = \dfrac{1}{x}$; $y = \dfrac{1}{x} + 2$ b) $y = \dfrac{1}{x}$; $y = -\dfrac{1}{x}$
   c) $y = \dfrac{1}{x}$; $y = \dfrac{1}{x - 3}$ d) $y = \dfrac{1}{x}$; $y = \dfrac{1}{x + 1}$

4 a) $y = a^x$; $y = a^x + 5$ b) $y = a^x$; $y = a^{x+1}$
   c) $y = a^x$; $y = a^{-x}$ d) $y = a^x$; $y = 3a^x$

5 a) $y = \sin x$; $y = 1 + \sin x$ b) $y = \sin x$; $y = \sin(x + 90°)$

   c) $y = \sin x$; $y = \sin 2x$ d) $y = \sin x$; $y = -\sin\left(\dfrac{x}{2}\right)$

6 Given $f(x) = x^2$, $x \in \mathbb{R}$, sketch the graph of each of the following functions on the same set of axes.
   a) $y = f(x)$ b) $y = -f(x)$
   c) $y = f(x - 2)$ d) $y = 3 + f(x - 2)$

7 The function f is defined by $f(x) = \dfrac{1}{x}$, $x \in \mathbb{R}$, $x > 0$.
   a) Sketch the graph of $y = f(x)$.
   b) On the same set of axes sketch the graphs of
     i) $y = f(x + 1)$ ii) $y = f(x + 1) - 2$

In each of questions **8** to **13**, a diagram is given for a graph of a function $f(x)$, where $f(x) = 0$ for $x < 0$ or $x > 4$. On separate axes, sketch the graphs of the functions listed below each diagram.

8

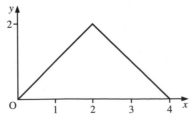

   a) $f(x) + 2$ b) $2f(x)$ c) $f(x + 2)$

9

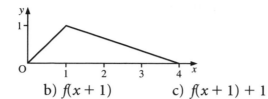

a) $f(x) + 1$ b) $f(x + 1)$ c) $f(x + 1) + 1$

10

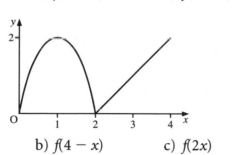

a) $f(x) - 2$ b) $f(4 - x)$ c) $f(2x)$

11

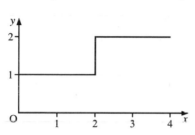

a) $f(x) + 2$ b) $f\left(\dfrac{x}{2}\right)$ c) $f(4x)$

12

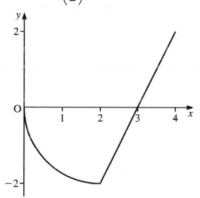

a) $f(x) - 2$ b) $-f(x)$ c) $f(-x)$

13

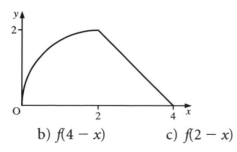

a) $f(x - 2)$ b) $f(4 - x)$ c) $f(2 - x)$

14 The function f is given by $f(x) = 3x - 2$, $x \in \mathbb{R}$.
Sketch the graph of f. Find a combination of geometrical
transformations which, when applied to the graph of f, will
give the graph of $g(x) = 6x + 1$.

15 The functions f and g are defined for all real numbers by
$f(x) = -x^2$ and $g(x) = x^2 + 2x + 8$.

a) Express $g(x)$ in the form $(x + a)^2 + b$, where a and b are constants.

b) Describe two transformations in detail, and the order in which they should be applied, whereby the graph of g may be obtained from the graph of f.

16 The function f is defined by $f(x) = x^2$, $x \in \mathbb{R}$. The graph of $g(x)$ is obtained by reflecting the graph of $f(x)$ in the x-axis, and the graph of $h(x)$ is obtained by translating the graph of $g(x)$ by $+2$ units parallel to the y-axis.

a) Sketch the graphs of $f(x)$, $g(x)$ and $h(x)$ on the same set of axes.

b) Find the equations of $g(x)$ and $h(x)$.

17 The function f is defined by $f(x) = \dfrac{3}{x}$, $x \in \mathbb{R}$, $x > 0$. The graph of $g(x)$ is obtained by translating the graph of $f(x)$ by -4 units parallel to the x-axis, and the graph of $h(x)$ is obtained by reflecting the graph of $g(x)$ in the x-axis.

a) Sketch the graphs of $f(x)$, $g(x)$ and $h(x)$ on the same set of axes.

b) Find the equations of $g(x)$ and $h(x)$.

18 Find an expression for the image of the function $f(x)$ under a translation $\begin{pmatrix} p \\ q \end{pmatrix}$.

3.2 Mappings

Consider two non-empty sets A and B. A **mapping** from A to B is a rule which associates each element of A with an element of B.

You can represent a mapping by a mapping diagram.
This diagram shows mappings from the set $A = \{-2, -1, 0, 1, 2\}$ to the set $B = \{0, 1, 2, 3, 4, 5, 6\}$.

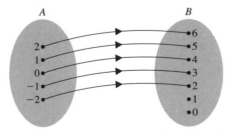

Notice that each element of A maps to one and only one element of B. This is called a **one-to-one mapping**. It doesn't matter that no element of A maps to either of the elements 0 or 1 in B.

In this diagram two elements of A map to one element of B.
This is called a **two-to-one mapping**, or a **many-to-one mapping**.

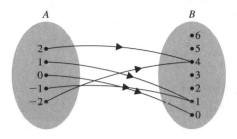

A one-to-one mapping or a many-to-one mapping is called a
function, usually denoted by f.

For example, in the first diagram the rule is 'add 4'.
Using functional notation, you can write:

$$f(x) = x + 4 \quad \text{or} \quad f: x \mapsto x + 4$$

The rule in the second diagram is

$$f(x) = x^2 \quad \text{or} \quad f: x \mapsto x^2$$

> A one-to-many mapping is *not* a function.

> $f: x \mapsto x + 4$ means that f is a function which maps x to $x + 4$.

Example 1

For each of the following mappings f, determine whether f is
one-to-one.

a) $f(x) = x^2, x \in \mathbb{R}$

b) $f(x) = \dfrac{x}{2} + 1, x \in \mathbb{R}$

....

a) Consider $f(x) = x^2$. Since $f(-1) = (-1)^2 = 1$ and
   $f(1) = (1)^2 = 1$, the mapping $f(x) = x^2$ is not one-to-one.
   You can also see this from the graph of the mapping.

b) The graph of $f(x) = \dfrac{x}{2} + 1$ is a straight line.
   It is clear from the graph that this mapping is one-to-one.

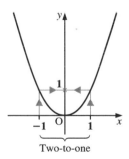

Two-to-one

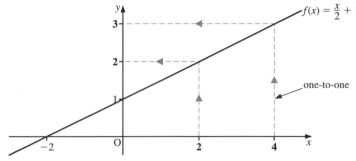

Range of a function

To identify the **range** of a function, it is very useful to have the graph of the function. For example, if the function $f(x) = 2x$ is defined for all real values of x then the graph of f is as shown.

> The range of the function is the set of all images of the function.

In other words, the range is 'that part of the y-axis which is used up by the function'. Therefore, the range of the function is the set of all real values. You write this as

$\{f(x) : f(x) \in \mathbb{R}\}$ or simply $f(x) \in \mathbb{R}$.

A function may be defined on a restricted domain. Consider the function f defined by

$f(x) = 2x, \; -1 < x < 4$

Now the range of the function is the set of real values from -2 to 8, excluding -2 and 8, since -1 and 4 are excluded in the domain. You write this as $-2 < f(x) < 8$.

Graph of $f(x) = 2x$

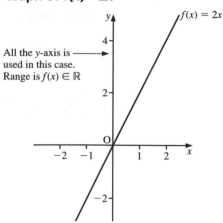

All the y-axis is used in this case.
Range is $f(x) \in \mathbb{R}$

Graph of f(x), $-1 < x < 4$

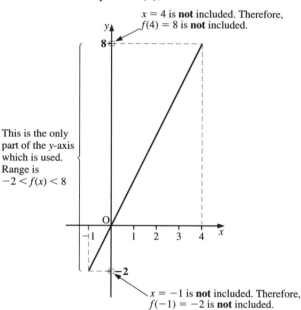

$x = 4$ is **not** included. Therefore, $f(4) = 8$ is **not** included.

This is the only part of the y-axis which is used.
Range is $-2 < f(x) < 8$

$x = -1$ is **not** included. Therefore, $f(-1) = -2$ is **not** included.

Example 2

Find the range of each of the following functions.
a) $f(x) = 2x - 1$, for $x \geq 0$
b) $f(x) = \dfrac{x}{4}$, for $x < 1$
c) $f(x) = x^2$, for $1 \leq x < 3$

....

a) The graph of $f(x) = 2x - 1$, for $x \geq 0$, is shown.
You can see that when $x \geq 0$, $f(x) \geq -1$.
The range of the function is $f(x) \geq -1$.

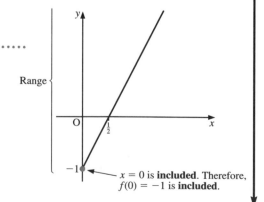

$x = 0$ is **included**. Therefore, $f(0) = -1$ is **included**.

b) The graph of $f(x) = \dfrac{x}{4}$, for $x < 1$, is shown.

You can see that if $x < 1$ then $f(x) < \frac{1}{4}$.

The range of the function is $f(x) < \frac{1}{4}$.

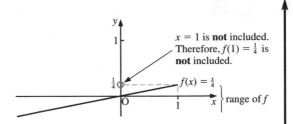

$x = 1$ is **not** included.
Therefore, $f(1) = \frac{1}{4}$ is
not included.

$f(x) = \frac{x}{4}$

range of f

c) The graph of $f(x) = x^2$, for $1 \leqslant x < 3$, is shown.

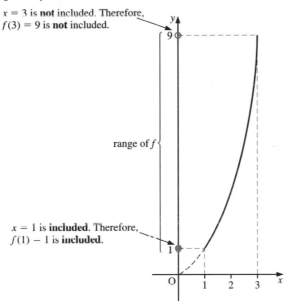

$x = 3$ is **not** included. Therefore,
$f(3) = 9$ is **not** included.

range of f

$x = 1$ is **included**. Therefore,
$f(1) - 1$ is **included**.

You can see that if $1 \leqslant x < 3$ then $1 \leqslant f(x) < 9$.

The range of the function is $1 \leqslant f(x) < 9$.

Example 3

The function f is defined as

$$f(x) = \begin{cases} x + 2 & \text{for} \quad 0 \leqslant x \leqslant 2 \\ x^2 & \text{for} \quad 2 \leqslant x \leqslant 4 \end{cases}$$

Draw a sketch graph of the function and state the range of f.

....

The sketch graph of f is shown.

The range is $2 \leqslant f(x) \leqslant 16$.

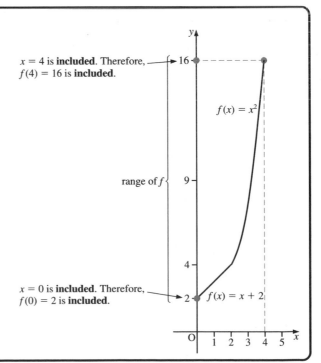

$x = 4$ is **included**. Therefore,
$f(4) = 16$ is **included**.

$f(x) = x^2$

range of f

$x = 0$ is **included**. Therefore,
$f(0) = 2$ is **included**.

$f(x) = x + 2$

Example 4

The function f is defined by $f(x) = x + \dfrac{3}{x}$, for $x \geqslant 2$.

a) Evaluate $f(2)$.

b) Find the value of x for which $f(x) = 4$.

....

a) $f(2) = 2 + \frac{3}{2} = \frac{7}{2}$

b) If $f(x) = 4$, then

$$x + \frac{3}{x} = 4$$

$$x^2 - 4x + 3 = 0$$

$$\therefore (x-1)(x-3) = 0$$

Solving gives $x = 1$ or $x = 3$.

Since the domain of f is $\{x : x \geqslant 2\}$ the only value of x that is required is $x = 3$.

Exercise 3B
....

1 Determine which of these functions are one-to-one and which are two-to-one.

a) $f : x \mapsto x + 3, x \in \mathbb{R}$

b) $f : x \mapsto x^2 + 3, x \in \mathbb{R}$

c) $f : x \mapsto \dfrac{1}{x}, x \in \mathbb{R}, x \neq 0$

d) $f : x \mapsto (x-4)^2, x \in \mathbb{R}, 2 \leqslant x \leqslant 6$

e) $f : x \mapsto x^2 - 4x, x \in \mathbb{R}, 0 < x < 4$

f) $f : x \mapsto x^2 - 4x, x \in \mathbb{R}, 0 < x < 2$

g) $f : x \mapsto x^4 - 3, x \in \mathbb{R}, 3 \leqslant x \leqslant 6$

h) $f : x \mapsto \dfrac{2}{x-3}, x \in \mathbb{R}, -1 < x < 2$

i) $f : x \mapsto x^3 - x^2, x \in \mathbb{R}, 0 \leqslant x \leqslant 1$

j) $f : x \mapsto x^6, x \in \mathbb{R}, -2 < x < 0$

k) $f : x \mapsto x^6, x \in \mathbb{R}, -2 < x < 2$

l) $f : x \mapsto (x^4 + 1)^2 - 3, x \in \mathbb{R}$

2 Determine the range of each of these functions.

a) $f : x \mapsto x + 4, x \in \mathbb{R}, 0 < x < 5$

b) $f : x \mapsto x^2 + 7, x \in \mathbb{R}$

c) $f : x \mapsto 2x - 3, x \in \mathbb{R}, 2 < x \leqslant 6$

d) $f : x \mapsto \dfrac{1}{x^2 + 2}, x \in \mathbb{R}, 1 \leqslant x \leqslant 4$

e) $f : x \mapsto (x^2 + 3)^2, x \in \mathbb{R}$

f) $f: x \mapsto 5x^3 - 1, x \in \mathbb{R}, 1 < x < 3$

g) $f: x \mapsto x^2 - 6x, x \in \mathbb{R}, 0 \leqslant x \leqslant 6$

h) $f: x \mapsto \dfrac{1}{x + 1}, x \in \mathbb{R}, 1 \leqslant x < 9$

i) $f: x \mapsto 3\sqrt{x} - 4, x \in \mathbb{R}, 0 < x < \infty$

j) $f: x \mapsto \sqrt{3x - 2}, x \in \mathbb{R}, 2 \leqslant x \leqslant 9$

k) $f: x \mapsto x^4 + x^2, x \in \mathbb{R}, 0 < x \leqslant 2$

l) $f: x \mapsto \dfrac{1}{3 + x^4}, x \in \mathbb{R}$

3 Sketch the graph of each of these functions and state its range.

a) $f(x) = \begin{cases} 3x + 4 & \text{for} \quad 0 \leqslant x \leqslant 4 \\ x^2 & \text{for} \quad 4 \leqslant x \leqslant 6 \end{cases}$

b) $g(x) = \begin{cases} x^2 & \text{for} \quad 0 \leqslant x \leqslant 3 \\ 12 - x & \text{for} \quad 3 \leqslant x \leqslant 12 \end{cases}$

c) $h(x) = \begin{cases} x + 3 & \text{for} \quad -3 \leqslant x \leqslant 0 \\ x^2 + 3 & \text{for} \quad 0 \leqslant x \leqslant -2 \end{cases}$

d) $f(x) = \begin{cases} -3(x + 2) & \text{for} \quad -3 \leqslant x \leqslant -2 \\ 4 - x^2 & \text{for} \quad -2 \leqslant x \leqslant 2 \\ 3(x - 2) & \text{for} \quad 2 \leqslant x \leqslant 3 \end{cases}$

e) $g(x) = \begin{cases} (x + 2)^2 & \text{for} \quad -1 \leqslant x \leqslant 0 \\ 4 & \text{for} \quad 0 \leqslant x \leqslant 3 \\ 7 \quad x & \text{for} \quad 3 \leqslant x \leqslant 6 \end{cases}$

f) $h(x) = \begin{cases} x^3 & \text{for} \quad 0 \leqslant x \leqslant 2 \\ 2x + 4 & \text{for} \quad 2 \leqslant x \leqslant 6 \\ 16 - (x - 6)^2 & \text{for} \quad 6 \leqslant x \leqslant 10 \end{cases}$

4 The function f is defined by $f(x) = 5x - 3, x \in \mathbb{R}$.
   a) Find the values of x for which $f(x) = 7$.
   b) Solve the equation $f(x) = x$.

5 The function f is defined by $f(x) = x^2 - 6x, x \in \mathbb{R}$.
   a) Find the values of x for which $f(x) = 16$.
   b) Solve the equation $f(x) = 5 - 2x$.

6 The function f is defined by $f(x) = 2 + \dfrac{5}{x}$, for $x \neq 0$.
   a) Evaluate $f(-2)$.
   b) Solve the equation $f(x) = 2x + 7$.

7 The functions f and g are defined by $f(x) = \dfrac{x}{x-2}$, $g(x) = \dfrac{2}{x-3}$

for $x \neq 2$ and $x \neq 3$.

a) Evaluate $f(6)$.

b) Solve the equation $g(x) = 6$.

c) Solve the equation $f(x) = g(x)$.

3.3 Composite functions

Consider the two functions $f(x) = 2x + 5$ and $g(x) = x - 3$, where the domain of f is $\{1, 2, 3, 4\}$ and the domain of g is the range of f. This can be shown on a mapping diagram.

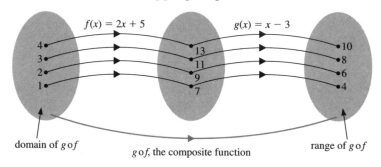

The function shown on the diagram with domain $\{1, 2, 3, 4\}$ and range $\{4, 6, 8, 10\}$ is called the **composite function**, gf or $g \circ f$.

You can work out a single 'rule' for the composite function $gf(x)$ in terms of x:

$$gf(x) = g(2x + 5)$$
$$= (2x + 5) - 3$$
$$\therefore \; gf(x) = 2x + 2$$

> Notice that f is written nearest to the variable x since f is the first function to operate on the set $\{1, 2, 3, 4\}$.

The composite function gf is defined by $gf(x) = 2x + 2$ with domain $\{1, 2, 3, 4\}$ and range $\{4, 6, 8, 10\}$.

If the function g were to operate first on the set $\{1, 2, 3, 4\}$ and then f were to operate on the range of g, the mapping diagram would look like this:

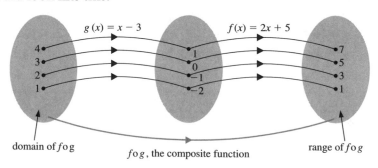

In this case, the composite function which has domain $\{1, 2, 3, 4\}$ and range $\{1, 3, 5, 7\}$ is written as fg or $f \circ g$.

The rule for $(f \circ g)(x)$ is given by

$$f(g(x)) = f(x - 3)$$
$$= 2(x - 3) + 5$$
$$= 2x - 6 + 5$$
$$\therefore \ f(g(x)) = 2x - 1$$

Notice that $fg(x) \neq gf(x)$.

The composite function $f \circ g$ is defined by $f(g(x)) = 2x - 1$ with domain $\{1, 2, 3, 4\}$ and range $\{1, 3, 5, 7\}$.

Example 1

The functions f and g are defined by $f(x) = 3x - 5, x \in \mathbb{R}$ and $g(x) = 3 - 2x, x \in \mathbb{R}$.

a) Evaluate

   i) $f(2)$ ii) $(f \circ g)(3)$

b) The composition function h is defined by $h = g \circ f$.
   Find $h(x)$.

....

a) i) $\qquad\qquad f(x) = 3x - 5$

    So: $\qquad f(2) = 3(2) - 5$

        $\therefore \ f(2) = 1$

   ii) To find $(f \circ g)(3)$, first evaluate $g(3)$.

      Since $\qquad g(x) = 3 - 2x$

             $g(3) = 3 - 2(3)$

        $\therefore \ g(3) = -3$

      Therefore,

$$(f \circ g)(3) = f(-3)$$
$$= 3(-3) - 5$$
$$\therefore \ (f \circ g)(3) = -14$$

b) $h = g \circ f$

    So: $\qquad h(x) = (g \circ f)(x)$

$$= g(3x - 5)$$
$$= 3 - 2(3x - 5)$$
$$= 3 - 6x + 10$$
$$\therefore \ h(x) = 13 - 6x$$

Example 2

The functions f and g are defined by

$$f(x) = x^2, 0 \leqslant x \leqslant 4$$

and $\quad g(x) = x + 3, x \in \mathbb{R}$

Find the composite function $(g \circ f)(x)$ and state the range of this function.

· ·

$$(g \circ f)(x) = g(x^2)$$
$$= x^2 + 3$$
$$\therefore (g \circ f)(x) = x^2 + 3$$

Since f is the first function to operate in the composite function $g \circ f$, you need the range of f, as this will be the domain of g.

The diagram shows the graph of $f(x) = x^2$, for $0 \leqslant x \leqslant 4$.

You can see that the range of f is $0 \leqslant f(x) \leqslant 16$. Therefore, the domain of g is $0 \leqslant x \leqslant 16$.

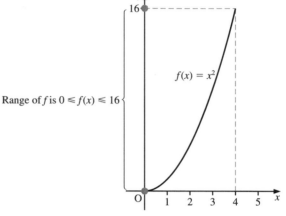

Range of f is $0 \leqslant f(x) \leqslant 16$

Graph of $f(x) = x^2$, $0 \leqslant x \leqslant 4$

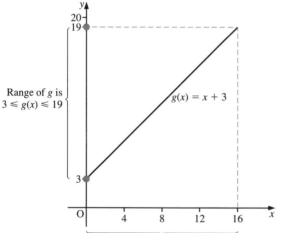

Graph of $g(x)$, $0 \leqslant x \leqslant 16$

From the graph of g you can see that the range of g (when its domain is $0 \leqslant x \leqslant 16$) is $3 \leqslant g(x) \leqslant 19$. Therefore, the composite function gf has range $3 \leqslant (g \circ f)(x) \leqslant 19$.

Alternatively, since you know that $(g \circ f)(x) = x^2 + 3$ and it has domain $0 \leqslant x \leqslant 4$, you can sketch the graph of $(g \circ f)(x)$, as shown.

The range is $3 \leqslant (g \circ f)(x) \leqslant 19$, as before.

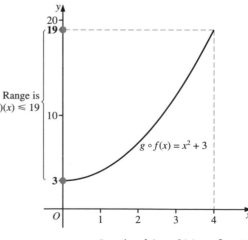

Graph of $(g \circ f)(x) = x^2 + 3$

Exercise 3C
....

Throughout this exercise, the domain of each function is the set of real numbers unless specifically stated otherwise.

1 Given $f(x) = 2x + 1$, $g(x) = x^2$ and $h(x) = \dfrac{1}{x}$ evaluate each of these composite functions.

a) $f(3)$
b) $g(2)$
c) $(h \circ g)(2)$
d) $(f \circ g)(-3)$
e) $(g \circ f)(1)$
f) $(g \circ h)(-2)$
g) $(h \circ f)(4)$
h) $(f \circ f)(5)$
i) $(g \circ g)(-3)$
j) $(h \circ h)(12)$
k) $(f \circ g \circ h)(2)$
l) $(h \circ f \circ g)(4)$

2 Given $f: x \mapsto 3x - 1$, $g: x \mapsto x^2$ and $h: x \mapsto \dfrac{2}{x}$, write down and simplify expressions for each of these composite functions.

a) $(f \circ g)(x)$
b) $(g \circ f)(x)$
c) $(f \circ h)(x)$
d) $(h \circ g)(x)$
e) $(g \circ g)(x)$
f) $(f \circ f)(x)$

3 Functions f and g are defined by
$$f: x \mapsto x^2 + 3 \quad g: x \mapsto x + 5$$
a) Write down and simplify expressions for
   i) $(f \circ g)(x)$ ii) $(g \circ f)(x)$
b) Hence solve the equation $(f \circ g)(x) = (g \circ f)(x)$.

4 Functions h and k are defined by
$$h: x \mapsto \dfrac{3}{x} \quad k: x \mapsto x + 5$$
a) Write down an expression for $(h \circ k)(x)$, and hence solve the equation $(h \circ k)(x) = 1$.
b) Write down an expression for $(k \circ h)(x)$, and hence solve the equation $(k \circ h)(x) = 6$.

5 Given $f(x) = x^2$ and $g(x) = 2x + 5$, solve these equations.
a) $(f \circ g)(x) = 9$
b) $(g \circ g)(x) = 21$

6 Given
$$f(x) = x^2, x \in \mathbb{R}, 1 \leqslant x \leqslant 5 \quad \text{and} \quad g(x) = 2x + 5, x \in \mathbb{R}$$
find an expression for the composite function $(g \circ f)(x)$.
State the domain and range of $(g \circ f)(x)$.

7 Functions p and q are defined by
$$p: x \mapsto 3x^2 + 1, x \in \mathbb{R}, 0 \leqslant x \leqslant 2$$
and
$$q: x \mapsto x^2 - 2, x \in \mathbb{R}$$
Find the composite function $(q \circ p)(x)$ and state its range.

8 Given

$$f: x \mapsto x^2 + 4, x \in \mathbb{R}$$

and

$$g: x \mapsto \frac{1}{x-3}, x \in \mathbb{R}, x \geqslant 4$$

find an expression for the composite function $(g \circ f)(x)$ and state its range.

9 Given

$$f(x) = \sqrt{x+1}, x \in \mathbb{R}, x > 0 \quad \text{and} \quad g(x) = x^2, x \in \mathbb{R}$$

a) find an expression for $(f \circ g)(x)$ and state its range

b) find an expression for $(g \circ f)(x)$ and state its range.

10 Functions h and k are defined by $h(x) = 3x + 5$, and $k(x) = 2 - x$.

a) Write down and simplify expressions for $(h \circ h)(x)$ and $(k \circ k)(x)$.

b) Hence solve the equation $(h \circ h)(x) = (k \circ k)(x)$.

11 Functions f and g are defined by

$$f: x \mapsto x + 1, x \in \mathbb{R} \quad \text{and} \quad g: x \mapsto x^2 - 3, x \in \mathbb{R}$$

a) Show that $(f \circ g)(x) + (g \circ f)(x) = 2x^2 + 2x - 4$.

b) Hence solve the equation $(f \circ g)(x) + (g \circ f)(x) = 0$.

3.4 Inverse functions

The mapping diagram represents the function f defined by $f(x) = x + 3$ with domain $\{1, 2, 3\}$. The range of f is $\{4, 5, 6\}$. The **inverse function** of f, denoted by f^{-1}, would have domain $\{4, 5, 6\}$ and range $\{1, 2, 3\}$, so that:

$$f^{-1}(4) = 1 \quad f^{-1}(5) = 2 \quad \text{and} \quad f^{-1}(6) = 3$$

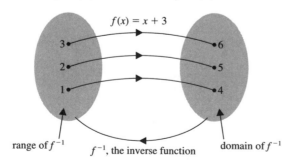

In this case, the inverse function, f^{-1}, is given by

$$f^{-1}(x) = x - 3$$

However, in some examples it is not quite so easy to identify a formula for f^{-1}. Consider the function $y = f(x)$.

You want the function f^{-1} such that $f^{-1}(y) = x$. In other words, x is to be expressed as a function of y. A useful technique for finding the formula for an inverse function is to let $y = f(x)$ and rearrange the equation to make x the subject.

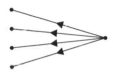

If $y = x + 3$, then rearranging for x gives $x = y - 3$. (In other words, if you are given y, the corresponding x value can be found using $x = y - 3$.) Therefore, the inverse function is $f^{-1}(x) = x - 3$.

Here is a one-to-one function. The inverse function is also a one-to-one function.

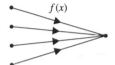

Here is a many-to-one function. The inverse is a one-to-many **mapping**.

A one-to-many mapping is not a function.

> Only one-to-one functions can have inverse functions.

If a function f has an inverse f^{-1}, then

* the composite function $f \circ f^{-1}$ is given by $(f \circ f^{-1})(x) = x$
* the composite function $f^{-1} \circ f$ is given by $(f^{-1} \circ f)(x) = x$.

Example 1

The function f is defined by $f(x) = 5x + 4, x \in \mathbb{R}$.
Find $f^{-1}(x)$ and verify that $(f \circ f^{-1})(x) = x$.

. . . .

To find $f^{-1}(x)$, let $y = 5x + 4$.

Rearranging for x gives $x = \dfrac{y - 4}{5}$.

Therefore, the inverse function is given by $f^{-1}(x) = \dfrac{x - 4}{5}$.

The composite function $(f \circ f^{-1})(x)$ is

$$(f \circ f^{-1})(x) = f\left(\frac{x - 4}{5}\right)$$

$$= 5\left(\frac{x - 4}{5}\right) + 4 = x$$

Therefore, $(f \circ f^{-1})(x) = x$, as required.

Example 2

Two functions f and g are defined by
$$f(x) = 7x + 1, x \in \mathbb{R} \quad \text{and} \quad g(x) = \frac{x}{3} - 1, x \in \mathbb{R}$$
Find the inverse functions f^{-1} and g^{-1} and verify that
$(f \circ g)^{-1} = g^{-1} \circ f^{-1}$.

··

Let $y = 7x + 1$.
$$\therefore \ x = \frac{y - 1}{7}$$
Therefore, $f^{-1}(x) = \frac{x - 1}{7}$.

Let $y = \frac{x}{3} - 1$.
$$\therefore \ x = 3y + 3$$
Therefore, $g^{-1}(x) = 3x + 3$.

To show that $(f \circ g)^{-1} = g^{-1} \circ f^{-1}$, first look at the LHS.
You need the composite function $fg(x)$, which is given by

$$(f \circ g)(x) = f\left(\frac{x}{3} - 1\right)$$
$$= 7\left(\frac{x}{3} - 1\right) + 1$$
$$= \frac{7x}{3} - 7 + 1$$
$$\therefore \ (f \circ g)(x) = \frac{7x}{3} - 6$$

To find the inverse of $(f \circ g)(x) = \frac{7x}{3} - 6$, let $y = \frac{7x}{3} - 6$.
Rearranging for x gives
$$x = \frac{3y + 18}{7}$$
Therefore,
$$(f \circ g)^{-1}(x) = \frac{3x + 18}{7}$$

Next, look at the RHS. You need the composite function
$g^{-1}(f^{-1}(x))$, which is given by

$$g^{-1}(f^{-1}(x)) = g^{-1}\left(\frac{x - 1}{7}\right)$$
$$= 3\left(\frac{x - 1}{7}\right) + 3$$
$$= \frac{3x - 3}{7} + 3$$
$$= \frac{3x - 3 + 21}{7}$$

Therefore $g^{-1}(f^{-1}(x)) = \frac{3x + 18}{7} = (fg)^{-1}(x)$ as required.

Remember:
$fg(x)$ is shorthand for $f \circ g(x)$.

Note:
$g^{-1} \circ f^{-1} = g^{-1}(f^{-1})$.

Graph of an inverse function

If you plot the graphs of the function $f(x) = x + 3$ and its inverse $f^{-1}(x) = x - 3$ on the same set of axes you get the lines shown.

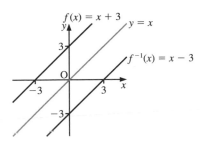

The graph of the inverse function is a reflection of the graph of f in the line $y = x$. This is because for every point (x, y) on the graph of the function f there is a point (y, x) on the graph of the function f^{-1}.

Example 3

The function f is defined by $f(x) = 3x - 6$ for all real values of x. Find the inverse function f^{-1}. Sketch the graphs of f and f^{-1} on the same set of axes and hence find the coordinates of the point of intersection of the graphs of f and f^{-1}.

....

Let $y = 3x - 6$. Rearranging for x gives $x = \dfrac{y + 6}{3}$.

Therefore, $f^{-1}(x) = \dfrac{x + 6}{3}$.

The graphs of f and f^{-1} are shown in the diagram.

The line $y = x$ is a line of symmetry.

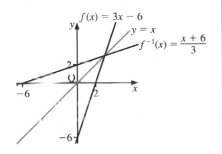

The point of intersection of the graphs of f and f^{-1} is also the point of intersection of the line $y = x$ with each of the graphs of f and f^{-1}. To find the coordinates of this intersection point you must solve simultaneously the equations

$$y = x \quad \text{and} \quad y = 3x - 6$$

Eliminating y gives

$$x = 3x - 6$$
$$2x = 6$$
$$\therefore \ x = 3$$

Substituting $x = 3$ into $y = x$ gives $y = 3$.

The coordinates of the point of intersection of the graphs of f and f^{-1} are $(3, 3)$.

Exercise 3D

....

1 Find the inverse of each of these functions.

a) $f: x \mapsto 3x + 2$

b) $f: x \mapsto 5x - 1$

c) $f: x \mapsto 4 - 3x$

d) $f: x \mapsto \dfrac{2}{x}, x \neq 0$

e) $f: x \mapsto \dfrac{3}{x - 1}, x \neq 1$

f) $f: x \mapsto \dfrac{5}{2 - 3x}, x \neq \frac{2}{3}$

g) $f: x \mapsto \dfrac{x}{2+x}, x \neq -2$ h) $f: x \mapsto \dfrac{2x}{5-x}, x \neq 5$

i) $f: x \mapsto \dfrac{3x}{2x+1}, x \neq -\frac{1}{2}$

2 Find the inverse of each of these functions, and state the domain on which each inverse is defined.

a) $f(x) = x^2, x \in \mathbb{R}, x > 2$

b) $f(x) = \dfrac{1}{2+x}, x \in \mathbb{R}, x > 0$

c) $f(x) = \sqrt{x-2}, x \in \mathbb{R}, x > 3$

d) $f(x) = 3x^2 - 1, x \in \mathbb{R}, 1 < x < 4$

e) $f(x) = \sqrt{2x+3}, x \in \mathbb{R}, x \geqslant 11$

f) $f(x) = \dfrac{1}{x} - 3, x \in \mathbb{R}, 2 < x < 5$

g) $f(x) = (x+2)^2 + 3, x \in \mathbb{R}, x \geqslant -2$

h) $f(x) = x^3 + 1, x \in \mathbb{R}$

3 Given $f: x \mapsto 3x - 4, x \in \mathbb{R}$,

a) find an expression for the inverse function $f^{-1}(x)$

b) sketch the graphs of $f(x)$ and $f^{-1}(x)$ on the same set of axes

c) solve the equation $f(x) = f^{-1}(x)$.

4 a) Sketch the graph of the function defined by
$$f(x) = 10 - 2x, x \in \mathbb{R}, x \geqslant 0$$

b) Find an expression for the inverse function $f^{-1}(x)$, and sketch the graph of $f^{-1}(x)$ on the same set of axes.

c) Calculate the value of x for which $f(x) = f^{-1}(x)$.

5 A function is defined by $f(x) = x^2 - 6, x \in \mathbb{R}, x > 0$.

a) Find an expression for the inverse function $f^{-1}(x)$.

b) Sketch the graphs of $f(x)$ and $f^{-1}(x)$ on the same set of axes.

c) Calculate the value of x for which $f(x) = f^{-1}(x)$.

6 a) Sketch the graph of the function defined by
$$f: x \mapsto (x-2)^2, x \in \mathbb{R}, x \geqslant 2$$

b) Find an expression for the inverse function $f^{-1}(x)$, and sketch the graph of $f^{-1}(x)$ on the same set of axes.

c) Calculate the value of x for which $f(x) = f^{-1}(x)$.

7 The functions f and g are defined by

$$f: x \mapsto 2x - 5,\, x \in \mathbb{R} \quad \text{and} \quad g: x \mapsto 7 - 4x,\, x \in \mathbb{R}$$

a) Solve the equation $f(x) = g(x)$.

b) Write down expressions for $f^{-1}(x)$ and $g^{-1}(x)$.

c) Solve the equation $f^{-1}(x) = g^{-1}(x)$, and comment on your answer.

8 The function h with domain $\{x : x \geqslant 0\}$ is defined by

$$h(x) = \frac{4}{x + 3}.$$

a) Sketch the graph of h and state its range.

b) Find an expression for $h^{-1}(x)$.

c) Calculate the value of x for which $h(x) = h^{-1}(x)$.

9 Functions f and g are defined by

$$f: x \mapsto 3x + 1,\, x \in \mathbb{R} \quad \text{and} \quad g: x \mapsto x - 2,\, x \in \mathbb{R}$$

a) Write down and simplify an expression for the composite function $(f \circ g)(x)$.

b) Find expressions for each of these inverse functions.

  i) $f^{-1}(x)$ ii) $g^{-1}(x)$ iii) $(f \circ g)^{-1}(x)$

c) Verify that $(f \circ g)^{-1}(x) = g^{-1}(f^{-1}(x))$.

10 The function g is defined by $g(x) = 2x^2 - 3,\, x \in \mathbb{R},\, x \geqslant 0$.

a) State the range of g and sketch its graph.

b) Explain why the inverse function g^{-1} exists and sketch its graph.

11 Functions f and g are defined by

$$f: x \mapsto 2x + 3,\, x \in \mathbb{R} \quad \text{and} \quad g: x \mapsto \frac{1}{x - 1},\, x \in \mathbb{R},\, x \neq 1$$

a) Find an expression for the inverse function $f^{-1}(x)$.

b) Find an expression for the composite function $(g \circ f)(x)$.

c) Solve the equation $f^{-1}(x) = (g \circ f)(x) - 1$.

12 Functions g and h are defined by

$$g(x) = \frac{5}{x - 3},\, x \in \mathbb{R},\, x \neq 3$$

and

$$h(x) = x^2 + 4,\, x \in \mathbb{R},\, x > 0$$

Find

a) an expression for the inverse function $g^{-1}(x)$

b) an expression for the composite function $(g \circ h)(x)$

c) the solutions to the equations $3g^{-1}(x) = 10g(h(x)) + 9$.

Summary

You should know how to ...

▶ Understand the concept of a function.

   ▷ A function is a one-to-one or a many-to-one mapping.
   ▷ The domain is the set of permitted values of x (x-values)
   ▷ The range is the set of all images (y-values)

▶ Understand and use composite functions.
   ▷ $(f \circ g)(x)$ means do g first then f.

▶ Understand and use inverse functions.
   ▷ $f^{-1}f(x) = x$
   ▷ Graphically, you obtain the inverse by reflecting $y = f(x)$ in the line $y = x$.
   ▷ Algebraically, you obtain the inverse by swapping x and y, and then rearranging.

▶ Recognise transformations of functions.
   ▷ $y = f(x) + b$ is a translation by $+b$ units parallel to the y-axis
   ▷ $y = f(x - a)$ is a translation by $+a$ units parallel to the x-axis
   ▷ $y = pf(x)$ is a stretch, scale factor p, parallel to the y-axis
   ▷ $y = f\left(\dfrac{x}{q}\right)$ is a stretch, scale factor q, parallel to the x-axis
   ▷ $y = -f(x)$ is a reflection in the x-axis
   ▷ $y = f(-x)$ is a reflection in the y-axis

Revision exercise 3

1 The diagram shows parts of the graphs of $y = x^2$ and $y = 5 - 3(x - 4)^2$.

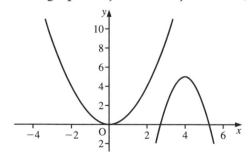

The graph of $y = x^2$ may be transformed into the graph of $y = 5 - 3(x - 4)^2$ by these transformations:

A reflection in the line $y = 0$ followed by

a vertical stretch with scale factor k followed by

a horizontal translation of p units   followed by

a vertical translation of q units.

Write down the value of

a) k b) p c) q

© IBO [2001]

2 The diagrams show how the graph of $y = x^2$ is transformed to the graph of $y = f(x)$ in three steps.

For each diagram give the equation of the curve.

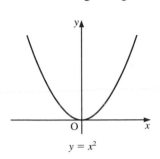

$y = x^2$

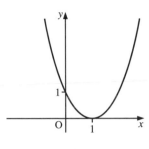

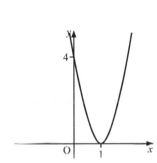

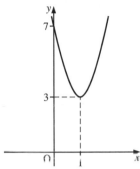

© *IBO* [*2000*]

3 The diagram shows the graph of $y = f(x)$.
It has minimum and maximum points at $(0, 0)$ and $(1, \frac{1}{2})$.

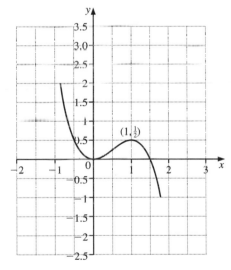

a) On a copy of the diagram, draw the graph of $y = f(x - 1) + \frac{3}{2}$.

b) What are the coordinates of the minimum and maximum points of $y = f(x - 1) + \frac{3}{2}$?

© *IBO* [*2001*]

4 Two functions f, g are defined as follows:

$$f: x \mapsto 3x + 5 \quad g: x \mapsto 2(1 - x)$$

Find

a) $f^{-1}(2)$ b) $(g \circ f)(-4)$

© *IBO* [*2000*]

5 The function f is defined by $f: x \mapsto \sqrt{3 - 2x}, \; x \leqslant \frac{3}{2}$

Find the value of $f^{-1}(5)$. © *IBO*[2001]

6 The graph of the function $f(x) = 3x - 4$ intersects the x-axis at A and the y-axis at B.

a) Find the coordinates of
   i) A   ii) B.

b) Let O denote the origin. Find the area of triangle OAB. © *IBO*[2003]

7 The function f is given by $f(x) = x^2 - 6x + 13$, for $x \geqslant 3$.

a) Write $f(x)$ in the form $(x - a)^2 + b$.

b) Find the inverse function f^{-1}.

c) State the domain of f^{-1}. © *IBO*[2003]

8 Two functions f and g are defined as follows:

$$f: x \mapsto x^2 - 4 \quad g: x \mapsto \frac{1}{3x}$$

Solve the equation $g \circ f(x) = g(x)$. © *IBO*[2003]

9 The function f is defined by $f(x) = x^3 - 3$.

a) Find $f(0)$.

b) Hence write down the solution of $f^{-1}(x) = 0$.

c) On the same axes, sketch the graphs of $y = f(x)$ and $y = f^{-1}(x)$ for $-5 < x < 5$.

d) State the number of solutions of the equation $f(x) = f^{-1}(x)$ for $0 < x < 5$.

10 The sketch shows part of the graph of $y = f(x)$ which passes through the points A$(-1, 3)$, B$(0, 2)$, C$(1, 0)$, D$(2, 1)$ and E$(3, 5)$.

A second function is defined by $g(x) = 2f(x - 1)$.

a) Calculate $g(0)$, $g(1)$, $g(2)$ and $g(3)$.

b) On the same axes, sketch the graph of the function $g(x)$.

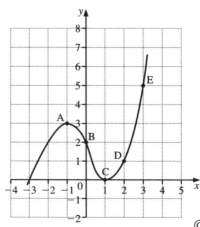

© *IBO*[2002]

4 Differentiation 1

Differentiation is part of the area of mathematics known as **calculus.** The differential calculus was first developed in the 17th century by Newton and Leibniz.

> Some of the examples in this chapter use the properties of surds and indices. You may find it helpful first to work through these topics in Chapter 10.

4.1 Gradient of a curve

The gradient of a straight line that passes through the points $A(x_1, y_1)$ and $B(x_2, y_2)$ is given by

$$m_{AB} = \frac{y_2 - y_1}{x_2 - x_1}$$

The gradient of a straight line is the same at all points on the line. However, this is not true on a curve.

Consider the curve shown in the diagram.

> For example, the line that passes through the points A(1, 2) and B(3, 8) has gradient
> $m_{AB} = \frac{8-2}{3-1} = \frac{6}{2} = 3$

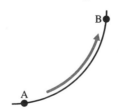

As you move along the curve from point A to point B, the gradient of the curve changes – it becomes steeper.

> The gradient of a curve at a point P is the gradient of the tangent line to the curve at point P.

For example, to find the gradient of the curve $y = x^2$ at the point P(3, 9), you need the gradient of the tangent line to the curve at point P.

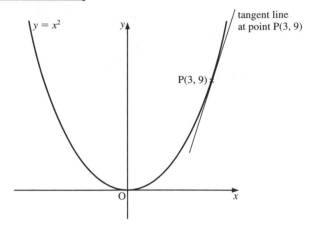

Drawing a tangent line by eye is not a very accurate way to find the gradient of a curve. Mathematicians had to develop an algebraic method for finding the gradient of any curve.

Differentiation from first principles

Consider the point $P(x, f(x))$ on the curve shown.

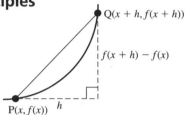

Imagine that you want to find the gradient of the curve at point P.

Let Q be a different point on the curve with coordinates $(x + h, f(x + h))$, where h is a small quantity. The straight line PQ is called a chord of the curve.

As the distance h becomes smaller, point Q moves closer to P and the chord PQ approaches the position of the tangent at P.

The gradient of PQ is given by

$$m_{PQ} = \frac{f(x + h) - f(x)}{(x + h) - x}$$

$$\therefore \ m_{PQ} = \frac{f(x + h) - f(x)}{h}$$

As h tends to zero ($h \to 0$), m_{PQ} approaches the value of the gradient of the tangent line at P. This limiting value is written as

$$\lim_{h \to 0} \left(\frac{f(x + h) - f(x)}{h} \right)$$

> The limiting value is called the differential coefficient or **first derivative** of y with respect to x and is denoted by $f'(x)$.
> When the notation '$y =$ some function of x' is used the first derivative is denoted by $\dfrac{dy}{dx}$.

The process of finding this limiting value is called **differentiation**.

Example 1

Given that $f(x) = x^2$, find $f'(x)$.

....

$$f'(x) = \lim_{h \to 0}\left(\frac{f(x + h) - f(x)}{h}\right)$$

Consider the points $P(x, f(x))$ and $Q(x + h, f(x + h))$ on the curve $f(x) = x^2$.

Since the coordinates of P and Q satisfy the equation $f(x) = x^2$, the coordinates can be written as $P(x, x^2)$ and $Q(x + h, (x + h)^2)$. Therefore

$$f'(x) = \lim_{h \to 0}\left(\frac{(x + h)^2 - x^2}{h}\right)$$

$$= \lim_{h \to 0}\left(\frac{x^2 + 2xh + h^2 - x^2}{h}\right)$$

$$= \lim_{h \to 0}\left(\frac{2xh + h^2}{h}\right)$$

$$\therefore \ f'(x) = \lim_{h \to 0}(2x + h)$$

As $h \to 0$, the expression $2x + h \to 2x$. Therefore,

$$f'(x) = 2x$$

The expression $f'(x) = 2x$ is called the derived expression or the gradient function or the first derivative of $f(x)$ with respect to x.

> The terms differential coefficient, gradient function, derived expression and first derivative all mean the same thing.

Investigating the slopes of curves

❖ Using your GDC plot the graph of $y = x^2$ for $-3 \leqslant x \leqslant 3$.
Investigate the slope of the curve at the points with x-coordinates $-2, -1, 0, 1, 2$.

You know from example 1 that the gradient function of x^2 is $2x$. Therefore the exact slope of the curve at each of these points is twice the value of the x-coordinate.
This gives the following table of results.

x value	-2	-1	0	1	2
Slope	-4	-2	0	2	4

❖ Using your GDC plot the graph of $y = 3x + 1$ for $-4 \leqslant x \leqslant 4$.
Investigate the slope of the graph at different points.
What do you notice?

Exercise 4A

Differentiate each of these functions from first principles.

1 $y = x^3$　　　　　　**2** $y = 2x^2 + 3$　　　　**3** $y = 1 - x^2$

4 $y = x^3 - 6x$　　　**5** $y = \dfrac{1}{x}$　　　　　　**6** $y = \sqrt{x}$

4.2　The derivative of $y = ax^n$

If you differentiate functions of the form $y = ax^n$, you will find this general result:

> If $y = ax^n$, then
>
> $$\dfrac{dy}{dx} = anx^{n-1}$$
>
> for all rational n.

Notice that if $y = a$, a constant, you can write this as $y = ax^0$ and

$$\dfrac{dy}{dx} = (a \times 0)x^{-1} = 0$$

In other words:

> The derivative of a constant is always zero.

If you think about this result geometrically, $y = a$ is a horizontal line which has a gradient of zero.

Remember:

$x^0 = 1$

You may need to read section 10.1 before tackling this section.

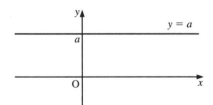

Example 1

Find $\dfrac{dy}{dx}$ for each of these functions.

a) $y = x^3$　　　　　　b) $y = 6x^4$　　　　　c) $y = \dfrac{1}{x^5}$

d) $y = \dfrac{3}{4x^2}$　　　　e) $y = x^{\frac{1}{3}}$　　　　f) $y = \dfrac{1}{\sqrt{x}}$

....

a) When $y = x^3$,

$$\dfrac{dy}{dx} = 3x^{3-1}$$

$$\therefore \dfrac{dy}{dx} = 3x^2$$

b) When $y = 6x^4$,

$$\dfrac{dy}{dx} = 6 \times (4x^{4-1})$$

$$\therefore \dfrac{dy}{dx} = 24x^3$$

c) When $y = \dfrac{1}{x^5} = x^{-5}$,

$$\dfrac{dy}{dx} = -5x^{-5-1}$$

$$= -5x^{-6}$$

$$\therefore \dfrac{dy}{dx} = -\dfrac{5}{x^6}$$

d) When $y = \dfrac{3}{4x^2} = \dfrac{3}{4}x^{-2}$

$$\dfrac{dy}{dx} = \dfrac{3}{4} \times -2x^{-2-1}$$

$$= -\dfrac{3}{2}x^{-3}$$

$$\therefore \dfrac{dy}{dx} = -\dfrac{3}{2x^3}$$

e) When $y = x^{\frac{1}{3}}$,

$$\dfrac{dy}{dx} = \dfrac{1}{3}x^{\frac{1}{3}-1}$$

$$= \dfrac{1}{3}x^{-\frac{2}{3}}$$

$$\therefore \dfrac{dy}{dx} = \dfrac{1}{3x^{\frac{2}{3}}}$$

f) When $y = \dfrac{1}{\sqrt{x}} = \dfrac{1}{x^{\frac{1}{2}}} = x^{-\frac{1}{2}}$,

$$\dfrac{dy}{dx} = -\dfrac{1}{2}x^{-\frac{1}{2}-1}$$

$$\therefore \dfrac{dy}{dx} = -\dfrac{1}{2x^{\frac{3}{2}}} \left(= -\dfrac{1}{2\sqrt{x^3}} \right)$$

Exercise 4B

1 Find $\dfrac{dy}{dx}$ for each of these functions.

a) $y = x^4$ b) $y = x^6$ c) $y = 6x^2$ d) $y = -5x^3$

e) $y = 3x$ f) $y = 2x^6$ g) $y = -7x^2$ h) $y = 2$

i) $y = \frac{1}{2}x^4$ j) $y = \frac{2}{3}x^6$ k) $y = -\frac{1}{4}x^3$ l) $y = \frac{2}{5}x$

2 Differentiate each of these expressions with respect to x.

a) x^{-2} b) x^{-4} c) $2x^{-3}$ d) $4x^{-1}$

e) $\dfrac{1}{x^3}$ f) $-\dfrac{1}{x^2}$ g) $\dfrac{3}{x^3}$ h) $-\dfrac{2}{x}$

i) $\dfrac{3}{2x^2}$ j) $\dfrac{9}{2x^3}$ k) $-\dfrac{3}{4x^4}$ l) $\dfrac{2}{5x}$

3 Find $f'(x)$ for each of these functions.

a) $f(x) = x^{\frac{1}{2}}$ b) $f(x) = 6x^{\frac{1}{3}}$ c) $f(x) = x^{-\frac{2}{3}}$

d) $f(x) = -10x^{-\frac{1}{5}}$ e) $f(x) = 7\sqrt{x}$ f) $f(x) = \sqrt[3]{x}$

g) $f(x) = \dfrac{4}{5\sqrt{x}}$ h) $f(x) = -\dfrac{6}{\sqrt[3]{x}}$ i) $f(x) = \dfrac{5}{2\sqrt{x}}$

4.3   Sum or difference of two functions

To find the derivative of a function that has more than one term, you need to differentiate each function in turn.

$$y = f(x) \pm g(x) \quad \text{then} \quad \frac{dy}{dx} = f'(x) \pm g'(x)$$

This applies to the sum or difference of any number of functions.

Example 1

Find $f'(x)$ for each of these functions.

a) $f(x) = 4x^2 + 1$ b) $f(x) = 2x^3 + \sqrt{x}$ c) $f(x) = x + \dfrac{1}{x}$

d) $f(x) = x^2 + 6x^{\frac{1}{3}} - 3$ e) $f(x) = \dfrac{2}{\sqrt{x}} + \dfrac{3}{x^2} - 1$

....

a) When $f(x) = 4x^2 + 1$,
$$f'(x) = 8x$$

b) When $f(x) = 2x^3 + \sqrt{x} = 2x^3 + x^{\frac{1}{2}}$,
$$f'(x) = 6x^2 + \tfrac{1}{2}x^{-\frac{1}{2}}$$
$$= 6x^2 + \frac{1}{2x^{\frac{1}{2}}}$$
$$\therefore f'(x) = 6x^2 + \frac{1}{2\sqrt{x}}$$

c) When $f(x) = x + \dfrac{1}{x} = x + x^{-1}$,
$$f'(x) = 1 - x^{-2}$$
$$\therefore f'(x) = 1 - \frac{1}{x^2}$$

d) When $f(x) = x^2 + 6x^{\frac{1}{3}} - 3$,
$$f'(x) = 2x + 2x^{-\frac{2}{3}}$$
$$= 2x + \frac{2}{x^{\frac{2}{3}}}$$
$$\therefore f'(x) = 2x + \frac{2}{\sqrt[3]{x^2}}$$

e) When $f(x) = \dfrac{2}{\sqrt{x}} + \dfrac{3}{x^2} - 1 = 2x^{-\frac{1}{2}} + 3x^{-2} - 1$,
$$f'(x) = -x^{-\frac{3}{2}} - 6x^{-3}$$
$$= -\frac{1}{x^{\frac{3}{2}}} - \frac{6}{x^3}$$
$$\therefore f'(x) = -\frac{1}{\sqrt{x^3}} - \frac{6}{x^3}$$

Remember:
$\sqrt{x} = x^{\frac{1}{2}}$

A function may not be given in the form ax^n. In this case, you must manipulate the expression for y and write it as a sum of functions, each in the form ax^n.

Example 2

Find $\dfrac{dy}{dx}$ for each of these functions.

a) $y = (x + 3)^2$ 　　　 b) $y = \sqrt{x}\,(x^2 - 1)$ 　　　 c) $y = \dfrac{x^3 + 6}{x}$

....

a) To differentiate $y = (x + 3)^2$, first expand the bracket:
$$y = (x + 3)(x + 3)$$
$$= x^2 + 6x + 9$$
$$\therefore \ \frac{dy}{dx} = 2x + 6$$

b) $y = \sqrt{x}\,(x^2 - 1) = x^{\frac{1}{2}}(x^2 - 1)$

So: 　　　　　$y = x^{\frac{5}{2}} - x^{\frac{1}{2}}$

$$\therefore \ \frac{dy}{dx} = \frac{5}{2}x^{\frac{3}{2}} - \frac{1}{2}x^{-\frac{1}{2}}$$

You can factorise this expression for $\dfrac{dy}{dx}$ by taking out $\dfrac{1}{2}x^{-\frac{1}{2}}$.

This gives
$$\frac{dy}{dx} = \frac{1}{2}x^{-\frac{1}{2}}(5x^2 - 1)$$

$$= \frac{1}{2x^{\frac{1}{2}}}(5x^2 - 1)$$

$$\therefore \ \frac{dy}{dx} = \frac{1}{2\sqrt{x}}(5x^2 - 1)$$

c) When $y = \dfrac{x^3 + 6}{x} = \dfrac{x^3}{x} + \dfrac{6}{x} = x^2 + \dfrac{6}{x} = x^2 + 6x^{-1}$,

$$\frac{dy}{dx} = 2x - 6x^{-2}$$

$$= 2x^{-2}(x^3 - 3)$$

$$\therefore \ \frac{dy}{dx} = \frac{2(x^3 - 3)}{x^2}$$

> Expand the bracket.

> $x^{-\frac{1}{2}}$ is the lowest power of x in the expression.

> Factorise: x^{-2} is the lowest power of x in the expression.

Examples 2 b) and c) show how you can manipulate the derivative to obtain a more mathematically tidy result. Note that any form of the correct derivative is acceptable. However, it is useful to understand such manipulation since many examination questions ask for the derivative in a particular form.

Exercise 4C

1 Find $\dfrac{dy}{dx}$ for each of these functions.

a) $y = x^2 + 2x$ b) $y = 3x^2 - 5x$ c) $y = x^2 + 1$

d) $y = 5 - 4x^3$ e) $y = x^2 + 2x + 3$ f) $y = x^7 + 3x^4$

g) $y = x^4 - 3x^2 + 2$ h) $y = x + \dfrac{1}{x}$ i) $y = 5x^2 - \dfrac{2}{x^3}$

2 Differentiate each of these expressions with respect to x.

a) $3x - 5x^3$ b) $2 - \dfrac{3}{x}$ c) $\dfrac{4}{x} - 2x^3$

d) $2\sqrt{x} + 1$ e) $\sqrt{x} + \dfrac{1}{\sqrt{x}}$ f) $4x^{-2} - 3x$

g) $3x^{\frac{1}{3}} - 4x^{-\frac{1}{3}}$ h) $4x^{\frac{1}{2}} + 2x - 1$ i) $6x^{\frac{2}{3}} - 4x^{\frac{5}{2}}$

3 Find $f'(x)$ for each of these functions.

a) $f(x) = 4x - 7$ b) $f(x) = 2\sqrt{x} + \dfrac{5}{2x}$

c) $f(x) = 4x^2 + 7x - 3$ d) $f(x) = 5x^{\frac{2}{5}} - 2x^{\frac{5}{2}}$

e) $f(x) = 6\sqrt{x} - \dfrac{3}{2x^2}$ f) $f(x) = 2x^{-7} - 5x^{-3} + x$

g) $f(x) = \dfrac{5}{x^2} - \dfrac{2}{x} + 3$ h) $f(x) = 9x^{\frac{4}{3}} + 3$

i) $f(x) = 2x^{-4} - 4x^{-2}$ j) $f(x) = \dfrac{3}{\sqrt{x}} - 2\sqrt{x}$

4 Find $\dfrac{dy}{dx}$ for each of these functions.

a) $y = x^2(x + 3)$ b) $y = x(2 - x)$

c) $y = x^3(4 - x^2)$ d) $y = x^{\frac{1}{3}}(2x - 5)$

e) $y = (x + 3)(x - 4)$ f) $y = (x + 4)^2$

g) $y = (x + 5)(2x - 1)$ h) $y = 2(x - 3)^2$

i) $y = (x + 8)(x - 2)$ j) $y = x^3(x + 1)$

5 Differentiate each of these expressions with respect to x.

a) $3x(x - 4)$ b) $x^3(3x^2 - 1)$ c) $\sqrt{x}(x^2 + 3)$

d) $x^{-\frac{1}{2}}(x + 1)$ e) $\dfrac{x^2 + 7}{x}$ f) $\dfrac{x + 5}{x^2}$

g) $\dfrac{3x^2 + 2}{x}$ h) $\dfrac{6x^3 - 7}{x^2}$ i) $\dfrac{2x + 3}{5x}$

6 Find $f'(x)$ for each of these functions.

a) $f(x) = x^3(3x - 1)$

b) $f(x) = 2x^2(x - 1)^2$

c) $f(x) = \dfrac{\sqrt{x} + 1}{\sqrt{x}}$

d) $f(x) = \dfrac{3x^3 + 5}{x^2}$

e) $f(x) = \dfrac{(x + 3)(x - 4)}{x}$

f) $f(x) = \dfrac{(2x - 5)(x - 4)}{x^3}$

g) $f(x) = \dfrac{(3x - 1)^2}{2x}$

h) $f(x) = \dfrac{5x + 3}{2\sqrt{x}}$

i) $f(x) = \dfrac{(x + 1)^2}{\sqrt{x}}$

j) $f(x) = \dfrac{x^2 + 5}{3\sqrt{x}}$

4.4   The second derivative

Differentiating a function gives you another function, the derived function or first derivative. You can differentiate this function to get the second derivative of the original function.

> The derivative of $\dfrac{dy}{dx}$, that is $\dfrac{d}{dx}\left(\dfrac{dy}{dx}\right)$, is denoted by $\dfrac{d^2y}{dx^2}$ and is called the **second derivative of y with respect to x.**

> The derivative of $f'(x)$ is denoted by $f''(x)$ and is called the **second derivative of $f(x)$ with respect to x.**

You will see how the second derivative can be used later, in section 4.7.

Example 1

Given that $f(x) = x + \dfrac{1}{x}$, find $f'(x)$ and $f''(x)$.

$$f(x) = x + \frac{1}{x} = x + x^{-1}.$$

Therefore,

$$f'(x) = 1 - x^{-2}$$

$$= 1 - \frac{1}{x^2}$$

Given that $f'(x) = 1 - x^{-2}$, then

$$f''(x) = 2x^{-3}$$

$$= \frac{2}{x^3}$$

Differentiate again.

Equations may contain derivatives of functions (these are called differential equations).

Example 2

If $y = 4x^3$, find $\dfrac{dy}{dx}$ and $\dfrac{d^2y}{dx^2}$. Hence show that y satisfies

$$3y\dfrac{d^2y}{dx^2} - 2\left(\dfrac{dy}{dx}\right)^2 \equiv 0$$

....

When $y = 4x^3$,

$$\dfrac{dy}{dx} = 12x^2 \quad \text{and} \quad \dfrac{d^2y}{dx^2} = 24x$$

Substituting into the LHS of

$$3y\dfrac{d^2y}{dx^2} - 2\left(\dfrac{dy}{dx}\right)^2 = 0$$

gives

$$3y\dfrac{d^2y}{dx^2} - 2\left(\dfrac{dy}{dx}\right)^2 \equiv 3(4x^3)(24x) - 2(12x^2)^2$$

$$\equiv 288x^4 - 288x^4 \equiv 0$$

as required.

Exercise 4D

1 Find $\dfrac{d^2y}{dx^2}$ for each of thesee functions.

a) $y = 3x^3 + 5x$
b) $y = x^2 - 4x^6$

c) $y = \dfrac{1}{x}$
d) $y = \dfrac{2}{x^2} - x$

e) $y = \dfrac{1}{\sqrt{x}} - \sqrt{x}$
f) $y = \sqrt[3]{x} - x^3$

g) $y = x^3(x^2 + 5)$
h) $y = (x^2 - 1)(2x + 3)$

i) $y = \dfrac{x^2 - 1}{x}$
j) $y = \dfrac{6x - 5}{x^2}$

2 Given that a, b, c and d are constants, find the values of $\dfrac{dy}{dx}$ and $\dfrac{d^2y}{dx^2}$ for each of these functions.

a) $y = ax^2 + bx + c$
b) $y = \dfrac{a}{x} + \dfrac{b}{x^2}$

c) $y = a\sqrt{x} + \dfrac{b}{\sqrt{x}}$
d) $y = (ax + b)(cx + d)$

e) $y = \dfrac{ax^3 + bx^2}{cx}$

4.5   Tangents and normals to a curve

As you saw on page 108, the gradient of a curve at a point P is the gradient of the tangent to the curve at the point P.

Example 1

Find the gradient of the curve $f(x) = x^2 + \dfrac{1}{x}$ at the point P(1, 2).

⋯⋯⋯⋯⋯⋯⋯⋯⋯⋯⋯⋯⋯⋯⋯⋯⋯⋯⋯⋯⋯⋯⋯⋯⋯⋯⋯⋯

To find the gradient of the curve, first find $f'(x)$:

$$f(x) = x^2 + \frac{1}{x}$$
$$= x^2 + x^{-1}$$
$$\therefore f'(x) = 2x - x^{-2}$$
$$= 2x - \frac{1}{x^2}$$

The gradient of the curve at the point P(1, 2) is given by $f'(1)$:

$$f'(1) = 2(1) - \frac{1}{(1)^2} = 1$$

The gradient of the curve at point P is 1.

You may be given the gradient of a curve at a point and asked to find the coordinates of the point.

Example 2

The gradient of the curve $y = 3x^2 + x - 3$ at the point P is 13. Find the coordinates of point P.

⋯⋯⋯⋯⋯⋯⋯⋯⋯⋯⋯⋯⋯⋯⋯⋯⋯⋯⋯⋯⋯⋯⋯⋯⋯⋯⋯⋯

When $y = 3x^2 + x - 3$, then $\dfrac{dy}{dx} = 6x + 1$.

$\dfrac{dy}{dx} = 13$ at point P. Therefore,

$$6x + 1 = 13$$
$$\therefore x = 2$$

To find the y-coordinate of point P substitute $x = 2$ into $y = 3x^2 + x - 3$:

$$y = 3(2)^2 + 2 - 3$$
$$\therefore y = 11$$

The coordinates of point P are (2, 11).

Example 3

The curve C is given by $y = ax^2 + b\sqrt{x}$, where a and b are constants. Given that the gradient of C at the point $(1, 1)$ is 5, find a and b.

Rewriting $y = ax^2 + b\sqrt{x}$ as $y = ax^2 + bx^{\frac{1}{2}}$ and then differentiating gives

$$\frac{dy}{dx} = 2ax + \frac{b}{2}x^{-\frac{1}{2}}$$

$$= 2ax + \frac{b}{2x^{\frac{1}{2}}}$$

Now the point $(1, 1)$ is on the curve C. So:

$$1 = a(1)^2 + b(1)^{\frac{1}{2}}$$

$$\therefore \ 1 = a + b \qquad\qquad [1]$$

The gradient of the curve when $x = 1$ is 5. So:

$$2a(1) + \frac{b}{2(1)^{\frac{1}{2}}} = 5$$

$$\therefore \ 4a + b = 10 \qquad\qquad [2]$$

Solving [1] and [2] simultaneously with the GDC or algebraically gives $a = 3$ and $b = -2$.

Exercise 4E

1 Find the gradient of each of these curves at the point given.

a) $y = x^2$, at $(3, 9)$

b) $y = 2x^3 - 4$, at $(2, 12)$

c) $y = \sqrt{x} + 2$, at $(9, 5)$

d) $y = \dfrac{1}{x}$, at $(3, \frac{1}{3})$

e) $y = 5 - x^2$, at $(-2, 1)$

f) $y = 3 - \dfrac{2}{x}$, at $(4, \frac{5}{2})$

g) $y = x + \dfrac{3}{x}$, at $(3, 4)$

h) $y = 2 - \dfrac{4}{x^2}$, at $(-2, 1)$

i) $y = 3x + 7$, at $(-3, -2)$

j) $y = \dfrac{x + 5}{x}$, at $(-1, -4)$

2 Find the coordinates of any points on each of these curves where the gradient is as stated.

a) $y = x^3$, grad 12

b) $y = 3x^2$, grad -6

c) $y = x^4 + 1$, grad 32

d) $y = \dfrac{4}{x}$, grad -16

e) $y = \dfrac{16}{x^2}$, grad 4

f) $y = x^3 + 2x - 1$, grad 29

g) $y = x^3 - x^2 + 3$, grad 0

h) $y = \sqrt{x} + 5$, grad 1

i) $y = \dfrac{4 - x}{x}$, grad -1

j) $y = \dfrac{x^2 + 3}{2x^2}$, grad 3

3 The curve C is defined by $y = ax^2 + b$, where a and b are constants. Given that the gradient of the curve at the point $(2, -2)$ is 3, find the values of a and b.

4 Given that the curve with equation $y = Ax^2 + Bx$ has gradient 7 at the point $(6, 8)$, find the values of the constants A and B.

5 A curve whose equation is $y = \dfrac{a}{x} + c$ passes through the point $(3, 9)$ with gradient 5. Find the values of the constants a and c.

6 Given that the curve with equation $y = a\sqrt{x} + b$ has gradient 3 at the point $(4, 6)$, find the values of the constants a and b.

7 A curve with equation $y = A\sqrt{x} + \dfrac{B}{\sqrt{x}}$, for constants A and B, passes through the point $(1, 6)$ with gradient -1. Find A and B.

Equation of a tangent

You can use differentiation to find the equation of the tangent line to a curve at a particular point.

Example 4

Find the equation of the tangent to the curve $y = x^2$ at the point P(3, 9).

....

The gradient of the tangent to the curve at point P is given by $\dfrac{dy}{dx}$ evaluated when $x = 3$.

$y = x^2$ so:
$$\frac{dy}{dx} = 2x$$
At the point P(3, 9),
$$\left.\frac{dy}{dx}\right|_{x=3} = 2(3) = 6$$

The tangent is a straight line, so its equation is given by $y = mx + c$. The gradient of the tangent at point P is 6, so the equation of the line is
$$y = 6x + c$$
The tangent passes through P(3, 9). Therefore,
$$9 = 6(3) + c$$
$$\therefore\ c = -9$$
The equation of the tangent is $y = 6x - 9$.

> $\left.\dfrac{dy}{dx}\right|_{x=3}$ means 'the value of the derived function when $x = 3$'.

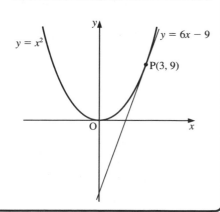

Note: The result in example 4 can easily be checked and illustrated using your GDC.

Plot the graphs of $y = x^2$ and $y = 6x - 9$ on the same set of axes to see that $y = 6x - 9$ is a tangent to the curve $y = x^2$.

Using your GDC show that the tangent and the curve intersect at the point $(3, 9)$.

Example 5

Find the equation of the tangent to the curve $f(x) = \dfrac{1}{x^2}$ at the point P$(-1, 1)$. Find the coordinates of the point where this tangent meets the curve again.

....

$f(x) = \dfrac{1}{x^2} = x^{-2}$. Therefore,

$$f'(x) = -2x^{-3} = -\frac{2}{x^3}$$

The gradient of the tangent to the curve at the point P$(-1, 1)$ is

$$f'(-1) = -\frac{2}{(-1)^3} = 2$$

The equation of the tangent is $y = 2x + c$.

The tangent passes through P$(-1, 1)$. So:

$$1 = 2(-1) + c$$

$$\therefore \ c = 3$$

The equation of the tangent to the curve at the point P is $y = 2x + 3$.

The tangent meets the curve again at the points whose x-coordinates satisfy

$$2x + 3 = \frac{1}{x^2}$$

$$2x^3 + 3x^2 - 1 = 0$$

$$\therefore \ (2x - 1)(x^2 + 2x + 1) = 0$$

> Factorise.

Solving this equation gives

$$2x - 1 = 0 \quad \text{or} \quad x^2 + 2x + 1 = 0$$

$$\therefore \quad x = \tfrac{1}{2} \quad \text{or} \quad (x + 1)(x + 1) = 0$$

$$\therefore \quad x = -1$$

> Factorise again.

When $x = \tfrac{1}{2}$, $f(\tfrac{1}{2}) = 4$.

When $x = -1$, $f(-1) = 1$. This is, in fact, just point P.

The tangent meets the curve again at the point with coordinates $(\tfrac{1}{2}, 4)$.

> $x = -1$ is a repeated root, which means the straight line, just touches the curve – it is a tangent (see page 108).

When two straight lines are perpendicular, the product of their gradients is -1.

Proof

Let PQ and RS be two straight lines which are perpendicular, as shown.

By inspection of the diagram, angle UWV $= \theta$.

Let the gradient of PQ be m. The gradient of RS, m_{RS}, is given by

$$m_{RS} = -\frac{WU}{UV} \qquad [1]$$

From triangle WUV,

$$\tan \theta = \frac{UV}{WU}$$

$$\therefore \ UV = WU \tan \theta \qquad [2]$$

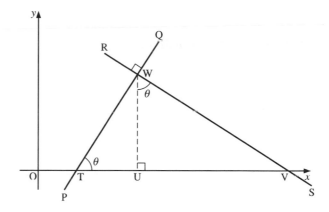

Substituting [2] into [1] gives

$$m_{RS} = -\frac{WU}{WU \tan \theta} = -\frac{1}{\tan \theta} = -\frac{1}{m}$$

Now

$$\text{Gradient of PQ} \times \text{Gradient of RS} = m \times \left(-\frac{1}{m}\right) = -1$$

Therefore, when two straight lines are perpendicular, the product of their gradients is -1.

Equation of a normal

The **normal** to a curve at a point P is the straight line through P which is perpendicular to the tangent at P.

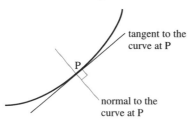

tangent to the curve at P

normal to the curve at P

Since the tangent and normal are perpendicular to each other, if the gradient of the tangent is m, then the gradient of the normal is $-\dfrac{1}{m}$.

Example 6

Find the equation of the normal to the curve $y = 3x^2 + 7x - 2$ at the point P where $x = -1$.

....

When $x = -1$,

$$y = 3(-1)^2 + 7(-1) - 2 = -6$$

Therefore, P has coordinates $(-1, -6)$.

When $y = 3x^2 + 7x - 2$, then $\dfrac{dy}{dx} = 6x + 7$.

At the point P$(-1, -6)$,

$$\left.\frac{dy}{dx}\right|_{x=-1} = 6(-1) + 7 = 1$$

The gradient of the tangent line at P is 1. Therefore, the

gradient of the normal at P is $-\dfrac{1}{(1)} = -1$.

The equation of the normal at P is

$$y = -x + c$$

The normal passes through P$(-1, -6)$. Therefore,

$$-6 = -(-1) + c$$

$$\therefore \ c = -7$$

The equation of the normal to the curve at the point P$(-1, -6)$ is $y = -x - 7$.

Exercise 4F

....

1 Find the equation of the tangent to each of the following curves at the point indicated by the given value of x.

   a) $y = x^2 + 3$, where $x = 2$ b) $y = 2x^3 - 1$, where $x = 1$

   c) $y = \dfrac{9}{x}$, where $x = -3$ d) $y = 6x - x^2$, where $x = 4$

   e) $y = 5 - \dfrac{8}{x^2}$, where $x = -2$ f) $y = 6\sqrt{x}$, where $x = 4$

   g) $y = x^2 - 10x + 30$, where $x = 5$

   h) $y = \dfrac{x + 4}{x}$, where $x = -2$ i) $y = x(x^2 - 3)$, where $x = 2$

   j) $y = \dfrac{x + 5}{\sqrt{x}}$, where $x = 25$

2 Find the equation of the normal to each of the following curves at the point indicated by the given value of x.

a) $y = x^2 - 3x$, where $x = 2$

b) $y = x^3 + 4$, where $x = -1$

c) $y = \dfrac{6}{x}$, where $x = 3$

d) $y = 2\sqrt{x}$, where $x = 9$

e) $y = 6 - \dfrac{1}{x^2}$, where $x = 1$

f) $y = x^3 + 2x^2 - 3$, where $x = -2$

3 The two tangents to the curve $y = x^2$ at the points where $y = 9$ intersect at the point P. Find the coordinates of P.

4 a) Find the values of x at which the gradient of the curve $y = x^3 - 6x^2 + 9x + 2$ is zero.

b) Hence find the equations of the tangents to the curve which are parallel to the x-axis.

5 Find the equations of the two normals to the curve $y = 8 + 15x + 3x^2 - x^3$ which have gradient $\frac{1}{9}$.

6 Show that there is no point on the curve $y = x^3 + 6x - 1$ at which the gradient of the tangent is 3.

7 a) Find the equation of the tangent at the point $(1, 2)$ on the curve $y = x^3 + 3x - 2$.

b) Use your GDC to find the coordinates of the point where this tangent meets the curve again.

8 a) Find the equation of the normal at the point $(2, 3)$ on the curve $y = 2x^3 - 12x^2 + 23x - 11$.

b) Use your GDC to find the coordinates of the points where this normal meets the curve again.

9 The tangent to the curve $y = ax^2 + 1$ at the point $(1, b)$ has gradient 6. Find the values of the constants a and b.

10 The normal to the curve $y = x^3 + cx$ at the point $(2, d)$ has gradient $\frac{1}{2}$. Find the values of the constants c and d.

4.6   Gradient as a rate of change

The gradient of a curve is given by

$$\frac{dy}{dx} = \frac{\text{difference in } y}{\text{difference in } x}$$

The derivative $\frac{dy}{dx}$ therefore represents the **rate of change of y with respect to x.**

It shows how changes in y are related to changes in x.

A common rate of change connects velocity, displacement and time.

Suppose s is the displacement of a body at time t, as shown in the graph relating s and t.

Suppose P and Q are any two points on the graph as shown. The length QR represents the distance travelled between the two points P and Q. The length PR represents the time taken to travel this distance.

The gradient of the line PQ represents the average velocity of the body during this time interval.

You can use the same technique as in section 4.1 to show that as Q moves closer to P the gradient of the line PQ gets closer to the gradient of the curve at P. Therefore the gradient of the tangent at P is equal to the **velocity** at that instant (point). So:

> The velocity v at time t is given by $v = \frac{ds}{dt}$.

> For example, if $\frac{dy}{dx} = 3$, then y is increasing 3 times as fast as x.

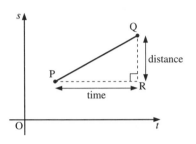

> See page 108.

Example 1

The displacement of a body at a time t seconds is given in metres by $s = t^2 + 3t$. Find

a) the velocity of the body at time t

b) the initial velocity of the body.

....

a) The velocity v is given by

$$v = \frac{ds}{dt} = 2t + 3$$

b) The initial velocity is the velocity when $t = 0$.
   Substituting $t = 0$ gives

$$v(0) = \frac{ds}{dt}\bigg|_{t=0} = 2(0) + 3$$
$$= 3$$

The initial velocity of the body is 3 m s^{-1}.

> When distance is in metres (m) and time in seconds (s), the units of velocity are metres per second (m s^{-1}).

Similarly, the rate of change of velocity with respect to time is called the **acceleration**.

> The acceleration a at time t is given by $a = \dfrac{dv}{dt}$.

Example 2

The displacement s metres of a body at time t seconds is given by $s = 2t^3 - t^2 + 2$. Find

a) the velocity after 1 second

b) the acceleration after 1 second

c) the time at which the acceleration is zero.

....

a) If $s = 2t^3 - t^2 + 2$ then the velocity is given by

$$v = \frac{ds}{dt} = 6t^2 - 2t$$

When $t = 1$,

$$v(1) = \frac{ds}{dt}\bigg|_{t=1} = 6(1)^2 - 2(1)$$
$$= 4$$

The velocity is 4 m s^{-1} after 1 second.

b) The acceleration is given by

$$a = \frac{dv}{dt} = 12t - 2$$

When $t = 1$,

$$a(1) = \frac{dv}{dt}\bigg|_{t=1} = 12(1) - 2$$
$$= 10$$

The acceleration is 10 m s^{-2} after 1 second.

c) When the acceleration is zero,

$$12t - 2 = 0$$
$$\therefore\ t = \tfrac{2}{12} = \tfrac{1}{6}$$

The acceleration is zero when $t = \tfrac{1}{6}$ second.

> The units of acceleration are metres per second per second (m s^{-2}).

Exercise 4G

....

1 The displacement, s metres, of a body at a time t seconds is given by the formula $s = t^2 + 3t$.

a) Find an expression for the velocity of the body at time t.

b) Calculate the velocity of the body when $t = 2$.

2 A body moves such that its displacement, s metres, at time t seconds, is given by the formula $s = 2t^3 + 5t$.

   a) Find an expression for the velocity of the body at time t.

   b) Calculate the velocity of the body when $t = 3$.

3 Given $s = 3t^2 - 5t^3$, where s is displacement at time t, find

   a) an expression for the velocity, v

   b) an expression for the acceleration, a.

4 Given $s = 1 + t + t^3$, where s is displacement at time t, find

   a) an expression for the velocity, v

   b) an expression for the acceleration, a.

> In questions 3 and 4, the displacement is in metres and the time in seconds.

5 The velocity, v m s^{-1} of a body at a time t seconds is given by the formula $v = 3t^2 - 2t$.

   a) Find an expression for the acceleration of the body at time t.

   b) Calculate the acceleration of the body when $t = 6$.

6 A body moves in such a way that its velocity, v m s^{-1}, at time t seconds is given by the formula $v = t^3 - 3t^2 + 6t$.

   a) Find an expression for the acceleration of the body at time t.

   b) Calculate the initial acceleration of the body.

7 The displacement, s metres, of a body at a time t seconds is given by the formula $s = t^2 - 8t$.

   a) Find an expression for the velocity of the body at time t.

   b) Calculate the value of t when the body is at rest.

8 A body moves such that its displacement, s metres, at time t seconds is given by the formula $s = t^3 - 6t^2 + 9t + 5$.

   a) Calculate the times at which the body is at rest.

   b) Find the values of s at these times.

9 A body moves such that its displacement, s metres, at a time t seconds is given by the formula $s = 14 + 9t^2 - t^3$.
Find the maximum velocity of the body.

....

4.7   Maximum, minimum and point of inflexion

> A point on a curve at which the gradient is zero, where $\dfrac{dy}{dx} = 0$,
> is called a **stationary point**.

At a stationary point, the tangent to the curve is horizontal and the curve is 'flat'.

There are three types of stationary point and you must know how to distinguish one from another.

Minimum point

In this case, the gradient of the curve is negative to the left of point P. To the right of point P, the gradient of the curve is positive.

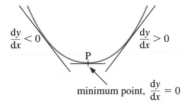

To the left of P	At point P	To the right of P
$\dfrac{\mathrm{d}y}{\mathrm{d}x} < 0$	$\dfrac{\mathrm{d}y}{\mathrm{d}x} = 0$	$\dfrac{\mathrm{d}y}{\mathrm{d}x} > 0$

Maximum point

In this case, the gradient of the curve is positive to the left of point P. To the right of point P, the gradient of the curve is negative.

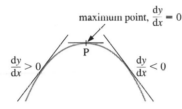

To the left of P	At point P	To the right of P
$\dfrac{\mathrm{d}y}{\mathrm{d}x} > 0$	$\dfrac{\mathrm{d}y}{\mathrm{d}x} = 0$	$\dfrac{\mathrm{d}y}{\mathrm{d}x} < 0$

> Maximum and minimum points are also called 'turning points', as this is where the graph 'turns' and the gradient changes from positive to negative or negative to positive.

Point of inflexion

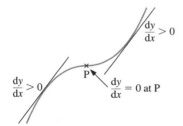 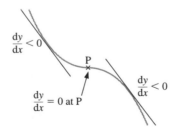

In this case, the gradient has the *same sign* each side of the stationary point. A point of inflexion which has zero gradient (such as this one) is called a **horizontal point of inflexion**. This distinguishes it from a point of inflexion which has non-zero gradient. Generally, no distinction is made between these two types and all are just called points of inflexion.

Example 1

Find the coordinates of the stationary points on the curve
$y = x^3 + 3x^2 + 1$ and determine their nature. Sketch the curve.

....

When $y = x^3 + 3x^2 + 1$ then $\dfrac{dy}{dx} = 3x^2 + 6x$.

At a stationary point, $\dfrac{dy}{dx} = 0$. Therefore:

$$3x^2 + 6x = 0$$

$$\therefore\ 3x(x + 2) = 0$$

Solving gives $x = 0$ or $x = -2$.

When $x = 0$: $y = (0)^3 + 3(0)^2 + 1 = 1$

When $x = -2$: $y = (-2)^3 + 3(-2)^2 + 1 = 5$

The coordinates of the stationary points are $(0, 1)$ and $(-2, 5)$.

To determine the nature of each stationary point, you must
examine the gradient each side of each point.

For the point $(0, 1)$:

x	-1	0	1
$\dfrac{dy}{dx}$	-3	0	9

    negative   positive

     \   —   /

The stationary point $(0, 1)$ is a minimum.

For the point $(-2, 5)$,

x	-3	-2	-1
$\dfrac{dy}{dx}$	9	0	-3

    positive   negative

     /   —   \

The stationary point $(-2, 5)$ is a maximum.

There are two stationary points on the curve: $(0, 1)$ a minimum
and $(-2, 5)$ a maximum.

To sketch the curve, plot the stationary points and notice that
the curve $y = x^3 + 3x^2 + 1$ cuts the y-axis at the point $(0, 1)$.

> Differentiate.

> Solve the quadratic for x.

> Find the corresponding
> y-coordinates.

> Note that when you choose x
> values on each side of $x = 0$,
> the chosen interval must not
> include any other stationary
> points.

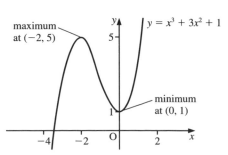

Second derivative and stationary points

The diagram shows the graph of some function $y = f(x)$ which possesses a maximum, a minimum and a point of inflexion. The graph of the derived function is shown below it.

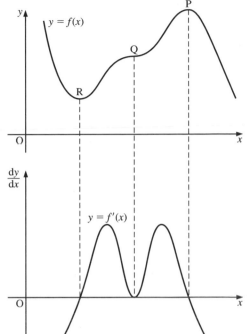

At all stationary points on $y = f(x)$ the gradient of the curve is zero. That is, $\dfrac{dy}{dx} = 0$.

❖ At the maximum point P, the gradient of the derived function is negative. That is,

$$\frac{d^2y}{dx^2} < 0 \quad \text{at a maximum}$$

The curve is said to be **concave down** at P.

❖ At the point of inflexion Q, the gradient of the derived function is zero. That is,

$$\frac{d^2y}{dx^2} = 0$$

However, the second derivative can sometimes also be zero at a maximum or a minimum. For this reason, you must also examine the sign of $\dfrac{dy}{dx}$ each side of the point.

$\dfrac{dy}{dx}$ gives the gradient of the function $y = f(x)$.

❖ At the minimum point R, the gradient of the derived function is positive. That is,

$$\frac{d^2y}{dx^2} > 0 \quad \text{at a minimum}$$

The curve is said to be **concave up** at R.

$\dfrac{d^2y}{dx^2}$ gives the gradient of the derived function $y = f'(x)$

This provides an easier method for classifying stationary points.

Example 2

Find the coordinates of the stationary points on the curve $f(x) = x^3 - 6x^2 - 15x + 1$, and using the second derivative determine their nature.

....

When $f(x) = x^3 - 6x^2 - 15x + 1$, then

$$f'(x) = 3x^2 - 12x - 15$$

Differentiate.

At stationary points, $f'(x) = 0$. That is,

$$3x^2 - 12x - 15 = 0$$

$$3(x^2 - 4x - 5) = 0$$

$$\therefore\ 3(x - 5)(x + 1) = 0$$

Solve the quadratic.

Solving gives $x = 5$ or $x = -1$.

When $x = 5$: $f(5) = (5)^3 - 6(5)^2 - 15(5) + 1 = -99$

When $x = -1$: $f(-1) = (-1)^3 - 6(-1)^2 - 15(-1) + 1 = 9$

The stationary points are $(5, -99)$ and $(-1, 9)$.

The second derivative is given by $f''(x) = 6x - 12$.
At the point $(5, -99)$,

$$f''(5) = 6(5) - 12 = 18 > 0$$

Therefore, the stationary point $(5, -99)$ is a minimum.
At the point $(-1, 9)$,

$$f''(-1) = 6(-1) - 12 = -18 < 0$$

Therefore, the stationary point $(-1, 9)$ is a maximum.

> Differentiate again.

If $\dfrac{d^2y}{dx^2} = 0$ at a stationary point, this does not necessarily mean

that it is a point of inflexion. You must examine the sign of $\dfrac{dy}{dx}$ on

each side of the point to see how the curve behaves.

Example 3

Find the coordinates of the stationary points on the curve
$y = x^4 - 4x^3$ and determine their nature. Sketch the curve.

....

When $y = x^4 - 4x^3$, then $\dfrac{dy}{dx} = 4x^3 - 12x^2$.

At a stationary point, $\dfrac{dy}{dx} = 0$.

So:

$$4x^3 - 12x^2 = 0$$
$$\therefore\ 4x^2(x - 3) = 0$$

Solving gives $x = 0$ or $x = 3$.

When $x = 0$: $\quad y = (0)^4 - 4(0)^3 = 0$
When $x = 3$: $\quad y = (3)^4 - 4(3)^3 = -27$

The coordinates of the stationary points are $(0, 0)$ and $(3, -27)$.
To determine the nature of each stationary point, check the sign
of $\dfrac{d^2y}{dx^2}$:

$$\frac{dy}{dx} = 4x^3 - 12x^2$$

$$\therefore\ \frac{d^2y}{dx^2} = 12x^2 - 24x$$

For the point $(0, 0)$,

$$\left.\frac{d^2y}{dx^2}\right|_{x=0} = 12(0)^2 - 24(0) = 0$$

This is not sufficient to conclude that the point $(0, 0)$ is a point of inflexion. You must examine the sign of $\dfrac{dy}{dx}$ each side of the point $(0, 0)$:

x	-1	0	1
$\dfrac{dy}{dx}$	-16	0	-8

   negative   negative

   $\searrow$ $-$ $\searrow$

The stationary point $(0, 0)$ *is* a point of inflexion, because the gradient of the curve has the same sign each side of $x = 0$.

For the point $(3, -27)$,

$$\left.\frac{d^2y}{dx^2}\right|_{x=3} = 12(3)^2 - 24(3) = 36 > 0$$

The stationary point $(3, -27)$ is a minimum.

There are two stationary points on the curve: $(0, 0)$ a point of inflexion and $(3, -27)$ a minimum.

To sketch the curve, plot the stationary points and notice that the curve crosses the x-axis when $y = 0$.

$$x^4 - 4x^3 = 0$$

$$\therefore \ x^3(x - 4) = 0$$

Solving gives $x = 0$ or $x = 4$.

The curve crosses the x-axis at $(0, 0)$ and $(4, 0)$.

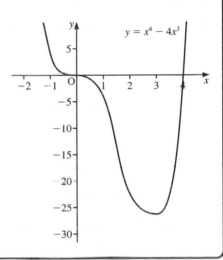

Example 4

a) Find the coordinates of the point of inflexion on the curve given by $y = x^3 - 6x^2 + x + 1$.

b) Using the GDC, plot the graph of $y = x^3 - 6x^2 + x + 1$ and estimate the x-coordinates of the stationary points on the curve.

....

a) For points of inflexion you need to look for $\dfrac{d^2y}{dx^2} = 0$ and the same gradient sign (both positive or both negative) either side of the point.

If $y = x^3 - 6x^2 + x + 1$ then

$$\frac{dy}{dx} = 3x^2 - 12x + 1$$

and

$$\frac{d^2y}{dx^2} = 6x - 12$$

> The point of inflexion is where the curve changes from concave down to concave up, or vice versa.

When $\dfrac{d^2y}{dx^2} = 0$,

$$6x - 12 = 0$$

$$\therefore \ x = 2$$

When $x = 2$, $y = (2)^3 - 6(2)^2 + 2 + 1 = -13$

Check the gradient either side of the point $(2, -13)$:

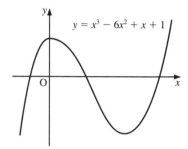

x	1	2	3
$\dfrac{dy}{dx}$	-8	-11	-8

On either side of the point $(2, -13)$ the gradient is negative.
Therefore the point $(2, -13)$ is a point of inflexion on the curve.

b) Using the GDC to plot the graph gives:

$y = x^3 - 6x^2 + x + 1$

Using the zoom function gives the x-coordinates of the
stationary points as $x = 0$ and $x = 4$.

Note that you can find the exact values of these

x-coordinates by putting $\dfrac{dy}{dx} = 0$:

$$3x^2 - 12x + 1 = 0$$

and solving to give $x = 2 + \tfrac{1}{3}\sqrt{33}$ and $x = 2 - \tfrac{1}{3}\sqrt{33}$.

> Use the quadratic formula
> (see page 30).

Increasing and decreasing functions

The function $f(x) = x$ increases as the values of x increase, and this
is true for all real values of x. The graph of the function shows this.

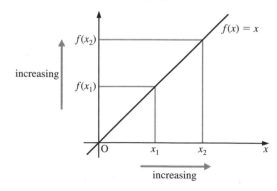

The function $g(x) = x^2$ behaves differently.

$g(x)$ increases as the values of x increase for $x > 0$.

$g(x)$ decreases as the values of x increase for $x < 0$.

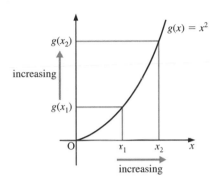
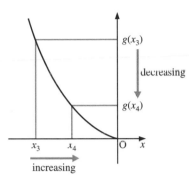

Generally, a function $f(x)$ is increasing when $x = a$ if $f'(a) > 0$. It is decreasing if $f'(a) < 0$. It is useful to be able to identify the turning points on the graph of a function when dealing with intervals of increase and decrease.

Example 5

Determine the range of values of x for which these functions increase and decrease.

a) $y = \dfrac{1}{x^2}$ $(x \neq 0)$

b) $y = \dfrac{1}{x^3}$ $(x \neq 0)$

. . . .

a) If $y = \dfrac{1}{x^2} = x^{-2}$ then $\dfrac{dy}{dx} = -2x^{-3} = -\dfrac{2}{x^3}$.

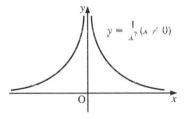

You can see that

$$\dfrac{dy}{dx} < 0 \text{ when } x > 0 \quad \text{and} \quad \dfrac{dy}{dx} > 0 \text{ when } x < 0$$

Therefore the function is decreasing for $x > 0$ and increasing for $x < 0$.

b) If $y = \dfrac{1}{x^3} = x^{-3}$ then $\dfrac{dy}{dx} = -3x^{-4} = -\dfrac{3}{x^4}$.

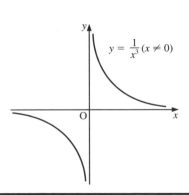

You can see that

$$\dfrac{dy}{dx} < 0 \text{ for all values of } x \, (x \neq 0)$$

Therefore the function is decreasing for all values of $x \, (x \neq 0)$.

Example 6

Determine the intervals of increase and decrease for each of these functions.

a) $f(x) = x^2 - 6x$

b) $f(x) = x^4 - 8x^2 - 3$

....

a) When $f(x) = x^2 - 6x$, then $f'(x) = 2x - 6$.
At a stationary point, $f'(x) = 0$. So:

$$2x - 6 = 0$$

$$\therefore \ x = 3$$

Now, $f''(x) = 2$ so $f''(x) > 0$ for all values of x.
Therefore, at $x = 3$, the curve has a minimum value.

From the graph it is clear that $f(x)$ is increasing for $x > 3$, and decreasing for $x < 3$.

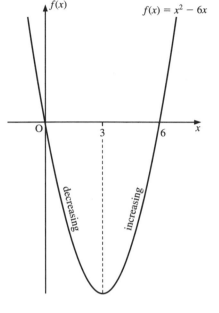

b) When $f(x) = x^4 - 8x^2 - 3$, then $f'(x) = 4x^3 - 16x$.
At a stationary point, $f'(x) = 0$. So:

$$4x^3 - 16x = 0$$

$$\therefore \ 4x(x - 2)(x + 2) = 0$$

which gives $x = -2$, $x = 0$ or $x = 2$.

Now, $f''(x) = 12x^2 - 16$.

When $x = -2$, $f''(-2) = 12(-2)^2 - 16 = 32 > 0$.
Therefore, at $x = -2$, the curve has a minimum value.

When $x = 0$, $f''(0) = 12(0)^2 - 16 = -16 < 0$.
Therefore, at $x = 0$, the curve has a maximum value.

When $x = 2$, $f''(2) = 12(2)^2 - 16 = 32 > 0$.
Therefore, at $x = 2$, the curve has a minimum value.

From the graph it is clear that $f(x)$ is increasing for $-2 < x < 0$ and $x > 2$, and that $f(x)$ is decreasing for $x < -2$ and $0 < x < 2$.

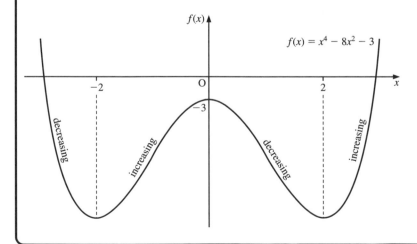

Exercise 4H

1 Find the coordinates of the points on each of these curves at which the gradient is zero.

a) $y = x^2 - 4x + 3$

b) $y = x^2 + 6x + 5$

c) $y = 6 - x^2$

d) $y = 3 - 5x + x^2$

e) $y = 2x^2 - 3x + 1$

f) $y = x^3 - 3x + 2$

g) $y = x^3 - 6x^2 - 36x$

h) $y = 6 + 9x - 3x^2 - x^3$

i) $y = 5 + 3x^2 - x^3$

j) $y = x^4 - 2x^2 + 3$

k) $y = x^4 - 32x + 3$

l) $y = 1 - 6x + 6x^2 + 2x^3 - 3x^4$

2 Find the coordinates of the stationary points on each of these curves, and determine their nature.

a) $y = x^2 - 2x + 5$

b) $y = x^2 + 4x + 2$

c) $y = 3 + x - x^2$

d) $y = (x - 4)(x - 2)$

e) $y = 3(x + 3)(2x - 1)$

f) $y = (x - 5)^2$

g) $y = x^3 + 6x^2 - 36x$

h) $y = x^3 - 5x^2 + 3x + 1$

i) $y = 3 + 15x - 6x^2 - x^3$

j) $y = x^4 - 8x^2 + 3$

k) $y = x^4 + 4x^3 + 1$

l) $y = x^4 - 14x^2 + 24x - 10$

3 Find the coordinates of the stationary points on these curves. In each case, determine their nature and sketch the curve.

a) $y = x^3 - 3x + 3$

b) $y = -x^3 + 6x^2 - 9x$

c) $y = x^3 - 9x^2 + 27x - 19$

d) $y = x^4 - 8x^2 - 9$

e) $y = 8x^3 - x^4$

f) $y = x^4 + x^3 - 3x^2 - 5x - 2$

> Use your GDC.

4 Find the coordinates of the stationary points on each of these curves, and determine their nature. Use your GDC to sketch the curve.

a) $y = x + \dfrac{1}{x}$

b) $y = x^2 + \dfrac{16}{x}$

c) $y = \dfrac{1}{x} - \dfrac{3}{x^2}$

d) $y = \dfrac{2}{x^3} - \dfrac{1}{x^2}$

e) $y = \dfrac{12x^2 - 1}{x^3}$

f) $y = \dfrac{2 - x^3}{x^4}$

5 Determine the intervals of increase and decrease for each of these functions.

a) $f(x) = x^2 - 8x + 3$

b) $f(x) = 7 - 2x - x^2$

c) $f(x) = x^3 - 3x^2 - 9x + 4$

d) $f(x) = 5 + 36x - 3x^2 - 2x^3$

e) $f(x) = x^4 + 4x^3 - 8x^2 - 48x + 20$

f) $f(x) = x + \dfrac{4}{x}$

6 Find the coordinates of the points of inflexion on each of the following curves:

a) $y = x^3 + 5x - 2$

b) $y = x^3 + 3x^2 + 3x - 2$

c) $y = 2x^3 + 10x - 2$

d) $y = x^3 + 6x^2 + 3x - 1$

e) $y = x^4 - 6x^2 + 9x$

f) $y = x^4 - 8x^3 + 18x^2 - 2$

7 By investigating the stationary points of $f(x) = x^3 + 3x^2 + 6x - 30$ and sketching the curve $y = f(x)$, show that the equation $f(x) = 0$ has only one real solution.

8 Show that the equation $x^4 - 4x^3 - 2x^2 + 12x + 12 = 0$ has no real solution.

9 Show that the equation $3x^4 + 4x^3 - 36x^2 + 64 = 0$ has precisely three real solutions.

4.8 Practical applications of maxima and minima

You can use differentiation to solve practical problems.
For example, it can be used to find the maximum (or minimum) possible value of an area or volume.

Example 1

In the right-angled triangle ABC shown, the lengths AB and BC vary such that their sum is always 6 cm.

a) If the length of AB is x cm, write down, in terms of x, the length BC.

b) Find the maximum area of triangle ABC.

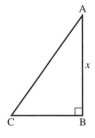

a) $\qquad$ AB + BC = 6

$\qquad$ x + BC = 6

$\qquad$ $\therefore$ BC = 6 - x

b) The area, A, of triangle ABC is given by
$$A = \tfrac{1}{2} \times BC \times AB$$
$$= \tfrac{1}{2}(6 - x)x$$
$$\therefore\ A = 3x - \frac{x^2}{2} \qquad [1]$$

> **Remember:**
>
> area of a triangle
> $= \tfrac{1}{2} \times$ base $\times$ height

You can see that A reaches a maximum value because the coefficient of the x^2 term is negative.
(That is, the graph of x against A is $\cap$ shaped.)
The maximum value of A occurs when $\dfrac{dA}{dx} = 0$.

$$\frac{dA}{dx} = 3 - x$$

> Differentiate [1] to find $\dfrac{dA}{dx}$.

When $\dfrac{dA}{dx} = 0$: $\qquad 3 - x = 0$

$\qquad\qquad\qquad \therefore\ x = 3$

The area of the triangle is a maximum when $x = 3$.

Therefore, the maximum area of triangle ABC is found by substituting $x = 3$ into [1], which gives:

$$A_{max} = 3(3) - \frac{(3)^2}{2} = \frac{9}{2} = 4.5$$

The maximum area of triangle ABC is 4.5 cm².

Example 2

A closed, right circular cylinder of base radius r cm and height h cm has a volume of 54π cm³. Show that S, the total surface area of the cylinder, is given by

$$S = \frac{108\pi}{r} + 2\pi r^2$$

Hence find the radius and height which make the surface area a minimum.

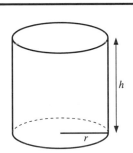

....

The volume, V, is

$$V = \pi r^2 h \qquad\qquad [1]$$

The total surface area, S, is

$$S = 2\pi rh + 2\pi r^2 \qquad\qquad [2]$$

$V = 54\pi$. Therefore, from [1]

$$\pi r^2 h = 54\pi$$

$$\therefore\ h = \frac{54}{r^2}$$

Substituting $h = \dfrac{54}{r^2}$ into [2] gives

$$S = 2\pi r\left(\frac{54}{r^2}\right) + 2\pi r^2$$

$$\therefore\ S = \frac{108\pi}{r} + 2\pi r^2$$

as required.

Maximum/minimum surface area occurs when $\dfrac{dS}{dr} = 0$.

$$S = 108\pi r^{-1} + 2\pi r^2$$

$$\therefore\ \frac{dS}{dr} = -108\pi r^{-2} + 4\pi r = -\frac{108\pi}{r^2} + 4\pi r$$

When $\dfrac{dS}{dr} = 0$: $\quad -\dfrac{108\pi}{r^2} + 4\pi r = 0$

$$4\pi r^3 = 108\pi$$

$$\therefore\ r^3 = 27$$

The radius, r, is 3 cm.

When $r = 3$: $\quad h = \dfrac{54}{(3)^2} = 6$

The height is 6 cm.

You must check that this value of r corresponds to a minimum, using the second derivative:

$$\frac{d^2S}{dr^2} = 216\pi r^{-3} + 4\pi$$

When $r = 3$,

$$\frac{d^2S}{dr^2} = \frac{216\pi}{(3)^3} + 4\pi = 12\pi$$

which is positive. That is, $\dfrac{d^2S}{dr^2} > 0$ when $r = 3$,

so $r = 3$ corresponds to a minimum S.

So, the surface area is a minimum when the radius is 3 cm and the height is 6 cm.

Exercise 4I

....

In these questions you should use calculus to solve the problem, and then use your GDC to confirm your answer.

1 The profit, $\$y$, generated from the sale of x items of a certain luxury product is given by the formula $y = 600x + 15x^2 - x^3$. Calculate the value of x which gives a maximum profit, and determine that maximum profit.

2 The profit, y hundred euro, generated from the sale, x thousand, of a certain product is given by the formula $y = 72x + 3x^2 - 2x^3$. Calculate how many items should be sold in order to maximise the profit, and determine that maximum profit.

3 At a speed of x km h^{-1} a certain car will travel y km on each litre of diesel, where $y = 2 + x - \dfrac{x^2}{110}$.
Calculate the speed at which the car should travel in order to maximise the distance it can cover on a single tank of diesel.

4 At a speed of x km h^{-1} a motorbike can travel y km on a litre of fuel, where $y = 5 + \dfrac{x}{2} + \dfrac{x^2}{60} - \dfrac{x^3}{1800}$.
Calculate the maximum distance that the motorbike can travel on 30 litres of fuel.

5 A ball is thrown vertically upwards. At time t seconds after the instant of projection, its height, y metres above the point of projection, is given by the formula $y = 15t - 5t^2$.
Calculate the time at which the ball is at its maximum height, and find the value of y at that time.

6 An unpowered missile is launched vertically from the ground. At a time t seconds after its launch its height, y metres, above the ground is given by the formula $y = 80t - 5t^2$.
Calculate the maximum height reached by the missile.

7 A piece of string which is 40 cm long is cut into two lengths. Each length is laid out to form a square.
Given that the length of the sides of one of the squares is x cm, find expressions in terms of x for

a) the length of the sides of the other square

b) the total area enclosed by the two squares.

Given also that the sum of the two areas is a minimum,

c) calculate the value of x.

8 A stick of length 24 cm is cut into three pieces, two of which are of equal length. The two pieces of equal length are then each cut into four equal lengths and constructed into squares of side x cm. The remaining piece is cut and constructed into a rectangle of width 3 cm.
Find expressions, in terms of x, for

a) the length of the rectangle

b) the total area enclosed by the three shapes.

Given also that the sum of the three areas is a minimum,

c) calculate the value of x.

9 A strip of wire of length 150 cm is cut into two pieces. One piece is bent to form a square of side x cm, and the other piece is bent to form a rectangle which is twice as long as it is wide. Find expressions, in terms of x, for

a) the width of the rectangle

b) the length of the rectangle

c) the area of the rectangle.

Given also that the sum of the two areas enclosed is a minimum,

d) calculate the value of x.

10 A rectangular enclosure is formed from 1000 m of fencing. Given that each of the two opposite sides of the rectangle has length x metres,

a) find, in terms of x, the length of the other two sides.

Given also that the area enclosed is a maximum,

b) find the value of x, and hence calculate the area enclosed.

11 A rectangular pen is formed from 40 m of fencing with a long wall forming one side of the pen, as shown in the diagram on the right.

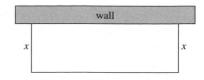

Given that the two opposite sides of the pen which touch the wall each have length x metres,

a) find, in terms of x, the length of each of the other two sides.

Given also that the area enclosed is a maximum,

b) find the value of x, and hence calculate the area enclosed.

12 A closed cuboidal box of square base has volume 8 m³. Given that the square base has sides of length x metres, find expressions, in terms of x, for

a) the height of the box

b) the surface area of the box.

Given also that the surface area of the box is a minimum,

c) find the value of x.

13 An *open* metal tank of square base has volume 108 m³. Given that the square base has sides of length x metres, find expressions, in terms of x, for

a) the height of the tank

b) the surface area of the tank.

Given also that the surface area is a minimum,

c) find the value of x.

14 A silver bar of volume 576 cm³ is cuboidal in shape, and has a length which is twice its breadth. Given that the breadth of the bar is x cm, find expressions, in terms of x, for

a) the length of the bar

b) the height of the bar

c) the surface area of the bar.

Given also that the surface area is a minimum,

d) find the value of x.

15 An *open* cuboidal tank of rectangular base is to be made with an external surface area of 36 m². The base is to be such that its length is three times its breadth.
Find the length of the base of the tank for the volume of the tank to be a maximum, and find this maximum volume.

16 A closed cuboidal plastic box is to be made with an external surface area of 216 cm². The base is to be such that its length is four times its breadth.
Find the length of the base of the box if the volume of the box is to be a maximum, and find this maximum volume.

17 A closed cuboidal box of square base and volume 36 cm³ is to be constructed and silver plated on the outside. Silver plating for the top and the base costs 40 cents per cm², and silver plating for the sides costs 30 cents per cm².
Given that the length of the sides of the base is to be x cm, find expressions, in terms of x, for

a) the height of the box

b) the cost of plating the top

c) the cost of plating a side

d) the total cost of plating the box.

Given also that this cost is to be a minimum,

e) find the value of x

f) calculate the cost of plating the box.

18 An *open* cuboidal fish tank of rectangular base and volume 2.5 m³ is to be made in such a way that its length is three times its breadth. Glass for the sides costs \$4 per m², and glass for the base costs \$15 per m². Given that the base has breadth x m, find expressions, in terms of x, for

a) the height of the tank

b) the cost of all of the glass for the sides

c) the cost of glass for the base.

Given also that the cost is to be a minimum,

d) find the value of x

e) calculate the cost of the glass for the tank.

19 *Open* cuboidal metal boxes of square base are to be made such that each box has a volume of 750 cm³. The metal sheeting used for the sides costs 2 cents per cm², and the metal sheeting used for the base costs 3 cents per cm². Calculate the dimensions of the boxes which should be made if the cost of the metal sheeting is to be a minimum.

20 An *open* cardboard box is to be made by cutting small squares of side x cm from each of the four corners of a larger square of card of side 10 cm, and folding along the dashed lines, as shown in the diagram.
Find the value of x such that the box has a maximum volume, and state the value of that maximum volume.

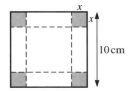

21 An *open* metal tray is to be made by cutting squares of side x cm from each of the four corners of a rectangular piece of metal measuring 8 cm by 5 cm, and folding the resulting shape as in question 20.
Find the value of x that will give the box a maximum volume.

22 A cylinder is to be made of circular cross-section with a specified volume. Prove that if the surface area is to be a minimum, then the height of the cylinder must be equal to the diameter of the cross-section of the cylinder.

Summary

You should know how to ...

▶ Differentiate a function.

▷ $y = x^n \ (n \in \mathbb{Q})$, $\dfrac{dy}{dx} = nx^{n-1}$

or $f(x) = x^n$, $f'(x) = nx^{n-1}$

▷ $\dfrac{dy}{dx}$, or $f'(x)$, is the derivative, and measures gradient

▷ $\dfrac{d^2y}{dx^2}$, or $f''(x)$, is the second derivative, and is given by $\dfrac{d}{dx}\left(\dfrac{dy}{dx}\right)$.

▶ Solve problems in kinematics.

▷ Distance $s = f(t)$

▷ Velocity $v = \dfrac{ds}{dt} = f'(t)$

▷ Acceleration $a = \dfrac{d^2s}{dt^2} = \dfrac{dv}{dt} = f''(t)$.

▶ Identify stationary points.

▷ At a stationary point, $\dfrac{dy}{dx} = 0$

▷ At a minimum, $\dfrac{d^2y}{dx^2} > 0$

▷ At a maximum, $\dfrac{d^2y}{dx^2} < 0$

▷ At a point of inflexion, $\dfrac{d^2y}{dx^2} = 0$.

Revision exercise 4

1 Differentiate these expressions with respect to x.

a) $3x^3 - 4x^2 + x + 1$

b) $\dfrac{1}{x^4} + \dfrac{5}{x^5}$

2 For each of these find $\dfrac{dy}{dx}$.

a) $y = \sqrt{x} - 4\sqrt[3]{x}$

b) $y = x^{-\frac{1}{3}} + 5x^{\frac{2}{3}}$

3 For each of these find $f'(x)$.

a) $f(x) = (x + 5)(2x - 3)$

b) $f(x) = \dfrac{x + 2}{x^3}$

c) $f(x) = \dfrac{x - 3}{x^2 - 3x}$

4 The diagram shows part of the graph of the curve $y = 2x^2 - 12x + 19$.

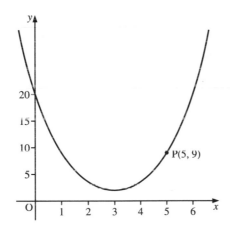

a) Find $\dfrac{dy}{dx}$.

A tangent is drawn to the curve at P(5, 9).

b) Calculate the gradient of this tangent.

c) Find the equation of this tangent.

© IBO [2003]

5 The diagram shows the graph with equation

$$y = 4x^3 - 24x^2 + 45x - 23$$

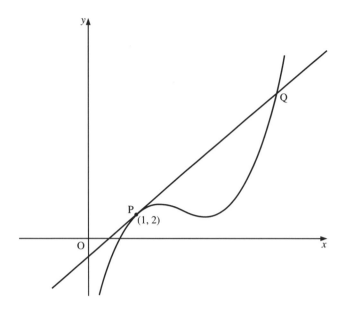

The tangent at P(1, 2) intersects the curve at Q.

a) Find $\dfrac{dy}{dx}$.

b) Find the gradient of the curve at P. Hence find the equation of the tangent at P in the form $y = ax + b$.

c) Show that the x-coordinate of the point Q is 4.   © IBO[1998]

6 Find the coordinates of the point on the graph of $y = x^2 - x$ at which the tangent is parallel to the line $y = 5x$.   © IBO[2000]

7 Find the equation of the normal to the curve with equation

$$y = x^3 + 1$$

at the point $(1, 2)$.   © IBO[1999]

8 The line $y = ax + 3$ is a tangent to the parabola $y = x^2 + x + b$ when $x = 1$. Find the values of the constants a and b.   © IBO[1997]

9 The graph of $y = x^3 - 10x^2 + 12x + 23$ has a maximum point between $x = -1$ and $x = 3$.
Find the coordinates of this maximum point.   © IBO[2002]

10 A particle moves in a straight line. Its displacement s metres from a fixed point O after t seconds is given by

$$s = 0.5t^3 - 13.5t + 3$$

Find

a) its velocity after 2 seconds

b) its acceleration after 2 seconds.

c) At what time is the particle momentarily at rest?

d) Find its acceleration at this time.

11 A stone is thrown vertically downwards from the top of a cliff. Its height h metres above the beach below at time t seconds is given by

$$h = (10 - \sqrt{5}t)(10 + \sqrt{5}t)$$

a) Show that is acceleration is constant and state its value.

b) Find the time when it hits the beach.

c) Find its velocity at this time.

12 A rocket is fired vertically upwards. Its height h metres after t seconds is given by $h - 98t - 4.9t^2$.

a) Find an expression in terms of t for its velocity.

b) Show that its height is a maximum when $t = 10$.

c) Find this maximum height.

5 Integration

5.1 Anti-differentiation

You know that if $y = x^2$, then $\dfrac{dy}{dx} = 2x$. Now suppose you are given $\dfrac{dy}{dx} = 2x$ and asked to find y in terms of x. This process is the reverse of differentiation and is called **integration**.

> Integration is a branch of calculus, as is differentiation.

In this particular case, you know that $y = x^2$ will satisfy $\dfrac{dy}{dx} = 2x$, but so will $y = x^2 + 1$ and $y = x^2 + 2$. In fact, $y = x^2 + c$, where c is a constant, will also satisfy $\dfrac{dy}{dx} = 2x$.

In other words, you do not know whether the original function contained a constant term or not. So you must write $y = x^2 + c$, where c is called the **constant of integration**.

> $y = x^2 + c$ is called the **integral** of $2x$ with respect to x.
> This is written as:
>
> $$\int 2x \, dx = x^2 + c$$
>
> integral sign
>
> indicating that the integration is with respect to the variable x

> This is called indefinite integration, because c could have any value.

To find $\int x^4 \, dx$, notice that x^5 differentiates to give $5x^4$, which is the required power of x. However, you do not want the constant 5. To get rid of it, multiply x^5 by $\frac{1}{5}$:

$\frac{1}{5}x^5$ differentiates to give x^4

So:

$$\int x^4 \, dx = \tfrac{1}{5}x^5 + c$$

> Don't forget to add the constant of integration.

In general:

If $\dfrac{dy}{dx} = ax^n$, then

$$y = \dfrac{ax^{n+1}}{n+1} + c \quad (n \neq -1)$$

That is:

$$\int ax^n \, dx = \dfrac{ax^{n+1}}{n+1} + c \quad (n \neq -1)$$

> One way of remembering this is 'add one to the power and divide by the new power'.

This can be shown to be true by differentiating y with respect to x:

$$y = \dfrac{ax^{n+1}}{n+1} + c$$

$$\dfrac{dy}{dx} = \dfrac{a(n+1)x^n}{(n+1)} = ax^n$$

Example 1

Find

a) $\displaystyle\int 3x^2 \, dx$ b) $\displaystyle\int \dfrac{1}{x^2} \, dx$ c) $\displaystyle\int 6\sqrt{x} \, dx$

- -

a) To find $\displaystyle\int 3x^2 \, dx$, notice that x^3 differentiates to give $3x^2$.
Therefore,

$$\int 3x^2 \, dx = x^3 + c$$

Alternatively, using the result above,

$$\int 3x^2 \, dx = \dfrac{3x^{2+1}}{(2+1)} + c$$

$$\therefore \int 3x^2 \, dx = x^3 + c$$

b)
$$\int \dfrac{1}{x^2} \, dx = \dfrac{x^{-2+1}}{(-2+1)} + c = \dfrac{x^{-1}}{-1} + c$$

$$\int \dfrac{1}{x^2} \, dx = -\dfrac{1}{x} + c$$

> $\dfrac{1}{x^2} = x^{-2}$

c)
$$\int 6\sqrt{x} \, dx = \dfrac{6x^{\frac{1}{2}+1}}{(\frac{1}{2}+1)} + c = \dfrac{6x^{\frac{3}{2}}}{(\frac{3}{2})} + c$$

$$\therefore \int 6\sqrt{x} \, dx = 4x^{\frac{3}{2}} + c = 4\sqrt{x^3} + c$$

> $6\sqrt{x} = 6x^{\frac{1}{2}}$

A general result is that:

$$\int af(x)\, dx = a\int f(x)\, dx$$

In section 4.3 you saw that

$$\frac{d}{dx}[f(x) \pm g(x)] = f'(x) \pm g'(x)$$

The integral behaves in exactly the same way. So:

$$\int [f(x) \pm g(x)]\, dx = \int f(x)\, dx \pm \int g(x)\, dx$$

This result applies to any number of functions.

Example 2

Find

a) $\int (x^2 + 6x - 3)\, dx$

b) $\int \left(1 - \frac{1}{x^2}\right) dx$

. . . .

a) $\quad \int (x^2 + 6x - 3)\, dx = \int x^2\, dx + \int 6x\, dx - \int 3\, dx$

$$= \frac{x^3}{3} + \frac{6x^2}{2} - 3x + c$$

$$\therefore \int (x^2 + 6x - 3)\, dx = \frac{x^3}{3} + 3x^2 - 3x + c$$

b) $\quad \int \left(1 - \frac{1}{x^2}\right) dx = \int 1\, dx - \int x^{-2}\, dx$

$$= x + x^{-1} + c$$

$$\therefore \int \left(1 - \frac{1}{x^2}\right) dx = x + \frac{1}{x} + c$$

Some expressions are not in the form ax^n or written as a sum of functions of this form. In these cases, you must manipulate the integrand into this form.

> The **integrand** is the expression that is to be integrated.

Example 3

Find

a) $\int (x - 4)^2\, dx$

b) $\int \left(\frac{x^3 - 3x}{x}\right) dx$

c) $\int \sqrt{x}(x + 1)\, dx$

. . . .

a) $\int (x-4)^2 \, dx = \int (x^2 - 8x + 16) \, dx$

$$= \frac{x^3}{3} - 4x^2 + 16x + c$$

b) $\int \left(\frac{x^3 - 3x}{x} \right) dx = \int \left(\frac{x^3}{x} - \frac{3x}{x} \right) dx$

$$= \int (x^2 - 3) \, dx$$

$\therefore \int \left(\frac{x^3 - 3x}{x} \right) dx = \frac{x^3}{3} - 3x + c$

c) $\int \sqrt{x} \, (x+1) \, dx = \int x^{\frac{1}{2}} (x+1) \, dx$

$$= \int \left(x^{\frac{3}{2}} + x^{\frac{1}{2}} \right) dx$$

$$= \frac{x^{\frac{5}{2}}}{\left(\frac{5}{2} \right)} + \frac{x^{\frac{3}{2}}}{\left(\frac{3}{2} \right)} + c = 2x^{\frac{3}{2}} \left(\frac{x}{5} + \frac{1}{3} \right) + c$$

$\therefore \int \sqrt{x} \, (x+1) \, dx = 2\sqrt{x^3} \left(\frac{x}{5} + \frac{1}{3} \right) + c$

$\qquad$ or $\qquad \frac{2}{15} \sqrt{x^3} \, (3x + 5) + c$

If you know the gradient of the tangent to a curve at a particular point, you can use integration to find the equation of the curve.

This is the reverse of finding the equation of the tangent by differentiation.

Example 4

The gradient of a curve at the point (x, y) is $12x^3 - \dfrac{1}{x^2}$ and the curve passes through the point $(1, 2)$.
Find the equation of the curve.

....

The gradient of a curve at any point is the derivative of the function at that point. So:

$$\frac{dy}{dx} = 12x^3 - \frac{1}{x^2} = 12x^3 - x^{-2}$$

$$\therefore y = \int (12x^3 - x^{-2}) \, dx$$

$$= \frac{12x^4}{4} - \frac{x^{-1}}{(-1)} + c$$

$$\therefore y = 3x^4 + \frac{1}{x} + c \qquad\qquad [1]$$

The curve passes through the point $(1, 2)$, so $x = 1$, $y = 2$ satisfies [1]. Therefore:

$$2 = 3(1)^4 + \frac{1}{1} + c$$

$$\therefore \ c = -2$$

The equation of the curve is $y = 3x^4 + \dfrac{1}{x} - 2$.

If you are given more information, you can find the value of the constant of integration.

Example 5

Given that

$$f''(x) = 2 - \frac{2}{\sqrt{x^3}} \quad \text{and} \quad f'(1) = 0$$

find $f'(x)$. Given further that $f(1) = 8$, find $f(x)$.

....

$$f''(x) = 2 - \frac{2}{\sqrt{x^3}} = 2 - 2x^{-\frac{3}{2}}$$

$$f'(x) = \int (2 - 2x^{-\frac{3}{2}}) \, dx = 2x - \frac{2x^{-\frac{1}{2}}}{(-\frac{1}{2})} + c_1$$

$$\therefore \ f'(x) = 2x + 4x^{-\frac{1}{2}} + c_1 \qquad [1]$$

Since $f'(1) = 0$, substituting into [1] gives $c_1 = -6$ and so

$$f'(x) = 2x + \frac{4}{\sqrt{x}} - 6 = 2x + 4x^{-\frac{1}{2}} - 6$$

Therefore,

$$f(x) = \int (2x + 4x^{-\frac{1}{2}} - 6) \, dx = x^2 + 8x^{\frac{1}{2}} - 6x + c_2$$

Since $f(1) = 8$,

$$8 = 1 + 8 - 6 + c_2$$

$$\therefore \ c_2 = 5$$

Therefore, $f(x) = x^2 + 8\sqrt{x} - 6x + 5$.

Exercise 5A

....

1 Integrate each of these expressions with respect to x.

a) x^3 b) x^4 c) $3x^2$ d) $12x^5$

e) $-4x$ f) $15x^4$ g) $2x^3$ h) 3

i) $\frac{1}{2}x^5$ j) $\frac{2}{3}x^3$ k) $-\frac{1}{3}x^2$ l) $\frac{2}{3}$

2 Find each of these integrals.

a) $\int x^{-2}\,dx$ b) $\int x^{-4}\,dx$ c) $\int 2x^{-3}\,dx$

d) $\int -6x^{-4}\,dx$ e) $\int \dfrac{1}{x^3}\,dx$ f) $\int -\dfrac{1}{x^5}\,dx$

g) $\int \dfrac{3}{x^2}\,dx$ h) $\int -\dfrac{2}{x^3}\,dx$ i) $\int \dfrac{4}{x^7}\,dx$

j) $\int \dfrac{3}{2x^4}\,dx$ k) $\int -\dfrac{5}{3x^2}\,dx$ l) $\int \dfrac{2}{3x^4}\,dx$

3 Integrate each of these functions with respect to x.

a) $f(x) = x^{\frac{1}{3}}$ b) $f(x) = 3x^{\frac{1}{2}}$ c) $f(x) = x^{-\frac{2}{3}}$

d) $f(x) = -4x^{-\frac{1}{5}}$ e) $f(x) = -3\sqrt{x}$ f) $f(x) = \sqrt[4]{x}$

g) $f(x) = \dfrac{4}{\sqrt[3]{x}}$ h) $f(x) = -\dfrac{2}{\sqrt[5]{x}}$ i) $f(x) = \dfrac{3}{7\sqrt{x}}$

j) $f(x) = \dfrac{6}{5\sqrt[3]{x}}$ k) $f(x) = \sqrt{x^3}$ l) $f(x) = \sqrt{9x}$

4 Integrate each of these expressions with respect to x.

a) $x^3 + 2x$ b) $3x^2 - 4x$ c) $x^3 - 1$

d) $6 + 3x^5$ e) $x^2 - 5x + 3$ f) $x^8 + 2x^5$

g) $x^4 - 3x + 2$ h) $x^2 - \dfrac{1}{x^2}$ i) $5x^4 - \dfrac{2}{x^3}$

j) $2x^6 + \dfrac{8}{x^5}$ k) $x^2 - \dfrac{3}{x^2}$ l) $\dfrac{5}{x^2} - 2 - 2x^3$

5 Find each of these integrals.

a) $\int (3\sqrt{x} - 4)\,dx$ b) $\int \left(\sqrt{x} + \dfrac{1}{\sqrt{x}}\right)\,dx$

c) $\int (3x^{\frac{1}{3}} - 2x^{\frac{1}{4}})\,dx$ d) $\int (5x^{-\frac{1}{2}} + 2x^{-\frac{1}{3}})\,dx$

e) $\int \left(4\sqrt{x} - \dfrac{2}{3x^2}\right)\,dx$ f) $\int (2\sqrt[3]{x} - \dfrac{6}{\sqrt{x}})\,dx$

6 Find $\int y\,dx$ for each of these functions.

a) $y = x(3 - x)$ b) $y = x^2(x + 5)$
c) $y = x^3(2 - x^2)$ d) $y = \sqrt{x}(x + 3)$
e) $y = 3\sqrt{x}(x^2 - x + 1)$ f) $y = x^{\frac{1}{3}}(2x + 3)$
g) $y = 3x^{\frac{1}{4}}(x - 2)$ h) $y = (x + 3)(x + 5)$
i) $y = (x - 2)^2$ j) $y = 2(x + 5)^2$
k) $y = x(x - 1)^2$ l) $y = (\sqrt{x} - 3)(\sqrt{x} + 5)$

7 Integrate each of these functions with respect to x.

a) $f(x) = 5x(x - 2)$ b) $f(x) = x^3(6x^2 - 1)$

c) $f(x) = \sqrt{x}(x^2 + 1)$ d) $f(x) = x^{-\frac{1}{3}}(2x + 3)$

e) $f(x) = \dfrac{x^2 + 5}{x^2}$ f) $f(x) = \dfrac{x - 4}{x^3}$

g) $f(x) = \dfrac{3x^2 + 5}{x^2}$ h) $f(x) = \dfrac{4x^3 - 3x^2}{2x}$

i) $f(x) = \dfrac{5x^2 - 4}{\sqrt{x}}$ j) $f(x) = \dfrac{\sqrt{x} - 1}{x}$

8 In each part of this question, use the information given to find an expression for y in terms of x.

a) $\dfrac{dy}{dx} = 3x^2 + 1$, $y = 12$ when $x = 2$

b) $\dfrac{dy}{dx} = 4x - 3$, $y = 6$ when $x = -1$

c) $\dfrac{dy}{dx} = 6x^2 - 4x$, $y = 24$ when $x = 3$

d) $\dfrac{dy}{dx} = 4 - 6x$, $y = -4$ when $x = -2$

e) $\dfrac{dy}{dx} = \dfrac{2}{x^2} - 1$, $x \neq 0$; $y = 5$ when $x = 1$

f) $\dfrac{dy}{dx} = -\dfrac{10}{x^3}$, $x \neq 0$; $y = 13$ when $x = \frac{1}{2}$

g) $\dfrac{dy}{dx} = \sqrt{x} - 5$, $x > 0$; $y = -18$ when $x = 9$

h) $\dfrac{dy}{dx} = x - \frac{1}{2}\sqrt{x}$, $x > 0$; $y = \frac{2}{3}$ when $x = 4$

i) $\dfrac{dy}{dx} = x^2(3 - x)$, $y = 2\frac{3}{4}$ when $x = -1$

j) $\dfrac{dy}{dx} = (2x - 3)^2$, $y = 13$ when $x = 3$

k) $\dfrac{dy}{dx} = \dfrac{\sqrt{x} - 1}{\sqrt{x}}$, $x > 0$; $y = 5\frac{1}{4}$ when $x = \frac{1}{4}$

l) $\dfrac{dy}{dx} = \dfrac{x - 2}{x^3}$, $x \neq 0$; $y = 3\frac{1}{2}$ when $x = -2$

9 The gradient of a curve at the point (x, y) on the curve is
given by $\dfrac{dy}{dx} = 3x^2 + 4$.

Given that the point $(1, 7)$ lies on the curve, determine the
equation of the curve.

Use your GDC to draw the curve and confirm that $(1, 7)$ lies
on this curve.

10 A curve passes through the point $(-2, 8)$ and its gradient
function is $4x^3 - 6x$.

Find the equation of the curve. Use your GDC to draw the
curve and confirm that the gradient at $x = -2$ is -20.

11 The gradient of a curve at the point (x, y) is $16x^3 + 2x + 1$.
Given that the curve passes through the point $(\frac{1}{2}, 3)$, find the
equation of the curve.

12 Find y as a function of x given that $\dfrac{dy}{dx} = \dfrac{5}{x^2} - 4 \ (x \neq 0)$, and
that $y = -12$ when $x = 5$.

13 A function $f(x)$ is such that $f'(x) = 3\sqrt{x} - 5$, $x \in \mathbb{R}$, $x \geqslant 0$.
Given that $f(4) = 3$, find an expression for $f(x)$.

14 A curve has an equation which satisfies $\dfrac{d^2y}{dx^2} = 6x - 4$.

The point P$(2, 11)$ lies on the curve, and the gradient of the
curve at the point P is 9. Determine the equation of the curve.

15 Find y as a function of x given that $\dfrac{d^2y}{dx^2} = 12x^2 - 6$, and that
when $x = 1$, $\dfrac{dy}{dx} = -2$ and $y = 1$.

16 Given that $\dfrac{d^2y}{dx^2} = 6x + \dfrac{4}{x^3} \ (x \neq 0)$, and that $y = 1$ when $x = 1$,
and that $y = 5$ when $x = 2$, find an expression for y in terms of x.

17 The curve with equation $y = ax^2 + bx + c$ passes through the
points P$(2, 6)$ and Q$(3, 16)$, and has a gradient of 7 at the
point P. Find the values of the constants a, b and c.

··

5.2   Variable acceleration

You saw in Chapter 4 that if s denotes the displacement of a body
at time t, then

the velocity v at time t is given by $v = \dfrac{ds}{dt}$

and the acceleration a at time t is given by $a = \dfrac{dv}{dt}$.

If you are given the velocity v in terms of the time t, you can find the displacement by integration. Similarly, if you are given the acceleration in terms of the time t, integration gives the velocity. To summarise:

$$s \quad \underset{\text{integrate}}{\overset{\text{differentiate}}{\rightleftarrows}} \quad v = \frac{ds}{dt} \quad \underset{\text{integrate}}{\overset{\text{differentiate}}{\rightleftarrows}} \quad a = \frac{dv}{dt}$$

Example 1

The acceleration a m s^{-2} of a particle moving in a straight line at time t seconds, is given by $a = t + 1$. Find formulae for:

a) the velocity v b) the displacement s,

given that $s = 0$ and $v = 8$ when $t = 0$.

....

a) Since $a = \dfrac{dv}{dt}$,

$$\frac{dv}{dt} = t + 1$$

$$v = \int t + 1 \, dt$$

$$\therefore \quad v = \frac{t^2}{2} + t + c$$

Since $v = 8$ when $t = 0$,

$$8 = c$$

Therefore

$$v = \frac{t^2}{2} + t + 8$$

b) Since $v = \dfrac{ds}{dt}$,

$$\frac{ds}{dt} = \frac{t^2}{2} + t + 8$$

$$s = \int \frac{t^2}{2} + t + 8 \, dt$$

$$\therefore \quad s = \frac{t^3}{6} + \frac{t^2}{2} + 8t + k$$

Since $s = 0$ when $t = 0$.

$$0 = k$$

Therefore

$$s = \frac{t^3}{6} + \frac{t^2}{2} + 8t$$

Exercise 5B

For questions **1–3**,

a) Find an expression, in terms of t, for the velocity of the particle at time t.

b) Find an expression, in terms of t, for the displacement of the particle at time t.

1 The acceleration, a m s^{-2}, of a particle at a time t seconds is given by the formula $a = 30t$. At time $t = 0$ the particle is moving through the origin with a velocity of 2 m s^{-1}.

2 The acceleration, a m s^{-2}, of a particle at a time t seconds is given by the formula $a = 48t^2 + 6$. At time $t = 0$, $s = 1$ and $v = 0$.

3 A particle moves such that its acceleration, a m s^{-2}, at a time t seconds is given by the formula $a = 20t^3 + 20$. At time $t = 1$, $s = 7$ and $v = 12$.

4 A particle moves such that its acceleration, a m s^{-2}, at a time t seconds is given by the formula $a = \dfrac{2}{\sqrt{t}} + 10t$, $t > 0$. At time $t = 4$, $v = 28$.

    a) Find an expression, in terms of t, for the velocity of the particle at time t.

    b) Hence find the value of v at $t = \frac{1}{4}$.

5 If $v = 5 - 2t$, and $s = 10$ m when $t = 1$ s, find:

    a) an expression, in terms of t, for the displacement of the particle at time t

    b) the value of t when the particle is at rest

    c) the distance the particle is from 0 when it comes to rest.

6 If $a = 6 - 2t$ and $v = 15$ m s^{-1} at $t = 2$ s, find:

    a) an expression, in terms of t, for the velocity of the particle at time t

    b) the value of t when the particle is at rest.

7 A particle moves such that its acceleration, a m s^{-2}, at a time t seconds is given by the formula $a = 2t - 4$. At time $t = 0$, $v = 3$ and $s = 2$. Find:

    a) an expression, in terms of t, for the velocity of the particle at time t

    b) an expression, in terms of t, for the displacement of the particle at time t

    c) the values of t when the particle is at rest

    d) the displacement of the particle at these times.

<u>8</u> A particle moves with constant acceleration, a m s^{-2}.
At time $t = 0$ it is moving through the origin with a velocity
of u m s^{-1}.

   a) Find an expression, in terms of t, for the velocity of the
particle at time t.

   b) Find an expression, in terms of t, for the displacement of
the particle at time t.

5.3 Area under a curve

Consider the area A under a curve $f(x)$ as shown in the diagram.

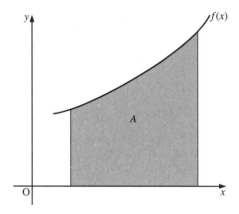

You can calculate an approximate value for the area A by splitting
the shaded region into rectangles, and summing the areas of these
rectangles. There are two cases to consider:

i)   ii)

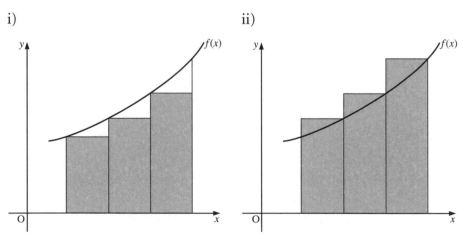

In case i) the approximation is less than A, whereas in case ii) the
approximation is greater than A. In both cases, as each rectangle is
made narrower (more rectangles are used) the approximation
approaches a limiting value, A.

Consider one such rectangle of width δx.
Let δA be the shaded area.

The area of rectangle ABEF is $y\delta x$.

The area of rectangle ABCD is $(y + \delta y)\delta x$.
So:

$$y\delta x < \delta A < (y + \delta y)\delta x$$

Since $\delta x > 0$, you can divide throughout by δx:

$$y < \frac{\delta A}{\delta x} < y + \delta y$$

If $\delta x \rightarrow 0$ (that is, you increase the number of rectangles) then
$\frac{\delta A}{\delta x} \rightarrow \frac{dA}{dx}$ and $\delta y \rightarrow 0$. Therefore,

$$\frac{dA}{dx} = y$$

Integrating each side with respect to x gives

$$\int \frac{dA}{dx}\,dx - \int y\,dx$$

$$\therefore\ A = \int y\,dx$$

This expression for the area will not give a definite value, only a function of x. This is because $\int y\,dx$ gives the area measured from an arbitrary origin to the point x. If you want to find the area under a curve between two values of x, you must evaluate the integral between the two given limits.

$A =$ (area up to the ordinate $x = b$) $-$ (area up to the ordinate $x = a$)
  $= A(b) - A(a)$

You write this as

$$A = \int_a^b y\,dx \quad \text{or} \quad A = \int_a^b f(x)\,dx$$

You call $\int_a^b f(x)\,dx$ a **definite integral** since it gives a definite answer.

❖ The dx indicates that the limits a and b are x limits.
❖ The constant a is called the **lower limit** of the integral.
❖ The constant b is called the **upper limit** of the integral.

For example, to evaluate the definite integral $\int_0^1 2x\,dx$, first integrate to obtain

$$\int_0^1 2x\,dx = \left[x^2 + c\right]_0^1$$

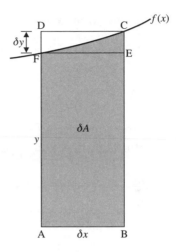

$\delta x \rightarrow 0$ means δx tends (gets closer) to zero.

$\dfrac{dA}{dx}$ is the **limit** of the ratio $\dfrac{\delta A}{\delta x}$

as $\delta x \rightarrow 0$.

Notice that square brackets are used here.

Substituting the values $x = 1$ and $x = 0$ gives

$$\int_0^1 2x\,dx = [(1)^2 + c] - [(0)^2 + c] = 1 - 0$$

$$\therefore \int_0^1 2x\,dx = 1$$

You subtract the expression containing the lower limit from the one containing the upper limit.

Notice that the constants of integration cancel. You can ignore the constants of integration when working with definite integrals.

Example 1

Evaluate these definite integrals.

a) $\displaystyle\int_0^2 4x^3\,dx$

b) $\displaystyle\int_{-1}^1 (3x^2 - 5)\,dx$

c) $\displaystyle\int_{-3}^{-2} \frac{1}{x^2}\,dx$

d) $\displaystyle\int_2^8 \left(x - \frac{3}{\sqrt{x}}\right)\,dx$

$\cdots\cdots\cdots\cdots\cdots\cdots\cdots\cdots\cdots\cdots\cdots\cdots\cdots\cdots\cdots\cdots\cdots\cdots\cdots\cdots\cdots\cdots$

a) $\displaystyle\int_0^2 4x^3\,dx = \left[x^4\right]_0^2$

$\qquad\qquad = (2)^4 - (0)^4$

$\therefore \displaystyle\int_0^2 4x^3\,dx = 16$

b) $\displaystyle\int_{-1}^1 (3x^2 - 5)\,dx = \left[x^3 - 5x\right]_{-1}^1$

$\qquad\qquad\qquad\qquad = (1 - 5) - (-1 + 5)$

$\qquad\qquad\qquad\qquad = -4 - 4$

$\therefore \displaystyle\int_{-1}^1 (3x^2 - 5)\,dx = -8$

c) $\displaystyle\int_{-3}^{-2} \frac{1}{x^2}\,dx = \int_{-3}^{-2} x^{-2}\,dx$

$\qquad\qquad\quad = \left[-x^{-1}\right]_{-3}^{-2}$

$\qquad\qquad\quad = \dfrac{1}{2} - \dfrac{1}{3}$

$\therefore \displaystyle\int_{-3}^{-2} \frac{1}{x^2}\,dx = \dfrac{1}{6}$

d) $\displaystyle\int_2^8 \left(x - \frac{3}{\sqrt{x}}\right)\,dx = \int_2^8 (x - 3x^{-\frac{1}{2}})\,dx$

$\qquad\qquad\qquad\quad = \left[\dfrac{x^2}{2} - 6x^{\frac{1}{2}}\right]_2^8$

$\qquad\qquad\qquad\quad = (32 - 6\sqrt{8}) - (2 - 6\sqrt{2})$

$\qquad\qquad\qquad\quad = (32 - 12\sqrt{2}) - (2 - 6\sqrt{2})$

$\therefore \displaystyle\int_2^8 \left(x - \frac{3}{\sqrt{x}}\right)\,dx = 30 - 6\sqrt{2}$

$\sqrt{8} = \sqrt{4 \times 2}$
$= \sqrt{4} \times \sqrt{2} = 2\sqrt{2}$

In the next example, integration is used to find the area between a curve and the x-axis.

Example 2

Find the area under the curve $y = x^2$ between $x = 1$ and $x = 3$.

....

Let A be the required area, then

$$A = \int_1^3 x^2 \, dx$$

$$= \left[\frac{x^3}{3} \right]_1^3 = 9 - \frac{1}{3}$$

$$\therefore A = \frac{26}{3}$$

The required area is $\frac{26}{3}$ or $8\frac{2}{3}$.

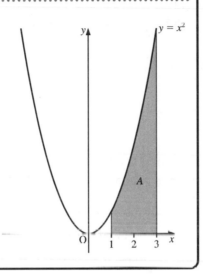

> You can use your GDC to find the area under a curve. The result will be approximate rather than exact.

The area between a curve and the x-axis may lie below the x-axis. In this case the value of the definite integral will be negative.

Example 3

Find the area between the curve $y = x^2 + 4x$ and the x-axis from

a) $x = -2$ to $x = 0$ b) $x = -2$ to $x = 2$

....

a) The required area is shaded in the diagram.

Let A be the required area, then

$$A = \int_{-2}^0 (x^2 + 4x) \, dx$$

$$= \left[\frac{x^3}{3} + 2x^2 \right]_{-2}^0$$

$$= 0 - \frac{16}{3}$$

$$\therefore A = -\frac{16}{3}$$

Therefore, the required area is $\frac{16}{3}$.

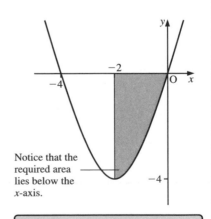

Notice that the required area lies below the x-axis.

> The minus sign tells you that the area is below the x-axis.

b) The required area has two parts, one part below the x-axis and one part above the x-axis.

Evaluating $\int_{-2}^{2} (x^2 + 4x)\, dx$ gives

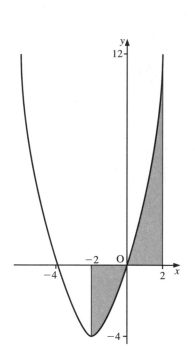

$$\int_{-2}^{2} (x^2 + 4x)\, dx = \left[\frac{x^3}{3} + 2x^2 \right]_{-2}^{2}$$

$$= \left(\frac{8}{3} + 8 \right) - \left(-\frac{8}{3} + 8 \right)$$

$$= \frac{32}{3} - \frac{16}{3} = \frac{16}{3}$$

It is obvious that this cannot be the total shaded area, since you know from part a) that the area between the curve and the x-axis from $x = -2$ to $x = 0$ is $\frac{16}{3}$.

Calculating each of the integrals $\int_{-2}^{0} (x^2 + 4x)\, dx$ and $\int_{0}^{2} (x^2 + 4x)\, dx$ will explain the mystery.

You know that

$$\int_{-2}^{0} (x^2 + 4x)\, dx = -\frac{16}{3}$$

and

$$\int_{0}^{2} (x^2 + 4x)\, dx = \left[\frac{x^3}{3} + 2x^2 \right]_{0}^{2} = \left(\frac{8}{3} + 8 \right) - (0) = \frac{32}{3}$$

Therefore, the required area, A, is given by

$$A = \frac{16}{3} + \frac{32}{3} = 16$$

Notice that

$$\int_{-2}^{0} (x^2 + 4x)\, dx + \int_{0}^{2} (x^2 + 4x)\, dx = -\frac{16}{3} + \frac{32}{3} = \frac{16}{3}$$

$$\therefore \int_{-2}^{0} (x^2 + 4x)\, dx + \int_{0}^{2} (x^2 + 4x)\, dx = \int_{-2}^{2} (x^2 + 4x)\, dx$$

> This example illustrates the importance of drawing a sketch in order to identify whether part of the required area lies below the x-axis or not.

When part of the curve lies above the x-axis and part below it, you must calculate the areas separately and add them together.

Example 4

The diagram shows a sketch of the curve given by
$y = x^3 - 4x^2 + 3x$. Find the area between the curve and the
x-axis from $x = 0$ to $x = 3$.

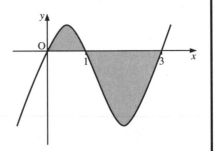

....

The required area is made up of two parts, A_1 and A_2, as shown.
Calculating A_1 gives

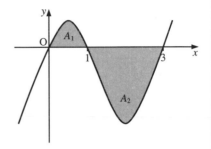

$$\int_0^1 (x^3 - 4x^2 + 3x)\,\mathrm{d}x = \left[\frac{x^4}{4} - \frac{4x^3}{3} + \frac{3x^2}{2}\right]_0^1$$

$$= \left(\frac{1}{4} - \frac{4}{3} + \frac{3}{2}\right) - (0)$$

$$\therefore\ A_1 = \frac{5}{12}$$

Calculating A_2 gives

$$\int_1^3 (x^3 - 4x^2 + 3x)\,\mathrm{d}x = \left|\frac{x^4}{4} - \frac{4x^3}{3} + \frac{3x^2}{2}\right|_1^3$$

$$= \left(\frac{81}{4} - 36 + \frac{27}{2}\right) - \left(\frac{1}{4} - \frac{4}{3} + \frac{3}{2}\right)$$

$$= -\frac{9}{4} - \frac{5}{12} = -\frac{8}{3}$$

$$\therefore\ A_2 = \frac{8}{3}$$

The required area A is given by

$$A = A_1 + A_2 = \frac{5}{12} + \frac{8}{3} = \frac{37}{12}$$

Exercise 5C
....

1 Work out each of these definite integrals.

a) $\displaystyle\int_0^2 x^2\,\mathrm{d}x$　　　b) $\displaystyle\int_0^3 4x^3\,\mathrm{d}x$　　　c) $\displaystyle\int_1^4 6x\,\mathrm{d}x$

d) $\displaystyle\int_2^3 (6x^2 - 1)\,\mathrm{d}x$　e) $\displaystyle\int_4^5 (4x + 3)\,\mathrm{d}x$　f) $\displaystyle\int_2^3 (4 - 3x^2)\,\mathrm{d}x$

g) $\displaystyle\int_2^8 \frac{1}{x^2}\,\mathrm{d}x$　　　h) $\displaystyle\int_1^2 \frac{4}{x^3}\,\mathrm{d}x$　　　i) $\displaystyle\int_4^9 \sqrt{x}\,\mathrm{d}x$

j) $\displaystyle\int_1^4 \left(3 - \frac{1}{\sqrt{x}}\right)\mathrm{d}x$　k) $\displaystyle\int_{\frac{1}{2}}^1 1 + \frac{1}{x^2}\,\mathrm{d}x$　l) $\displaystyle\int_1^8 \sqrt[3]{x}\,\mathrm{d}x$

2 Evaluate each of these.

a) $\int_{-1}^{3} 4x \, dx$

b) $\int_{-2}^{3} 6x^2 \, dx$

c) $\int_{-3}^{-1} 2x^3 \, dx$

d) $\int_{-2}^{-1} \dfrac{2}{x^3} \, dx$

e) $\int_{-4}^{5} (x-1) \, dx$

f) $\int_{-1}^{2} (2x - x^5) \, dx$

g) $\int_{-3}^{3} (x^3 + x) \, dx$

h) $\int_{-4}^{-1} \dfrac{16}{x^5} \, dx$

i) $\int_{1}^{5} \left(\dfrac{x^3 - 1}{x^2} \right) dx$

j) $\int_{1}^{16} \left(\dfrac{\sqrt{x} - 4}{\sqrt{x}} \right) dx$

k) $\int_{-1}^{1} x^3 (2x - 1) \, dx$

l) $\int_{-2}^{4} (x-2)^2 \, dx$

3 Work out the shaded area on each of these diagrams.

a)

b)
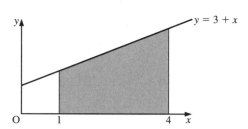

c)
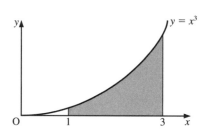

d)
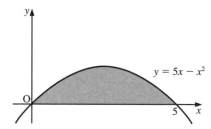

e)
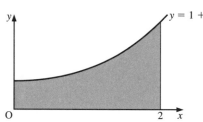

f)
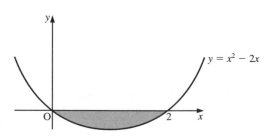

g)
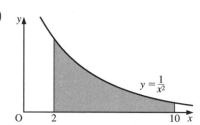

h)
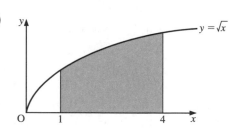
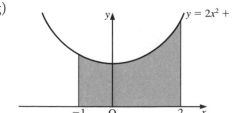

i)

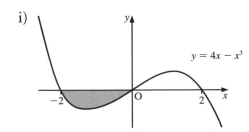

j)

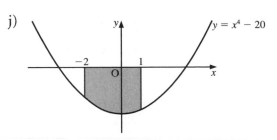

k)

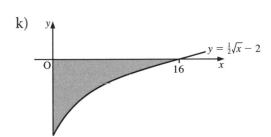

l)

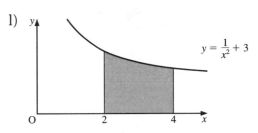

4 Find the areas enclosed by the x-axis and these curves and straight lines.

a) $y = x^2 + 3x$, $x = 2$, $x = 5$

b) $y = \frac{1}{8}x^3 + 2x$, $x = 2$, $x = 4$

c) $y = 2 - x^3$, $x = -3$, $x = -2$

d) $y = \dfrac{4}{x^3}$, $x = \frac{1}{4}$, $x = \frac{1}{2}$

e) $y = 6 - \dfrac{1}{\sqrt{x}}$, $x = 16$, $x = 25$

f) $y = (3x - 4)^2$, $x = 1$, $x = 3$

5 Use your GDC to sketch the graph of the region bounded by the curve $y = x^3 - 5$, the lines $x = 2$ and $x = 4$, and the x-axis. Find the area of the region.

6 Find the area enclosed above the x-axis and below the curve $y = 16 - x^2$.

7 Use your GDC to sketch the curve $y = (x - 2)(x - 3)$, showing where it crosses the x-axis. Hence find the area enclosed below the x-axis and above the curve.

8 Use your GDC to sketch the curve with equation $y = (x - 2)^2$. Calculate the area of the region bounded by the curve and the x- and y-axes.

9 Use your GDC to sketch the curve $y = 3x^2 - x^3$. Hence find the area of the region bounded by the curve and the x-axis.

10 Use your GDC to sketch the graph of the function $f(x) = \sqrt{x} - 3$ for $x > 0$. Calculate the area of the region bounded by the curve and the x- and y-axes.

11 Calculate the area of the region bounded by the x-axis and the function $f : x \mapsto 5 + \dfrac{1}{x^2}, x \in \mathbb{R}, 2 \leqslant x \leqslant 8$.

12 a) Use your GDC to sketch the curve $y = x(x + 1)(x - 3)$, showing where it cuts the x-axis.

b) Calculate the area of the region above the x-axis, bounded by the x-axis and the curve.

c) Calculate the area of the region below the x-axis, bounded by the x-axis and the curve.

13 a) Use your GDC to sketch the curve $y = x^2(x - 1)(x + 2)$.

b) Calculate the area of the region bounded by the positive x-axis and the curve.

c) Calculate the area of the region bounded by the negative x-axis and the curve.

....

Area between two curves

Consider two intersecting curves $f(x)$ and $g(x)$ as shown.

The shaded area, A, between the two curves is given by

$$A = \int_a^b g(x)\, dx - \int_a^b f(x)\, dx$$

$$\therefore A = \int_a^b (g(x) - f(x))\, dx$$

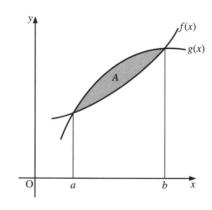

Example 5

Find the area enclosed between the curves $y = x^2 + 2x + 2$ and $y = -x^2 + 2x + 10$.

....

The x-coordinates of the points of intersection satisfy the equation

$$x^2 + 2x + 2 = -x^2 + 2x + 10$$

Simplifying gives

$$2x^2 - 8 = 0$$

$$x^2 = 4$$

$$\therefore x = \pm 2$$

> You must first find the points of intersection of the two curves.

> You can check this on your GDC.

The diagram shows the two curves.

The shaded area A is given by

$$A = \int_{-2}^{2} (-x^2 + 2x + 10)\,\mathrm{d}x - \int_{-2}^{2} (x^2 + 2x + 2)\,\mathrm{d}x$$

$$= \int_{-2}^{2} [(-x^2 + 2x + 10) - (x^2 + 2x + 2)]\,\mathrm{d}x$$

$$= \int_{-2}^{2} (-2x^2 + 8)\,\mathrm{d}x$$

$$= \left[-\frac{2x^3}{3} + 8x\right]_{-2}^{2} = \left(-\frac{16}{3} + 16\right) - \left(\frac{16}{3} - 16\right)$$

$$\therefore A = \frac{32}{3} + \frac{32}{3} = \frac{64}{3}$$

The area enclosed between the two curves is $\frac{64}{3}$.

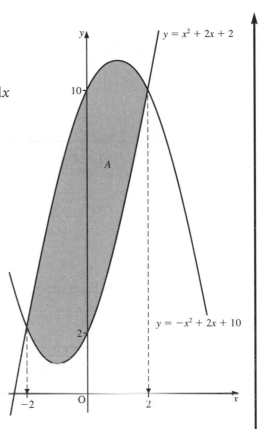

You may be asked to find the area between a curve and a straight line.

Example 6

Find the area enclosed between the curve $y = x^2 - 2x - 3$ and the line $y = x + 1$.

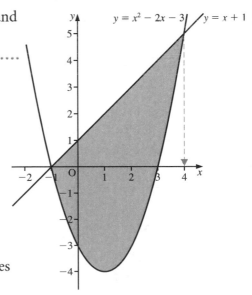

You must first find the coordinates of the points of intersection. The x-coordinates satisfy the equation

$$x^2 - 2x - 3 = x + 1$$

Simplifying and factorising gives

$$x^2 - 3x - 4 = 0$$

$$\therefore (x - 4)(x + 1) = 0$$

So $x = 4$ or $x = -1$.

Sketching the curve and the line on the same set of axes gives the diagram shown.

Notice that part of the required area lies below the x-axis.

There are two ways of finding the area required.
First, you can consider the two parts A_1 and A_2 together
with triangle PQR, as shown.

The area A_1 is found by calculating:

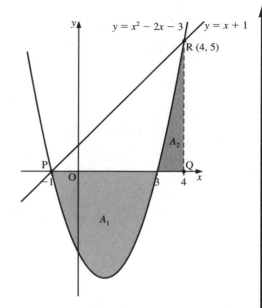

$$-A_1 = \int_{-1}^{3} (x^2 - 2x - 3)\, dx = \left[\frac{x^3}{3} - x^2 - 3x\right]_{-1}^{3}$$

$$= (9 - 9 - 9) - \left(-\frac{1}{3} - 1 + 3\right)$$

$$= -9 - \frac{5}{3}$$

$$\therefore A_1 = \frac{32}{3}$$

Calculating A_2 gives

$$A_2 = \int_{3}^{4} (x^2 - 2x - 3)\, dx = \left[\frac{x^3}{3} - x^2 - 3x\right]_{3}^{4}$$

$$= -\frac{20}{3} - (-9)$$

$$\therefore A_2 = \frac{7}{3}$$

The area, A_3, or triangle PQR is

$$A_3 = \frac{1}{2}(5)(5) = \frac{25}{2}$$

So: $\quad A = A_1 + (A_3 - A_2)$

$$= \frac{32}{3} + \left(\frac{25}{2} - \frac{7}{3}\right) = \frac{125}{6}$$

The area between the curve and the line is $\frac{125}{6} = 20\frac{5}{6}$.

The second method for finding the area between the two
curves uses the result

$$A = \int_{a}^{b} [(g(x) - f(x)]\, dx$$

This gives

$$A = \int_{-1}^{4} (x + 1)\, dx - \int_{-1}^{4} (x^2 - 2x - 3)\, dx$$

$$= \int_{-1}^{4} [(x + 1) - (x^2 - 2x - 3)]\, dx$$

$$= \int_{-1}^{4} (-x^2 + 3x + 4)\, dx$$

$$= \left[-\frac{x^3}{3} + \frac{3x^2}{2} + 4x\right]_{-1}^{4} = \left(-\frac{64}{3} + 24 + 16\right) - \left(\frac{1}{3} + \frac{3}{2} - 4\right)$$

Notice how much easier it is
to use the result.

$$= \frac{56}{3} - \left(-\frac{13}{6}\right)$$

$$\therefore A = \frac{125}{6} \text{ as before.}$$

Exercise 5D

1 The line $y = 3x + 1$ meets the curve $y = x^2 + 3$ at the points P and Q.

   a) Calculate the coordinates of P and Q.

   b) Sketch the line and the curve on the same set of axes.

   c) Calculate the area of the finite region bounded by the line and the curve.

2 The curve $y = x^2 - 2x + 3$ meets the line $y = 9 - x$ at the points A and B.

   a) Find the coordinates of A and B.

   b) Sketch the line and the curve on the same set of axes.

   c) Calculate the area of the finite region bounded by the line and the curve.

3 The curve $y = x^2 + 16$ meets the curve $y = x(12 - x)$ at the points C and D.

   a) Find the coordinates of C and D.

   b) Sketch the two curves on the same set of axes.

   c) Calculate the area bounded by the two curves.

4 a) Use your GDC to sketch the curves $y = x^2 - 5x$ and $y = 3 - x^2$, and find their points of intersection.

   b) Find the area of the region bounded by the two curves.

5 a) Use your GDC to sketch the graphs of the line $y = \frac{1}{3}x$ and the curve $y = \sqrt{x}$ for positive values of x, and find the coordinates of their points of intersection.

   b) Find the area of the region bounded by the line and the curve.

6 Find the area enclosed between the curves $y = 2x^2 - 7$ and $y = 5 - x^2$.

7 Find the area enclosed between the curves $y = (x - 1)^2$ and $y = 8 - (x - 1)^2$.

8 Calculate the area bounded by the y-axis, the line $y = 8$ and the curve $y = x^3$.

9 Calculate the area bounded by the curve $y = \frac{16}{x^2}$, the line $x = 4$ and the line $y = 16$.

10 The curve $y = x^2 - 2x$ cuts the x-axis at the points O and P, and meets the line $y = 2x$ at the point Q, as in the diagram.

   a) Calculate the coordinates of P and Q.

   b) Find the area of the shaded region.

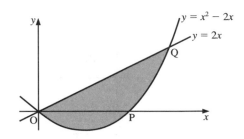

11 The curve $y = 3x - x^2$ cuts the x-axis at the points O and A, and meets the line $y = -3x$ at the point B, as in the diagram on the right.

a) Calculate the coordinate of A and B.

b) Find the area of the shaded region.

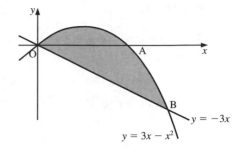

12 The curve $y = x^2 - 1$ cuts the x-axis at the points P and Q, and meets the line $y = x + 1$ at the points P and R, as in the diagram on the right.

a) Calculate the coordinates of P, Q and R.

b) Find the area of the shaded region.

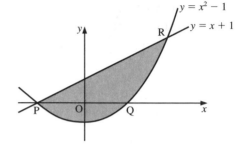

13 a) Sketch the curve of $y = f(x)$, where $f(x) = x(x + 1)(x - 2)$.

b) Hence evaluate the following.

i) $\displaystyle\int_0^2 f(x)\,dx$ ii) $\displaystyle\int_{-1}^0 f(x)\,dx$ iii) $\displaystyle\int_{-1}^2 f(x)\,dx$

....

5.4   Volume of revolution about the x-axis

Consider the area under the curve $y = x^2$ between $x = 1$ and $x = 2$, as shown.

Now imagine that this area is rotated through 2π radians about the x-axis. It will form a three-dimensional solid. The volume of this solid can be calculated using integration.

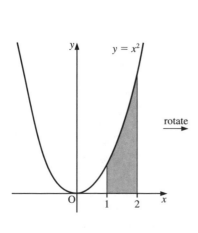

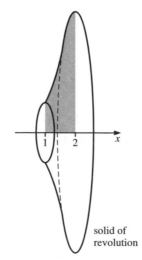

solid of revolution

Think of a small strip of width δx under the curve $f(x)$, as shown. When this small area is rotated through 2π radians about the x-axis, it forms a disc radius y and thickness δx. The volume, δV, of the disc is given by

$$\delta V = \pi y^2 \delta x$$

To find the volume, V, of the whole solid, you must add the volumes of all such discs from $x = a$ to $x = b$. Therefore,

$$V = \sum_{x=a}^{b} \pi y^2 \, \delta x$$

This is an approximate value for the volume of the solid. The more, smaller discs you divide the solid into, the closer this value is to the actual volume.

As $\delta x \to 0$, the summation approaches a limiting value. So:

$$V = \lim_{\delta x \to 0} \pi \sum_{x=a}^{b} y^2 \, \delta x$$

In other words,

$$V = \pi \int_{a}^{b} y^2 \, dx$$

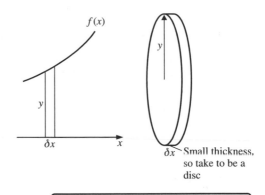

Σ means 'the sum of'.

The volume, V, of the solid of revolution formed when the curve $y = f(x)$ is rotated through 2π radians between the limits of $x = a$ and $x = b$ about the x-axis is given by

$$V = \pi \int_{a}^{b} y^2 \, dx$$

Example 1

Find the volume of the solid formed when the area between the curve $y = x^2 + 2$ and the x-axis from $x - 1$ to $x = 3$ is rotated through 2π radians about the x-axis.

....

The volume V is given by

$$V = \int_{1}^{3} y^2 \, dx$$

Now $y^2 = (x^2 + 2)^2 = x^4 + 4x^2 + 4$. Therefore,

$$V = \pi \int_{1}^{3} (x^4 + 4x^2 + 4) \, dx$$

$$= \pi \left[\frac{x^5}{5} + \frac{4x^3}{3} + 4x \right]_{1}^{3} = \pi \left(\frac{483}{5} - \frac{83}{15} \right)$$

$$\therefore \quad V = \frac{1366\pi}{15}$$

The volume of the solid formed is $\dfrac{1366\pi}{15}$.

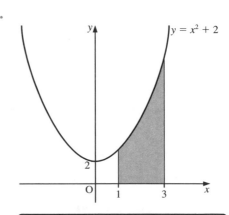

Always draw a sketch to see what is happening.

You can also rotate the area between two curves about the x-axis to form a solid. In this case the volume of the solid is given by

$$V = \pi \int_a^b (g(x))^2 \, \mathrm{d}x - \pi \int_a^b (f(x))^2 \, \mathrm{d}x$$

or

$$V = \pi \int_a^b [(g(x))^2 - (f(x))^2] \, \mathrm{d}x$$

> Compare this with the formula for the area between two curves, page 164.

Example 2

The area enclosed between the curve $y = 4 - x^2$ and the line $y = 4 - 2x$ is rotated through 2π radians about the x-axis. Find the volume of the solid generated.

> You say that the volume is *generated* by rotating the area.

....

The diagram shows the area to be rotated.

$$V = \pi \int_0^2 (4 - x^2)^2 \, \mathrm{d}x - \pi \int_0^2 (4 - 2x)^2 \, \mathrm{d}x$$

$$= \pi \int_0^2 [(4 - x^2)^2 - (4 - 2x)^2] \, \mathrm{d}x$$

$$= \pi \int_0^2 x^4 - 8x^2 + 16 - (4x^2 - 16x + 16) \, \mathrm{d}x$$

$$= \pi \int_0^2 (x^4 - 12x^2 + 16x) \, \mathrm{d}x$$

$$= \pi \left[\frac{x^5}{5} - 4x^3 + 8x^2 \right]_0^2$$

$$\therefore \ V = \frac{32\pi}{5}$$

The volume of the solid of revolution is $\dfrac{32\pi}{5}$.

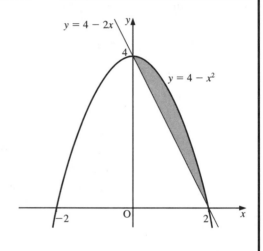

Exercise 5E
....

Throughout this exercise leave your answers as **multiples of π**.

1 Find the volume generated when each of the areas, bounded by the following curves and the x-axis, is rotated through $360°$ about the x-axis between the given lines.

a) $y = x$; $x = 0$ and $x = 6$

b) $y = x^2$; $x = 0$ and $x = 5$

c) $y = \sqrt{x}$; $x = 0$ and $x = 4$

d) $y = \dfrac{1}{x^2}$; $x = 1$ and $x = 2$

e) $y = 3\sqrt{x}$; $x = 2$ and $x = 4$

f) $y = 2x + 1$; $x = 1$ and $x = 3$

g) $y = 5 - x$; $x = 2$ and $x = 5$

h) $y = x^2 + 1$; $x = 0$ and $x = 3$

i) $y = \sqrt{x^2 + 3x}$; $x = 2$ and $x = 6$

j) $y = \sqrt{3x^2 + 8}$; $x = 1$ and $x = 3$

2 The curve $y = x^2$ meets the line $y = 4$ at the points P and Q.

   a) Find the coordinates of P and Q.

   b) Calculate the volume generated when the region bounded by the curve and the line is rotated through 360° about the x-axis.

3 The curve $y = x^2 + 1$ meets the line $y = 2$ at the points A and B.

   a) Find the coordinates of A and B.

   The region bounded by the curve and the line is rotated through 360° about the x-axis.

   b) Calculate the volume of the solid generated.

4 The region bounded by the lines $y = x + 1$, $y = 3$ and the y-axis is rotated through 360° about the x-axis. Calculate the volume of the solid generated.

5 Calculate the volume generated when the region bounded by the curve $y = \dfrac{4}{x}$ and the lines $x = 1$ and $y = 1$ is rotated through 360° about the x-axis.

6 The line $y = 3x$ meets the curve $y = x^2$ at the points O and P.

   a) Calculate the coordinates of P.

   b) Find the volume of the solid generated when the area enclosed by the line and the curve is rotated through 360° about the x-axis.

7 a) On one set of axes sketch the graphs of the curves $y = x(1 - x)$ and $y = 2x(1 - x)$.

   b) Calculate the volume generated when the finite region bounded by the two curves is rotated through 360° about the x-axis.

8 The curve $y = x^2$ meets the curve $y = 8 - x^2$ at the points P and Q.

   a) Find the coordinates of P and Q.

   b) Calculate the volume generated when the region bounded by the two curves is rotated through 180° about the x-axis.

Summary

You should know how to ...

► Integrate a polynomial.

▷ $\displaystyle\int ax^n \, dx = \frac{ax^{n+1}}{n+1} + c \qquad (n \neq 1)$

► Use integration to solve kinematics problems.

▷

$$s \quad \underset{\text{integrate}}{\overset{\text{differentiate}}{\underset{\longleftarrow}{\longrightarrow}}} \quad v = \frac{ds}{dt} \quad \underset{\text{integrate}}{\overset{\text{differentiate}}{\underset{\longleftarrow}{\longrightarrow}}} \quad a = \frac{dv}{dt}$$

► Use integration to solve problems with area and volume.

▷ The area enclosed by the curve $y = f(x)$, the lines $x = a$ and $x = b$ and the x-axis is given by

$$A = \int_a^b f(x) \, dx$$

▷ The area between the curves $y = f(x)$ and $y = g(x)$ is given by

$$A = \int_a^b \left(f(x) - g(x) \right) dx$$

▷ The volume of revolution about the x-axis is given by

$$V = \pi \int_a^b y^2 \, dx$$

Revision exercise 5

1 Integrate

a) $\sqrt{x^3}$

b) $\dfrac{x^4 - 1}{x^3}$

2 It is given that $\dfrac{dy}{dx} = x^3 + 2x - 1$ and that $y = 13$ when $x = 2$.

Find y in terms of x.

© IBO[2003]

3 A curve with equation $y = f(x)$ passes through the point $(1, 1)$.
Its gradient function is $f'(x) = -2x + 3$.

Find the equation of the curve.

© IBO[2000]

4 The diagram shows part of the graph of $y = x^4 - 8x$.

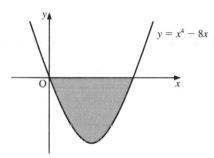

Find the area of the shaded region.

5 The definite integral $\displaystyle\int_1^2 \dfrac{6}{x^2}\, dx$ represents an area.

a) Find this area.

b) Sketch a diagram to show this area.

6 a) Find $\displaystyle\int (1 - x)(x^2 - 9)\, dx$.

The diagram shows part of the graph of the function

$$f(x) = (1 - x)(x^2 - 9).$$

The area of the shaded region is given by $\displaystyle\int_a^b f(x)\, dx$.

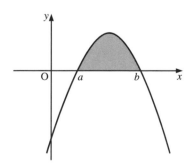

b) Find the values of a and b.

7 A solid is formed by rotating the area between the curve
$y = 4 - x^2$ and the x-axis through 360° about the x-axis.

Show that its volume is $\dfrac{512\pi}{15}$.

8 The area in question **4** is rotated through 360° about the x-axis.
Find the volume of the solid that is generated.

9 The diagram shows part of the graph of the function
$f(x) = x(x - 2)(x + 3)$.

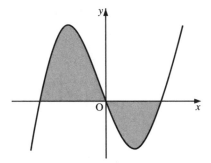

a) Find the total area of the two shaded regions.

b) Find the value of $\displaystyle\int_{-3}^{2} f(x)\, dx$.

Explain why your answers to parts a) and b) are different.

10 The diagram shows part of the curve $y = x^2$.
The shaded area is rotated through 360° about the x-axis.

The volume of the solid formed is $\dfrac{\pi}{5}$.

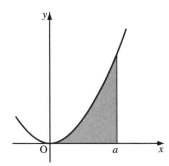

Find the value of a.

11 The function f is such that $f''(x) = 2x - 2$. When the graph of
f is drawn it has a minimum point at $(3, -7)$.

a) Show that $f'(x) = x^2 - 2x - 3$ and hence find $f(x)$.

b) Find $f(0), f(-1)$ and $f'(-1)$.

c) Hence sketch the graph of f, labelling it with the information
obtained in part b).

© IBO [1999]

6 Sequences and series

6.1   Sequences

A **sequence** is a set of numbers in a particular order where each number is derived from a particular rule.

For example, 3, 6, 9, 12, ... is a sequence.

The first term is 3, the second term is 6, the nth term is given by $3n$.

You write the nth term as u_n. In this example,

$$u_n = 3n \quad \text{for } n \geqslant 1$$

> When $n = 1$, $u_1 = 3 \times 1 = 3$
> When $n = 2$, $u_2 = 3 \times 2 = 6$
> and so on.

> u_n can be called the general term of the sequence.

Example 1

Write down the first three terms of the sequence whose nth term is given by

$$u_n = n^2 + 6n \quad \text{for } n \geqslant 1$$

....

Let $n = 1$, then $\quad u_1 = (1)^2 + 6(1) = 7$

Let $n = 2$, then $\quad u_2 = (2)^2 + 6(2) = 16$

Let $n = 3$, then $\quad u_3 = (3)^2 + 6(3) = 27$

The first three terms of the sequence are 7, 16 and 27.

If you are given the first few terms of a sequence you can work out the general term.

Example 2

Write down an expression, in terms of n, for the nth term of the sequence 5, 9, 13, 17, ...

....

The difference between each of the consecutive terms is 4.
Consider the sequence defined by $u_n = 4n$:

$\quad$ 4, 8, 12, 16, ...

Each of these numbers is one less than the numbers in the given sequence.
So the sequence 5, 9, 13, 17, ... is defined by $u_n = 4n + 1$.

You can also define a sequence by giving an expression that relates one general term of the sequence to another.

This relationship between the terms occurs throughout the sequence and is therefore called a **recurrence relation**.

> You also need to know the first one or two terms of the sequence.

Example 3

A sequence is defined by $u_{n+1} = 2u_n + 1$, where the first term is $u_1 = 3$. Write down the first four terms of the sequence.

....

Let $n = 2$, then $u_2 = 2u_1 + 1$. Since $u_1 = 3$,

$$u_2 = 2(3) + 1$$

$$\therefore u_2 = 7$$

Let $n = 3$, then $u_3 = 2u_2 + 1$. Since $u_2 = 7$,

$$u_3 = 2(7) + 1$$

$$\therefore u_3 = 15$$

Let $n = 4$, then $u_4 = 2u_3 + 1$. Since $u_3 = 15$,

$$u_4 = 2(15) + 1$$

$$\therefore u_4 = 31$$

The first four terms of the sequence are 3, 7, 15 and 31.

> $u_{n+1} = 2u_n + 1$ is a recurrence relation.

> You can check your answers using a GDC. Input the first term, 3. Then input '2 × answer + 1'. If you keep pressing 'Enter', you will get successive terms of the sequence.

If you know the first few terms of a sequence you can work out the recurrence relation.

Example 4

Write down a recurrence relation between the terms of the sequence 5, 14, 41, 122, ...

....

$$u_1 = 5$$

You can see that:

$$14 = 3(5) - 1$$

$$41 = 3(14) - 1$$

$$122 = 3(41) - 1$$

$$\therefore u_{n+1} = 3u_n - 1$$

The recurrence relation is $u_{n+1} = 3u_n - 1$, where the first term is $u_1 = 5$.

> You must always state the first term.

Some recurrence relations are more complicated.

Example 5

A sequence is defined by the recurrence relation
$$u_{n+1} = 2u_n + u_{n-1} \quad \text{for } n > 1$$
Given that $u_5 = 99$, show that $u_7 - 2u_6 = 99$. Given further that $u_8 = 1393$, find the terms u_6 and u_7.

Let $n = 6$, then
$$u_7 = 2u_6 + u_5$$
$$= 2u_6 + 99$$
$$\therefore \quad u_7 - 2u_6 = 99 \qquad [1]$$
Let $n = 7$, then
$$u_8 = 2u_7 + u_6$$
$$\therefore \quad 1393 = 2u_7 + u_6 \qquad [2]$$
Solving [1] and [2] simultaneously using a GDC gives $u_6 = 239$ and $u_7 = 577$.

Exercise 6A

1 Write down the first four terms of these sequences. Check using your GDC.

a) $u_n = 2n + 1$
b) $u_n = 3n - 2$
c) $u_n = 5 - 2n$
d) $u_n = n^2 + 3$
e) $u_n = \dfrac{1}{n}$
f) $u_n = \dfrac{n}{n + 1}$
g) $u_n = \dfrac{1}{n^2 + 1}$
h) $u_n = 3 + \dfrac{1}{n(n + 1)}$
i) $u_n = n(n + 1)(n + 2)$

2 Write down an expression, in terms of n, for the nth term of each of these sequences.

a) $4, 8, 12, 16, 20, \ldots$
b) $5, 7, 9, 11, 13, \ldots$
c) $4, 9, 14, 19, 24, \ldots$
d) $8, 11, 14, 17, 20, \ldots$
e) $\frac{1}{2}, \frac{1}{3}, \frac{1}{4}, \frac{1}{5}, \frac{1}{6}, \ldots$
f) $\frac{1}{3}, \frac{1}{6}, \frac{1}{9}, \frac{1}{12}, \frac{1}{15}, \ldots$
g) $\frac{2}{5}, \frac{2}{8}, \frac{2}{11}, \frac{2}{14}, \frac{2}{17}, \ldots$
h) $\frac{1}{2}, \frac{2}{3}, \frac{3}{4}, \frac{4}{5}, \frac{5}{6}, \ldots$
i) $\frac{2}{1}, \frac{3}{4}, \frac{4}{7}, \frac{5}{10}, \frac{6}{13}, \ldots$

3 Write down an expression, in terms of n for the nth term of each of these sequences.

a) $2, 4, 8, 16, 32, \ldots$
b) $10, 20, 40, 80, 160, \ldots$
c) $5, 10, 20, 40, 80, \ldots$
d) $4, 12, 36, 108, 324, \ldots$
e) $2, -6, 18, -54, 162, \ldots$
f) $1, -\frac{1}{2}, \frac{1}{4}, -\frac{1}{8}, \frac{1}{16}, \ldots$
g) $1, 4, 9, 16, 25, \ldots$
h) $\frac{1}{4}, \frac{2}{9}, \frac{3}{16}, \frac{4}{25}, \frac{5}{36}, \ldots$
i) $-2, 6, -12, 20, -30, \ldots$

4 Write down the first four terms of each of these sequences. Check using your GDC.

a) $u_{n+1} = 2 + u_n, u_1 = 5$ b) $u_{n+1} = 6 + u_n, u_1 = 3$

c) $u_{n+1} = 3 - u_n, u_1 = 2$ d) $u_{n+1} = 1 + 2u_n, u_1 = 3$

e) $u_{n+1} = 10 - u_n, u_1 = 5$ f) $u_{n+1} = \dfrac{1}{u_n}, u_1 = 7$

5 Write down the recurrence relation between the terms of these sequences.

a) $4, 9, 24, 69, \ldots$ b) $5, 11, 23, 47, \ldots$

c) $2, 6, 26, 126, \ldots$ d) $1, 5, 21, 85, \ldots$

e) $3, 5, 11, 29, \ldots$ f) $10, 22, 58, 166, \ldots$

6 A Fibonacci sequence is defined by the recurrence relation $u_{n+1} = u_n + u_{n-1}$, where $u_1 = 1$ and $u_2 = 2$.

a) Write down the first seven terms of the sequence.

b) Given that $u_{15} = 987$ and $u_{12} = 233$, deduce the equations

$$u_{14} + u_{13} = 987 \quad \text{and} \quad u_{14} - u_{13} = 233$$

c) Hence find the value of u_{14}.

> Fibonacci sequences are named after the Italian mathematician Fibonacci of Pisa (1175–1250 AD). He was one of the first people to introduce the Hindu–Arabic number system into Europe and it is the number system that we use today.

7 The sequence $\{u_n\}$ is defined by the recurrence relation $u_{n+1} = u_n + 2u_{n-1}$.

a) Given that $u_{11} = 683$ and $u_8 = 85$, deduce the equations

$$u_{10} + 2u_9 = 683 \quad \text{and} \quad u_{10} - u_9 = 170$$

b) Hence find the value of u_9.

6.2   Series and sigma notation

A **series** is the sum of the terms of a sequence. You can write the sum of the first n terms of a sequence as S_n, where

$$S_n = u_1 + u_2 + u_3 + \ldots + u_n$$

This is an example of a finite series since there is a finite number of terms. It can be expressed more concisely using sigma (Σ) notation, as:

$$u_1 + u_2 + u_3 + \ldots + u_n = \sum_{r=1}^{n} u_r$$

For example, the finite series $7 + 11 + 15 + 19$ can be written as

$$\sum_{r=1}^{4} (4r + 3)$$

The infinite series $1 + 4 + 9 + 16 + \ldots$ can be written as

$$\sum_{r=1}^{\infty} r^2$$

> Σ (sigma) is the Greek capital 'S' and in mathematics means 'the sum of'.

> '+ ...' means that the series goes on and on – it is infinite. ∞ is the symbol for infinity.

Example 1

Find the sum of the first four terms of the sequence defined by
$$u_r = 3^{r+1} \quad \text{for } r \geqslant 1$$

S_4, the sum of the first four terms, is given by

$$S_4 = \sum_{r=1}^{4} 3^{r+1}$$

$$= 3^2 + 3^3 + 3^4 + 3^5$$

$$= 9 + 27 + 81 + 243$$

$$\therefore \; S_4 = 360$$

The sum of the first four terms is 360.

Example 2

Write each of these series in Σ notation.

a) $-1 + 2 + 7 + 14 + 23$ 　　　　b) $6 - 7 + 8 - 9 + \ldots$

a) The series whose terms are defined by $u_r = r^2$ is
$$1 + 4 + 9 + 16 + 25$$

You want the series whose rth term is given by $u_r = r^2 - 2$.
Therefore, the series can be written as

$$\sum_{r=1}^{5} (r^2 - 2)$$

b) Notice that this is an infinite series. If you ignore the
alternating plus and minus signs, you see that the terms
increase by 1. The rth term of the series $6 + 7 + 8 + 9 + \ldots$
is given by $u_r = r + 5$.

So the rth term of the given series is defined by
$$u_r = (-1)^{r+1}(r + 5)$$
and can be written as

$$\sum_{r=1}^{\infty} (-1)^{r+1}(r + 5)$$

The term $(-1)^{r+1}$ simply gives you the alternating sign.
When $(r + 1)$ is even, you get a positive term.
When $(r + 1)$ is odd, you get a negative term.
Notice that this series could also be written as

$$\sum_{r=5}^{\infty} (-1)^{r+1}(r + 1)$$

$u_1 = 1 - 2 = -1$
$u_2 = 4 - 2 = 2$
$u_3 = 9 - 2 = 7$
etc.

The first term is $u_1 = 1 + 5 = 6$

Example 3

Find $\displaystyle\sum_{r=3}^{6}(2r+1)$.

$$\sum_{r=3}^{6}(2r+1) = [2(3)+1] + [2(4)+1] + [2(5)+1] + [2(6)+1]$$
$$= 7 + 9 + 11 + 13$$
$$= 40$$

Exercise 6B

1 Write down all the terms in each of these series.

a) $\displaystyle\sum_{r=1}^{5} r^2$

b) $\displaystyle\sum_{r=1}^{6}(3r-1)$

c) $\displaystyle\sum_{r=1}^{4}(2r^2+3)$

d) $\displaystyle\sum_{r=3}^{6} r^3$

e) $\displaystyle\sum_{r=5}^{10} r(r-3)$

f) $\displaystyle\sum_{r=0}^{5}(2r+1)^2$

g) $\displaystyle\sum_{r=1}^{5}\frac{1}{r}$

h) $\displaystyle\sum_{r=4}^{8} r^r$

i) $\displaystyle\sum_{r=1}^{5} 3$

2 Write each of these series in Σ notation.

a) $1 + 2 + 3 + 4 + 5$

b) $1^3 + 2^3 + 3^3 + 4^3 + 5^3 + 6^3 + 7^3$

c) $7 + 10 + 13 + 16 + 19 + 22 + 25$

d) $\frac{1}{3} + \frac{1}{4} + \frac{1}{5} + \dots + \frac{1}{20}$

e) $5 \times 6 + 6 \times 7 + 7 \times 8 + \dots + 18 \times 19$

f) $3^4 + 4^4 + 5^4 + \dots + n^4$

g) $\dfrac{5}{5^2-1} + \dfrac{6}{6^2-1} + \dfrac{7}{7^2-1} + \dots + \dfrac{n}{n^2-1}$

h) $\dfrac{1}{2\times3} + \dfrac{2}{3\times4} + \dfrac{3}{4\times5} + \dots + \dfrac{n}{(n+1)(n+2)}$

3 Find $\displaystyle\sum_{r=4}^{7}(3r-8)$

4 Find $\displaystyle\sum_{r=2}^{6}(2r+5)$

5 Find $\displaystyle\sum_{r=1}^{5} r^2$

6.3 Arithmetic sequences

Consider the sequence of numbers $1, 3, 5, 7, \dots$.
You can obtain each term from the previous term by adding 2.
This sequence is an example of an arithmetic sequence (AS).

An **arithmetic sequence** is a sequence of numbers in which any term can be obtained from the previous term by adding a certain number called the **common difference**.

❖ The first term of an AS is denoted by u_1.
❖ Its common difference is denoted by d.

In the example above, $u_1 = 1$ and $d = 2$. You can write down an expression for the nth term in terms of u_1 and d.

Write $1, 3, 5, 7, \dots$ as $1, 1 + 2, 1 + (2 \times 2), 1 + (3 \times 2), \dots$
or $u_1, u_1 + d, u_1 + 2d, u_1 + 3d, \dots$

So: $u_n = u_1 + (n - 1)d$

The general expression for the nth term of an arithmetical sequence is

$$u_1 + (n - 1)d$$

where u_1 is the first term and d is the common difference.

Example 1

Write down the nth term for each of these arithmetic sequences.
a) $2, 5, 8, 11, \dots$ 　　　　　b) $10, 6, 2, -2, \dots$

....

a) The first term is $u_1 = 2$ and the common difference is $d = 5 - 2 = 3$.
The nth term is given by
$$u_1 + (n - 1)d = 2 + 3(n - 1)$$
$$= 2 + 3n - 3$$
$$= 3n - 1$$
b) The first term is $u_1 = 10$ and the common difference is $d = 6 - 10 = -4$.
The nth term is given by
$$u_1 + (n - 1)d = 10 + (n - 1)(-4)$$
$$= 10 - 4n + 4$$
$$= 14 - 4n$$

You find the common difference by subtracting consecutive terms.

If you know the first term, the last term and the common difference, you can find the number of terms in an arithmetic sequence.

Example 2

Find the number of terms in the arithmetic sequence

$$4, 4\tfrac{1}{2}, 5, 5\tfrac{1}{2}, \ldots \ 10$$

....

The first term is $u_1 = 4$ and the common difference is $d = \tfrac{1}{2}$.

Let the number of terms be n, then the nth term is 10. So:

$$10 = u_1 + (n - 1)d$$
$$10 = 4 + (n - 1) \times \tfrac{1}{2}$$
$$12 = n - 1$$
$$\therefore \ n = 13$$

The sum of an arithmetic sequence

Consider the sum of the first 10 natural numbers,

$$\sum_{r=1}^{10} r = 1 + 2 + 3 + \ldots + 10$$

Simply adding these up gives 55. However, a useful technique for calculating such a sum is to pair off terms.
Notice that the first and last terms added together give $1 + 10 = 11$.
The second and next to last terms added together give $2 + 9 = 11$.
In fact, by pairing off in this way you get:

$$1 + 10 = 11$$
$$2 + 9 = 11$$
$$3 + 8 = 11$$
$$4 + 7 = 11$$
$$5 + 6 = 11$$

This gives 5 pairings of 11 so the sum is $5 \times 11 = 55$.
In other words, you can calculate the sum by using

$$\frac{10}{2} \times (1 + 10) = 55$$

If you write S_n for the sum, then

$$S_n = \frac{n}{2}(u_1 + u_n)$$

where u_n is the last term.

As the last term is the nth term $u_1 + (n - 1)d$, you can write

$$S_n = \frac{n}{2}[u_1 + u_1 + (n - 1)d]$$
$$= \frac{n}{2}[2u_1 + (n - 1)d]$$

General formula for the sum of an arithmetic sequence

The sum of the first n terms of an arithmetic sequence is

$$S_n = u_1 + [u_1 + d] + \dots + [u_1 + (n-1)d] \quad [1]$$

Write the terms on the right in reverse order:

$$S_n = [u_1 + (n-1)d] + [u_1 + (n-2)d] + \dots + u_1 \quad [2]$$

Adding [1] and [2] gives

$$2S_n = \{u_1 + [u_1 + (n-1)d]\} + \{(u_1 + d) + [u_1 + (n-2)d]\}$$
$$+ \dots + \{[u_1 + (n-1)d] + u_1\}$$
$$= [2u_1 + (n-1)d] + [2u_1 + (n-1)d]$$
$$+ \dots + [2u_1 + (n-1)d]$$
$$= n[2u_1 + (n-1)d]$$

$$\therefore \; S_n = \frac{n}{2}[2u_1 + (n-1)d]$$

Alternatively,

$$S_n = \frac{n}{2}[u_1 + u_1 + (n-1)d]$$

$$= \frac{n}{2}(\text{1st term} + n\text{th term})$$

$$\therefore \; S_n = \frac{n}{2}(u_1 + u_n)$$

where u_n is the nth term.

> The sum of the first n terms of an arithmetic sequence is given by
> $$S_n = \frac{n}{2}[2u_1 + (n-1)d]$$
> or $S_n = \frac{n}{2}(u_1 + u_n)$
>
> where u_n is the last term.

Example 3

Find the sum of the first 20 terms of the arithmetic sequence

$$\tfrac{1}{2}, \tfrac{5}{2}, \tfrac{9}{2}, \dots$$

....

The first term is $u_1 = \tfrac{1}{2}$ and the common difference is given by $d = \tfrac{5}{2} - \tfrac{1}{2} = 2$.

Therefore the sum of the first 20 terms is given by

$$S_{20} = \tfrac{20}{2}[2(\tfrac{1}{2}) + (20-1) \times 2]$$
$$= 10(1 + 38)$$
$$= 390$$

Example 4

An AS has a first term of 2 and an nth term of 32.
Given that the sum of the first n terms is 357, find n and the common difference of the AS.

....

Since the nth term is 32,

$$u_1 + (n-1)d = 32$$

$u_1 = 2$, so:

$$2 + (n-1)d = 32$$

$$\therefore (n-1)d = 30 \qquad [1]$$

Since the sum of the first n terms is 357,

$$\frac{n}{2}[2u_1 + (n-1)d] = 357$$

$u_1 = 2$, so:

$$\frac{n}{2}[2(2) + (n-1)d] = 357$$

$$\therefore \ n[4 + (n-1)d] = 714 \qquad [2]$$

Substituting [1] into [2] gives

$$n(4 + 30) = 714$$

$$34n = 714$$

$$\therefore \ n = 21$$

Substituting $n = 21$ into [1] gives

$$(21 - 1)d = 30$$

$$\therefore \ d = \tfrac{3}{2}$$

The value of n is 21 and the common difference is $\tfrac{3}{2}$.

Remember:

AS is an abbreviation of 'arithmetic sequence'.

Example 5

The sum of the first five terms of an AS is $\frac{65}{2}$. Also, five times the 7th term is the same as six times the 2nd term.
Find the first term and common difference of the AS.

....

Since the sum of the first five terms is $\frac{65}{2}$,

$$\tfrac{5}{2}[2u_1 + (5-1)d] = \tfrac{65}{2}$$

$$\therefore \ 2u_1 + 4d = 13 \qquad [1]$$

You also know that five times the 7th term is the same as six times the 2nd term. That is,

$$5(u_1 + 6d) = 6(u_1 + d)$$

$$\therefore \ u_1 - 24d = 0 \qquad [2]$$

Solving [1] and [2] simultaneously gives $u_1 = 6$ and $d = \tfrac{1}{4}$.

The first term of the AS is 6 and the common difference is $\tfrac{1}{4}$.

A useful result is the sum of the first n natural numbers.

The sum of the first n natural numbers is

$$S_n = \frac{n}{2}(1 + n)$$

Example 6

The population of a type of insect is known to be 200 000 on 1st January in a particular year. Each month the population increases by 75 000. Find

a) the total population by the end of the same year
b) the month in which the population exceeds 600 000.

....

The growth can be modelled by an AS with $a = 200\,000$, $d = 75\,000$.

a) For the population at the end of the year, calculate the 12th term:

$$200\,000 + (12 - 1) \times 75\,000 = 1\,025\,000$$

b) The nth term of the sequence is given by
$200\,000 + (n - 1) \times 75\,000$. Putting

$$200\,000 + (n - 1) \times 75\,000 = 600\,000 \text{ gives}$$
$$(n - 1) = 5.3$$
$$\therefore \qquad\qquad n = 6.3$$

Therefore the population exceeds 600 000 by the seventh month, which is July.

Example 6

a) Show that the sum of the first n natural numbers can be written as:

$$\frac{n}{2}(1 + n)$$

b) Hence find

$$\sum_{r=1}^{n} (2r - 1)$$

....

a) The first n natural numbers form an arithmetic sequence with first term $u_1 = 1$ and common difference 1. So:

$$S_n = \frac{n}{2}(u_1 + u_n)$$

$$= \frac{n}{2}(1 + n)$$

Using sigma notation, this can be written as

$$\sum_{r=1}^{n} r = \frac{n}{2}(1 + n)$$

b) $\displaystyle\sum_{r=1}^{n}(2r-1) = \sum_{r=1}^{n}2r - \sum_{r=1}^{n}1$

$\qquad\qquad\quad = 2\sum_{r=1}^{n}r - \sum_{r=1}^{n}1$

$\qquad\qquad\quad = 2 \times \dfrac{n}{2}(1+n) - n$

$\qquad\qquad\quad = n(1+n) - n$

$\qquad\qquad\quad = n^2$

$$\sum_{r=1}^{n}1 = 1 \times n = n$$

Exercise 6C

1 Decide which of these sequences are arithmetic sequences. For those which are, write down the value of the common difference.

a) 8, 11, 14, 17, 20, 23

b) 83, 72, 61, 50, 39, 28

c) 1, 2, 4, 8, 16, 32

d) 1, 1.1, 1.11, 1.111, 1.1111, 1.111 11

e) 1, 1.1, 1.2, 1.3, 1.4, 1.5

f) $1, \frac{1}{2}, \frac{1}{3}, \frac{1}{4}, \frac{1}{5}$

g) $1, -2, 3, -4, 5, -6$

h) $-1, -2, -3, -4, -5, -6$

2 Write down the term indicated in square brackets for each of these arithmetic sequences.

a) 1, 5, 9, ... [10th term]

b) 7, 9, 11, ... [30th term]

c) 20, 17, 14, ... [16th term]

d) $-6, -11, -16, ...$ [12th term]

e) 81, 77, 73, ... [nth term]

f) $0.1, -0.2, -0.5, ...$ [25th term]

g) $\frac{1}{6}, \frac{1}{3}, \frac{1}{2}, ...$ [nth term]

h) $a, 3a, 5a, ...$ [nth term]

3 Find the sum, as far as the term indicated in square brackets, of each of these arithmetic sequences.

a) 1, 2, 3, ... [10th term] b) 5, 7, 9, ... [25th term]

c) 4, 9, 14, ... [18th term] d) 60, 55, 50, ... [12th term]

e) 9, 5, 1, ... [20th term] f) 7, 10, 13, ... [nth term]

g) $9, -1, -11, ...$ [25th term] h) $-2, -\frac{1}{2}, 1, ...$ [30th term]

4 Find the number of terms in each of these arithmetic sequences.

   a) $5, 6, 7, \ldots 15$ b) $10, 20, 30, \ldots 210$

   c) $5, 8, 11, \ldots 302$ d) $-8, -6, -4, \ldots 78$

   e) $97, 85, 73, \ldots 13$ f) $46, 42, 38, \ldots -26$

   g) $9, -11, -31, \ldots -571$ h) $2.1, 3.2, 4.3, \ldots 31.8$

5 Find the sum of each of these arithmetic sequences.

   a) $1, 2, 3, \ldots 100$ b) $6, 8, 10, \ldots 30$

   c) $9, 13, 17, \ldots 41$ d) $62, 60, 58, \ldots 38$

   e) $8, 3, -2, \ldots -42$ f) $1.3, 1.6, 1.9, \ldots 4.6$

   g) $3\frac{1}{3}, 4, 4\frac{2}{3}, \ldots 12\frac{2}{3}$ h) $9\frac{1}{5}, 8\frac{4}{5}, 8\frac{2}{5}, \ldots 3\frac{3}{5}$

6 In the arithmetic sequence, the 1st term is 13 and the 15th term is 111. Find the common difference and the sum of the first 20 terms.

7 In an arithmetic sequence, the 3rd term is 4 and the 8th term is 49. Find the 1st term, the common difference and the sum of the first ten terms.

8 The 5th term of an arithmetic sequence is 7 and the common difference is 4. Find the 1st term and the sum of the first ten positive terms.

9 The sum of the first ten terms of an arithmetic sequence is 95, and the sum of the first 20 terms of the same arithmetic sequence is 290.
   Calculate the first term and the common difference.

10 The 17th term of an arithmetic sequence is 22, and the sum of the first 17 terms is 102. Find the 1st term, the common difference and the sum of the first 30 terms.

11 An arithmetic sequence has 1st term 2 and common difference 5. Given that the sum of the first n terms of the sequence is 119, calculate the value of n.

12 An arithmetic sequence has 1st term 6 and common difference 12. Calculate how many terms should be taken in order that the total exceeds 500.

13 A child is collecting marbles. He collects six marbles on the first day of the month and stores them in a box. On the second day of the month he collects another ten marbles, and adds them to his box. He continues in this way, each day collecting four marbles more than he collected on the previous day. Find the day of the month on which the number of marbles in his box will first exceed 1000.

14 Peter is given an interest-free loan to buy a second-hand car. He repays the loan in monthly instalments. He repays $30 the first month, $32 the second month and the repayments continue to rise by $2 per month until the loan is repaid. Given the final monthly repayment is $100

a) calculate the number of months it takes Peter to repay the loan

b) find the amount, in dollars, of the loan.

15 The training programme of an athlete requires her to run laps of a track. Each day she must run three laps more than the day before. On the sixth day she runs 17 laps.

a) Calculate how many laps she runs

i) on the first day

ii) in total by the end of the sixth day.

c) After how many days of training will her total number of laps reach 100?

....

6.4 Geometric sequences

Consider the sequence of numbers 2, 6, 18, 54, … . You can find each term of the sequence by multiplying the previous term by 3. This is an example of a geometric series, also known as a geometric sequence (GS).

> A **geometric sequence** is a sequence of numbers in which any term can be obtained from the previous term by multiplying by a certain number called the **common ratio**.

❖ The first term of a GS is denoted by u_1.

❖ Its common ratio is denoted by r.

In this example, $u_1 = 2$ and $r = 3$. You can write down an expression for the nth term in terms of u_1 and r.

2	or u_1
2×3	$u_1 \times r$
2×3^2	$u_1 \times r^2$
2×3^3	$u_1 \times r^3$

> The general expression for the nth term of a geometrical sequence is $u_1 r^{n-1}$.

Example 1

Write down the nth term for each of these geometric sequences.

a) $3, 12, 48, \ldots$

b) $10, 5, 2.5, \ldots$

a) The first term is 3 and the common ratio is $r = \frac{12}{3} = 4$.
Therefore the nth term is
$$u_1 r^{n-1} = 3(4)^{n-1}$$

b) The first term is 10 and the common ratio is $r = \frac{5}{10} = \frac{1}{2}$.
Therefore the nth term is
$$u_1 r^{n-1} = 10(\tfrac{1}{2})^{n-1}$$

The sum of a geometric sequence

Consider the sum of the first 8 terms of a geometric sequence with $u_1 = 1, r = 3$.

$$S_8 = 1 + 3 + 9 + 27 + \ldots + 729 + 2187 \qquad [1]$$

Multiplying by the common ratio of 3 gives

$$3S_8 = 3 + 9 + 27 + 81 + \ldots + 2187 + 6561 \qquad [2]$$

Subtract [1] from [2]:

$$3S_8 - S_8 = -1 + 6561$$
$$2S_8 = 6560$$
$$\therefore \ S_8 = 3280$$

The sum of this geometric sequence is 3280.

This technique can be used to find the sum of any geometric sequence.

General formula for the sum of a geometric sequence

The sum of the first n terms is

$$S_n = u_1 + u_1 r + u_1 r^2 + \ldots + u_1 r^{n-1} \qquad [1]$$

Multiply throughout by r:

$$rS_n = u_1 r + u_1 r^2 + u_1 r^3 + \ldots + u_1 r^n \qquad [2]$$

Subtract [2] from [1]:

$$S_n - rS_n = (u_1 + u_1 r + u_1 r^2 + \ldots + u_1 r^{n-1})$$
$$- (u_1 r + u_1 r^2 + u_1 r^3 + \ldots + u_1 r^n)$$

$$S_n(1 - r) = u_1 - u_1 r^n$$

$$\therefore \quad S_n = \frac{u_1(1 - r^n)}{1 - r} = u_1 \left(\frac{1 - r^n}{1 - r} \right), \quad r \neq 1 \qquad [3]$$

Multiplying both the numerator and the denominator of [3] by -1 gives

$$S_n = u_1\left(\frac{r^n - 1}{r - 1}\right), \quad r \neq 1$$

which is an alternative form.

The sum of the first n terms of a geometric sequence is given by

$$S_n = u_1\left(\frac{1 - r^n}{1 - r}\right) = u_1\left(\frac{r^n - 1}{r - 1}\right)$$

Example 2

Given the geometric sequence $2, 6, 18, 54, \ldots$ find

a) the common ratio r

b) the 10th term

c) the sum of the first 10 terms.

. . . .

a) The first term is $u_1 = 2$ and the common ratio is $r = \frac{6}{2} = 3$.

b) The 10th term is

$$u_1 r^9 = 2(3)^9 = 39\,366$$

c) The sum of the first 10 terms is

$$S_{10} = u_1\left(\frac{r^{10} - 1}{r - 1}\right)$$

$$= 2\left(\frac{3^{10} - 1}{3 - 1}\right)$$

$$= 2\left(\frac{59\,048}{2}\right)$$

$$= 59\,048$$

Example 3

A GS has a 1st term of 1 and a common ratio of $\frac{1}{4}$.
Find the sum of the first four terms and show that the nth term is given by $4^{(1-n)}$.

. . . .

The sum of the first four terms is

$$S_4 = 1\left[\frac{1 - (\frac{1}{4})^4}{1 - (\frac{1}{4})}\right] = \frac{85}{64}$$

The sum of the first four terms is $\frac{85}{64}$.
The nth term is given by

$$u_n = ar^{n-1}$$

$$= 1(\tfrac{1}{4})^{n-1} = (4^{-1})^{n-1}$$

$$\therefore \; u_n = 4^{(1-n)}$$

Geometric sequences are useful when you want to calculate something that changes over time at a fixed rate, such as population numbers or invested money.

Example 4

An employee of a company starts on a salary of $20 000 per year with an annual increase of 4% of the previous year's salary.

a) Show that the amounts of annual salary form a geometric sequence.

b) Find
   i) how much the employee earns in the tenth year with the company
   ii) the total amount earned by the employee over the first ten years with the company.

. . . .

a) The starting salary is $20 000.
In the second year the employee will earn
$20 000 $\times$ 1.04 = $20 800.
In the third year the employee will earn

$$(\$20\,000 \times 1.04) \times 1.04 = \$20\,000 \times 1.04^2$$
$$= \$21\,632$$

In the fourth year the employee will earn

$$(\$20\,000 \times 1.04^2) \times 1.04 = \$20\,000 \times 1.04^3$$
$$= \$22\,497.28$$

The amounts form a GS with first term $u_1 = 20\,000$ and common ratio $r = 1.04$.

b) i) In the tenth year the salary will be given by the 10th term of the sequence:

$$u_1 r^9 = \$20\,000 \times (1.04)^9$$
$$= \$28\,466.24$$

  ii) The total amount earned over the first ten years with the company is given by the sum of the first 10 terms of the GS:

$$S_{10} = u_1\left(\frac{r^9 - 1}{r - 1}\right)$$

$$= \$20\,000\left(\frac{(1.04)^9 - 1}{1.04 - 1}\right)$$

$$= \$211\,655.91$$

Example 5

The sum of the 2nd and 3rd terms of a GS is 12. The sum of the 3rd and 4th terms is -36.
Find the first term and the common ratio.

....

Since the sum of the 2nd and 3rd terms is 12,

$$u_1r + u_1r^2 = 12$$
$$u_1r(1 + r) = 12$$
$$\therefore \ 1 + r = \frac{12}{u_1r} \qquad [1]$$

Since the sum of the 3rd and 4th terms is -36,

$$u_1r^2 + u_1r^3 = -36$$
$$\therefore \ u_1r^2(1 + r) = -36 \qquad [2]$$

Substituting [1] into [2] gives

$$u_1r^2\left(\frac{12}{u_1r}\right) = -36$$
$$12r = -36$$
$$\therefore \ r = -3$$

From [1]:

$$u_1 = \frac{12}{r(1 + r)}$$

Substituting $r = -3$ gives

$$u_1 = \frac{12}{(-3)(1 - 3)} = 2$$

The 1st term of the GS is 2 and the common ratio is -3.

Example 6

Show that there are two possible geometric sequences in each of which the 1st term is 8 and the sum of the first three terms is 14. For the GS with positive common ratio find, in terms of n, an expression for the sum of the first n terms.

....

Since the 1st term is 8 and the sum of the first three terms is 14,

$$8 + 8r + 8r^2 = 14$$
$$8r^2 + 8r - 6 = 0$$
$$\therefore \ 2(2r - 1)(2r + 3) = 0$$

Solving gives $r = \frac{1}{2}$ or $r = -\frac{3}{2}$.

Hence, there are two geometric sequences which have a first term of 8 and have the sum of their first three terms equal to 14, namely, one with a common ratio of $\frac{1}{2}$ and a second with a common ratio of $-\frac{3}{2}$.

To find the sum of the first n terms of the GS with positive common ratio, use

$$S_n = u_1\left(\frac{1-r^n}{1-r}\right) \quad \text{with } u_1 = 8 \text{ and } r = \tfrac{1}{2}$$

This gives

$$S_n = 8\left[\frac{1-(\tfrac{1}{2})^n}{1-(\tfrac{1}{2})}\right] = 16[1-(2^{-1})^n]$$

$$\therefore\ S_n = 16(1-2^{-n})$$

Exercise 6D

1 Decide which of the following series are geometric sequences. For those which are, write down the value of the common ratio.

a) $2, 6, 18, 54, 162, 486$

b) $3, -6, 12, -24, 48, -96$

c) $3, 9, 15, 21, 27, 33$

d) $1, 1.1, 1.11, 1.111, 1.1111, 1.111\,11$

c) $1, 2, -4, -8, 16, 32$

f) $1, \tfrac{1}{2}, \tfrac{1}{4}, \tfrac{1}{8}, \tfrac{1}{16}, \tfrac{1}{32}$

g) $3, -3, 3, -3, 3, -3$

h) $1, a, a^2, a^3, a^4, a^5$

2 Write down the term indicated in square brackets in each of these geometric sequences.

a) $2, 4, 8, \ldots$ [10th term]

b) $1, -3, 9, \ldots$ [7th term]

c) $5, 10, 20, \ldots$ [8th term]

d) $2, 3, 4\tfrac{1}{2}, \ldots$ [9th term]

e) $81, -54, 36, \ldots$ [8th term]

f) $2, \tfrac{2}{5}, \tfrac{2}{25}, \ldots$ [5th term]

g) $1, \tfrac{1}{2}, \tfrac{1}{4}, \ldots$ [12th term]

h) $1, -\tfrac{1}{3}, \tfrac{1}{9}, \ldots$ [6th term]

3 Find the sum, as far as the term indicated in square brackets, of each of these geometric sequences.

a) $3 + 6 + 12 + \ldots$ [10th term]

b) $3 - 6 + 12 - \ldots$ [10th term]

c) $3 - 6 + 12 - \ldots$ [11th term]

d) $5 + 10 + 20 + \ldots$ [8th term]

e) $-2 + 8 - 32 + \ldots$ [6th term]

f) $1 + 10 + 100 + \ldots$ [7th term]

g) $1 + \tfrac{1}{3} + \tfrac{1}{9} + \ldots$ [7th term]

h) $\tfrac{1}{2} + \tfrac{1}{4} + \tfrac{1}{8} + \ldots$ [nth term]

4 Find the number of terms in each of these geometric sequences.

a) $2, 10, 50, \ldots 1250$ b) $3, 6, 12, \ldots 768$

c) $2, 6, 18, \ldots 1458$ d) $1, -2, 4, \ldots 1024$

e) $4, -12, 36, \ldots -972$ f) $5, 20, 80, \ldots 5120$

g) $54, 18, 6, \ldots \frac{2}{27}$ h) $64, 32, 16, \ldots \frac{1}{8}$

5 Find the sum of each of these geometric sequences.

a) $3 + 6 + 12 + \ldots + 384$ b) $2 + 6 + 18 + \ldots + 1458$

c) $4 - 12 + 36 - \ldots - 972$ d) $7 - 14 + 28 - \ldots + 448$

e) $36 + 12 + 4 + \ldots + \frac{4}{27}$ f) $20 + 10 + 5 + \ldots + \frac{5}{16}$

g) $\frac{1}{3} - \frac{1}{9} + \frac{1}{27} - \ldots - \frac{1}{729}$ h) $1 + \frac{1}{2} + \frac{1}{4} + \ldots \frac{1}{2^n}$

6 A geometric sequence has 3rd term 75 and 4th term 375. Find the common ratio and the first term.

7 In a geometric sequence the 2nd term is -12 and the 5th term is 768. Find the common ratio and the first term.

8 The 4th term of a geometric sequence is 48, and the 6th term is 12. Find the possible values of the common ratio and the corresponding values of the 1st term.

9 A geometric sequence has 3rd term 7 and 5th term 847. Find the possible values of the common ratio, and the corresponding values of the 4th term.

10 Find the sum of the first ten terms of a geometric sequence which has 3rd term 20 and 8th term 640.

11 In a geometric sequence the 2nd term is 15 and the 5th term is -405. Find the sum of the first eight terms.

12 A geometric sequence has common ratio -3. Given that the sum of the first nine terms of the sequence is 703, find the 1st term.

13 Find the 1st term of the geometric sequence in which the common ratio is 2 and the sum of the first ten terms is 93.

14 The common ratio of a geometric sequence is -5 and the sum of the first seven terms of the sequence is 449. Find the first three terms.

15 A geometric sequence has 1st term $\frac{1}{11}$ and common ratio 2. Given that the sum of the first n terms is 93, calculate the value of n.

16 Find how many terms of the geometric sequence $5 - 10 + 20 - \ldots$ should be taken in order that the total equals 215.

17 In a geometric sequence the 1st term is 8 and the sum of the first three terms is 104. Calculate the possible values of the common ratio, and, in each case, write down the corresponding first three terms of the sequence.

18 Given that the 1st term of a geometric sequence is 5 and the sum of the first three terms is 105, find the possible values of the common ratio, and, in each case, write down the corresponding values of the first three terms of the sequence.

19 In a geometric sequence the sum of the 2nd and 3rd terms is 12, and the sum of the 3rd and 4th terms is 60.
Find the common ratio and the 1st term.

20 A geometric sequence is such that the sum of the 4th and 5th terms is -108, and the sum of the 5th and 6th terms is 324. Calculate the common ratio and the value of the 1st term.

21 Find the first five terms in the geometric sequence which is such that the sum of the 1st and 3rd terms is 50, and the sum of the 2nd and 4th terms is 150.

22 In a geometric sequence in which all the terms are positive and increasing, the difference between the 7th and 5th terms is 192, and the difference between the 4th and 2nd terms is 24. Find the common ratio and the 1st term.

23 A child tries to negotiate a new deal for her pocket money for the 30 days of the month of June. She wants to be paid 1p on the 1st of the month, 2p on the 2nd of the month, and, in general, (2^{n-1})p on the nth day of the month. Calculate how much she would get, in total, if this were accepted.

24 A man, who started work in 1990, planned an investment for his retirement in 2030 in the following way. On the first day of each year, from 1990 to 2029 inclusive, he is to place $100 in an investment account. The account pays 10% compound interest per annum, and interest is added on 31 December of each year of the investment.
Calculate the value of his investment on 1 January 2030.

25 A woman borrows $50 000 in order to buy a house. Compound interest at the rate of 12% per annum is charged on the loan. She agrees to pay back the loan in 25 equal instalments at yearly intervals, the first repayment being made exactly one year after the loan is taken out.
Calculate the value of each instalment.

6.5 Infinite geometric sequences

Consider the geometric sequence

$$1, \tfrac{1}{2}, \tfrac{1}{4}, \tfrac{1}{8}, \dots \left(\tfrac{1}{2}\right)^{n-1}$$

It is clear that $u_1 = 1$ and $r = \tfrac{1}{2}$. The sum of the first n terms is given by

$$S_n = \left[\frac{1 - \left(\tfrac{1}{2}\right)^n}{1 - \left(\tfrac{1}{2}\right)} \right] = 2\left[1 - \left(\frac{1}{2}\right)^n \right]$$

Look at S_n for $n = 2, 10, 20$ and 30:

n	2	10	20	30
S_n	1.5	1.998	1.999 998 093	1.999 999 998

> As n gets larger, S_n gets closer to 2.

As $n \to \infty$, the term $\left(\tfrac{1}{2}\right)^n \to 0$, therefore $S_n \to 2$.
So 2 is the **limit** of the series, written as

$$\lim_{n \to \infty} S_n = 2$$

> $n \to \infty$ means 'n tends to infinity', that is, it gets closer and closer to infinity.

This limit is called the **sum to infinity** of the geometric sequence.

General formula for the sum to infinity of a geometric sequence

The sum of the first n terms of a geometric sequence is given by

$$S_n = u_1\left(\frac{1 - r^n}{1 - r}\right)$$

If $-1 < r < 1$ then as $n \to \infty$, $r^n \to 0$. Therefore, as $n \to \infty$:

$$S_n \to u_1\left(\frac{1 - 0}{1 - r}\right) = \frac{u_1}{1 - r}$$

> Notice that the proof of this result hangs on the fact that $-1 < r < 1$.
> If this is not the case, the sum to infinity does not exist.

The sum to infinity of a geometric sequence is given by

$$S_\infty = \sum_{n=1}^{\infty} u_1 r^{n-1} = \frac{u_1}{1 - r}$$

where $-1 < r < 1$.

> Another way of writing $-1 < r < 1$ is $|r| < 1$, where $|r|$ means the modulus of r.

Example 1

Calculate the sum to infinity of the series $2 + \tfrac{1}{2} + \tfrac{1}{8} + \tfrac{1}{32} + \dots$

This is a geometric sequence with $u_1 = 2$ and $r = \tfrac{1}{4}$. Therefore,

$$S_\infty = \frac{2}{1 - \tfrac{1}{4}} = \frac{8}{3}$$

> **Remember:**
> A series is the sum of the terms of a sequence.

A rational number which is a recurring decimal can be written as the sum to infinity of a geometric sequence.

Example 2

Write the recurring decimal 0.3232 … as the sum of a GS.
Hence write this recurring decimal as a rational number.

....

$$0.323\,232\ldots = \frac{32}{100} + \frac{32}{10\,000} + \frac{32}{1\,000\,000} + \ldots$$

This is a geometric series with $u_1 = \frac{32}{100}$ and $r = \frac{1}{100}$.
Since $-1 < r < 1$ the sum to infinity exists and is given by

$$S_\infty = \frac{\left(\frac{32}{100}\right)}{\left(1 - \frac{1}{100}\right)} = \frac{32}{99}$$

The recurring decimal $0.\dot{3}\dot{2}$ can be written as $\frac{32}{99}$.

There may be two infinite geometric sequences with the same sum and common ratio.

Example 3

The sum to infinity of a geometric sequence is 7 and the sum of the first two terms is $\frac{48}{7}$.
Show that the common ratio, r, satisfies the equation
$$1 - 49r^2 = 0$$
Hence find the first term of the GS with positive common ratio.

....

Since the sum to infinity is 7, we have

$$\frac{u_1}{1 - r} = 7$$

$$\therefore \qquad u_1 = 7(1 - r) \qquad\qquad [1]$$

The sum of the first two terms is $\frac{48}{7}$, so:

$$u_1 + u_1 r = \frac{48}{7}$$

$$\therefore \quad u_1(1 + r) = \frac{48}{7} \qquad\qquad [2]$$

Substituting [1] into [2] gives

$$7(1 - r)(1 + r) = \frac{48}{7}$$

$$49(1 - r^2) = 48$$

$$\therefore \quad 1 - 49r^2 = 0$$

as required.

Solving gives $r = \frac{1}{7}$ or $r = -\frac{1}{7}$.

The positive common ratio is $r = \frac{1}{7}$. So from [1] the first term of the required GS is

$$u_1 = 7(1 - r)$$
$$= 7(1 - \tfrac{1}{7})$$
$$\therefore \ u_1 = 6$$

The first term of the GS with positive common ratio is 6.

Exercise 6E

1 Work out each of these sums.

a) $\displaystyle\sum_{r=0}^{\infty} \left(\frac{1}{2}\right)^r$ b) $\displaystyle\sum_{r=0}^{\infty} \left(\frac{1}{3}\right)^r$ c) $\displaystyle\sum_{r=1}^{\infty} \left(\frac{1}{5}\right)^r$

d) $\displaystyle\sum_{r=0}^{\infty} \left(-\frac{1}{4}\right)^r$ e) $\displaystyle\sum_{r=2}^{\infty} \left(-\frac{1}{8}\right)^r$ f) $\displaystyle\sum_{r=0}^{\infty} \left(\frac{1}{9}\right)^{r+1}$

g) $\displaystyle\sum_{r=1}^{\infty} (0.3)^{r+1}$ h) $\displaystyle\sum_{r=0}^{\infty} (-0.7)^{r+2}$ i) $\displaystyle\sum_{r=0}^{\infty} 4 \times \left(\frac{1}{3}\right)^r$

2 Express each of these recurring decimals as a fraction in its simplest form.

a) $0.\dot{5}$ b) $0.\dot{8}$ c) $0.\dot{7}\dot{2}$

d) $0.\dot{1}0\dot{2}$ e) $2.\dot{4}$ f) $3.2\dot{8}1\dot{4}$

3 Find the 1st term of a geometric sequence that has a common ratio of $\frac{2}{5}$ and a sum of infinity of 20.

4 Find the 3rd term of a geometric sequence that has a common ratio of $-\frac{1}{3}$ and a sum to infinity of 18.

5 A geometric sequence has a 1st term of 6 and a sum to infinity of 60. Find the common ratio.

6 Find the common ratio of a geometric sequence that has a 1st term of 6 and a sum to infinity of 4.

7 Find the common ratio of a geometric sequence which has a 2nd term of 6 and a sum to infinity of 24.

8 A geometric sequence has a 2nd term of 6 and a sum to infinity of 27. Write down the possible values of the first three terms.

9 Given that $\displaystyle\sum_{r=0}^{\infty} 5 \times a^r = 15$, find the value of a.

10 In an arithmetic sequence the 1st, 2nd and 5th terms form a geometric sequence.
Find the common ratio of the geometric sequence.

11 The 1st, 2nd and 3rd terms of a geometric sequence are the 1st, 7th and 9th terms of an arithmetic sequence.
Find the common ratio of the geometric sequence.

12 The 2nd, 4th and 5th terms of an arithmetic sequence are the first three terms of a geometric sequence.
Find the common ratio of the geometric sequence.

6.6 Binomial expansions

You may often be required to expand binomial expressions of the form $(a + b)^n$.

> A **binomial** expression has only two terms.

You can use ordinary algebraic multiplication to show that:

$$(1 + x) = 1 + x$$
$$(1 + x)^2 = (1 + x)(1 + x) = 1 + 2x + x^2$$
$$(1 + x)^3 = (1 + x)^2(1 + x) = 1 + 3x + 3x^2 + x^3$$
$$(1 + x)^4 = (1 + x)^3(1 + x) = 1 + 4x + 6x^2 + 4x^3 + x^4$$

and so on.

The coefficients of the terms in these expansions can be written as a triangle:

```
            1        1
        1       2        1
     1      3        3        1
   1     4       6        4        1
 ...    ...     ...      ...      ...      ...
```

> Blaise Pascal (1623–1662) was a French mathematician who first studied this number pattern.

This triangular array of numbers is known as **Pascal's triangle**.

The next row is obtained like this:

$$1 \quad (1 + 4) \quad (4 + 6) \quad (6 + 4) \quad (4 + 1) \quad 1$$
$$1 \quad\quad 5 \quad\quad 10 \quad\quad 10 \quad\quad 5 \quad\quad 1$$

The entry in the 5th row, 3rd position from the left is 10. Therefore, the coefficient of the x^2 term in the expansion of $(1 + x)^5$ is 10.

> Work out the next five rows of the triangle.

The entry in the 3rd row, 2nd position from the left is 3. Therefore, the coefficient of the x term in the expansion of $(1 + x)^3$ is 3.

A general expression for the coefficient of the $(r + 1)$th term in the expansion of $(1 + x)^n$ is:

$$\binom{n}{r} = \frac{n!}{r!(n-r)!}$$

where

$$n! = n(n-1)(n-2) \ldots 3 \times 2 \times 1 \quad \text{(called } n \text{ factorial)}$$

and where, by definition, $0! = 1$.

For example:
$4! = 4 \times 3 \times 2 \times 1 = 24$

The GDC will give you the value of $\binom{n}{r}$ without having to use this formula.

$\binom{n}{r}$ often appears on calculators as ^nC_r or nCr.

Using this result, you can write down a general formula for the expansion of $(1 + x)^n$:

$$(1 + x)^n = 1 + \binom{n}{1}x + \binom{n}{2}x^2 + \binom{n}{3}x^3 + \ldots + x^n$$

This is known as the **binomial expansion** of $(1 + x)^n$, for positive integer n.

Example 1

Use your GDC to evaluate

a) $\binom{5}{1}$ b) $\binom{6}{4}$ c) $\binom{10}{3}$ d) $\binom{8}{8}$

....

Using the GDC gives

a) $\binom{5}{1} = 5$ b) $\binom{6}{4} = 15$ c) $\binom{10}{3} = 120$ d) $\binom{8}{8} = 1$

Example 2

Find the coefficients of the x^2 and x^3 terms in the expansion of $(1 + x)^7$.

....

The coefficient of the x^2 term is

$$\binom{7}{2} = 21$$

$n = 7, r = 2$

The coefficient of the x^2 term in the expansion of $(1 + x)^7$ is 21.
The coefficient of the x^3 term is

$$\binom{7}{3} = 35$$

$n = 7, r = 3$

The coefficient of the x^3 term in the expansion of $(1 + x)^7$ is 35.

Example 3

Write down the expansions of   a) $(1 + x)^8$ b) $(1 - 2x)^4$.

a) Use Pascal's triangle or the GDC to generate the coefficients:

$$(1 + x)^8 = 1 + 8x + 28x^2 + 56x^3 + 70x^4 + 56x^5 + 28x^6$$
$$+ 8x^7 + x^8$$

b) $(1 - 2x)^4 = 1 + 4(-2x) + 6(-2x)^2 + 4(-2x)^3 + (-2x)^4$
$$= 1 - 8x + 24x^2 - 32x^3 + 16x^4$$

Exercise 6F

1 Use your GDC to work out each of the following.

a) $\binom{5}{3}$ b) $\binom{6}{2}$ c) $\binom{9}{7}$ d) $\binom{6}{5}$

e) $\binom{5}{5}$ f) $\binom{12}{2}$ g) $\binom{7}{3}$ h) $\binom{100}{99}$

2 Expand

a) $(1 + x)^4$ b) $(1 + x)^5$ c) $(1 + 3x)^4$ d) $(1 - x)^3$

e) $(1 - 2x)^4$ f) $(1 - 5x)^3$ g) $(1 + \frac{1}{2}x)^4$ h) $(1 - \frac{1}{5}x)^2$

3 Find the coefficient of the term indicated in square brackets in the expansion of each of these expressions.

a) $(1 + x)^7$ $[x^4]$ b) $(1 + x)^9$ $[x^2]$ c) $(1 + 2x)^5$ $[x^3]$

d) $(1 + 5x)^8$ $[x^2]$ e) $(1 - 3x)^6$ $[x^3]$ f) $(1 - 6x)^7$ $[x]$

g) $(1 - 4x)^4$ $[x^2]$ h) $(1 + 2x)^5$ $[x^4]$ i) $(1 - \frac{1}{2}x)^3$ $[x^2]$

Expansion of $(a + x)^n$

Consider these expansions

$$(a + x)^1 = a + x$$
$$(a + x)^2 = a^2 + 2ax + x^2$$
$$(a + x)^3 = a^3 + 3a^2x + 3ax^2 + x^3$$

Notice that the coefficients are the Pascal coefficients.

The general formula for the expansion of $(a + x)^n$ is

$$(a + x)^n = a^n + \binom{n}{1}a^{n-1}x + \binom{n}{2}a^{n-2}x^2 + \binom{n}{3}a^{n-3}x^3$$
$$+ \dots + x^n$$

This is often known as the **binomial theorem**.

Example 4

Expand these expressions in ascending powers of x.

a) $(3 + x)^4$

b) $(2 - x)^6$

c) $(2 + 5x)^3$

··

a) Using Pascal coefficients or the GDC to generate $1, 4, 6, 4, 1$ gives

$$(3 + x)^4 = 3^4 + 4(3)^3(x) + 6(3)^2(x)^2 + 4(3)(x)^3 + x^4$$

$$\therefore (3 + x)^4 = 81 + 108x + 54x^2 + 12x^3 + x^4$$

$a = 3, n = 4$

b) Using Pascal coefficients or the GDC to generate $1, 6, 15, 20, 15, 6, 1$ gives

$$(2 - x)^6 = 2^6 + 6(2)^5(-x) + 15(2)^4(-x)^2 + 20(2)^3(-x)^3$$
$$+ 15(2)^2(-x)^4 + 6(2)(-x)^5 + (-x)^6$$

$$\therefore (2 - x)^6 = 64 - 192x + 240x^2 - 160x^3 + 60x^4$$
$$- 12x^5 + x^6$$

$a = 2, n = 6, 'x' = -x$

c) Using Pascal coefficients or the GDC to generate $1, 3, 3, 1$ gives

$$(2 + 5x)^3 = 2^3 + 3(2)^2(5x) + 3(2)(5x)^2 + (5x)^3$$

$$\therefore (2 + 5x)^3 = 8 + 60x + 150x^2 + 125x^3$$

$a = 2, n = 3, 'x' = 5x$

You can find particular terms in a binomial expansion without working out the whole expansion.

Example 5

Find the coefficient of the x^4 term in the expansion of $(3 - 2x)^{10}$.

··

Rather than expand the whole expression, the x^4 term is given by the Pascal coefficient $\binom{10}{4}$ multiplied by the term $3^6(-2x)^4$.

This gives

$$\binom{10}{4}3^6(-2x)^4 = 210 \times 729 \times 16x^4$$

$$= 2\,449\,440x^4$$

So the coefficient of the x^4 term is $2\,449\,440$.

Example 6

Expand $(x + 2y)^5$.

. . . .

Using the general formula gives

$$(x + 2y)^5 = x^5 + \binom{5}{1}x^4(2y) + \binom{5}{2}x^3(2y)^2 + \binom{5}{3}x^2(2y)^3$$
$$+ \binom{5}{4}x(2y)^4 + (2y)^5$$

$$\therefore (x + 2y)^5 = x^5 + 10x^4y + 40x^3y^2 + 80x^2y^3 + 80xy^4 + 32y^5$$

Example 7

Calculate the value of the constant a if the coefficient of the x^3 term in the expansion of $(a + 2x)^4$ is 160.

. . . .

The x^3 term in the expansion of $(a + 2x)^4$ is

$$\binom{4}{3}a(2x)^3 = 32ax^3$$
$$32a = 160$$
$$\therefore a = 5$$

Approximations

You can often use a binomial expansion to provide a good approximation to decimal quantities raised to a high power.

Example 8

Expand $(1 + 4x)^{14}$ in ascending powers of x, up to and including the 4th term.
Hence evaluate $(1.0004)^{14}$, correct to four decimal places.

. . . .

Using the binomial expansion gives

$$(1 + 4x)^{14} = 1 + \binom{14}{1}(4x) + \binom{14}{2}(4x)^2 + \binom{14}{3}(4x)^3 + \dots$$

$$\therefore (1 + 4x)^{14} = 1 + 56x + 1456x^2 + 23\,296x^3 + \dots$$

Since

$$(1.0004)^{14} = [1 + 4(0.0001)]^{14}$$

you can use this expansion with $x = 0.0001$:

$$(1.0004)^{14} = 1 + 56(0.0001) + 1456(0.0001)^2 + 23\,296(0.0001)^3$$
$$= 1.005\,614\,583$$
$$= 1.0056 \quad \text{(to 4 d.p.)}$$

The next term in the expansion is

$$\binom{14}{4}(4x)^4 = 256\,256x^4$$

This term would contribute the value

$$256\,256(0.0001)^4 = 2.562\,56 \times 10^{-11}$$

to the approximation of $(1.0004)^{14}$. In other words, it would not affect the answer of 1.0056, which is given to four decimal places.

Exercise 6G

1 Expand
  a) $(2 + x)^3$ b) $(3 + x)^4$ c) $(6 - 5x)^3$ d) $(2 + \frac{1}{2}x)^4$
  e) $(3x + 2y)^3$ f) $(2x - y)^5$ g) $(2x + 5y)^3$ h) $(3x - 4y)^4$

2 Find the coefficient of the term indicated in square brackets in the expansion of each of these expressions.
  a) $(2 + 3x)^5$ $[x^3]$ b) $(5 + 2x)^8$ $[x^6]$ c) $(3 + 2x)^7$ $[x^5]$
  d) $(7 - 4x)^5$ $[x^4]$ e) $(2 - 7x)^4$ $[x]$ f) $(5 + 2x)^6$ $[x^3]$
  g) $(\frac{1}{3} + \frac{3}{2}x)^6$ $[x^3]$ h) $(\frac{2}{3} - \frac{2}{5}x)^3$ $[x]$

3 Expand each of these in ascending powers of x, up to and including the term in x^3.
  a) $(1 - 3x)^5$ b) $(1 + 2x)^{10}$ c) $(1 - 5x)^7$ d) $(2 - 3x)^5$
  e) $(4 - x)^5$ f) $(2 + 3x)^6$ g) $(1 + \frac{1}{3}x)^9$ h) $(4 + \frac{1}{4}x)^6$

4 Expand
  a) $(1 + x^3)^4$ b) $(1 + 3x^2)^3$ c) $(3 - 2x^3)^3$
  d) $(1 + x + x^2)^2$ e) $(1 + 2x - x^2)^3$ f) $(2 + 3x - x^2)^2$
  g) $(2 + x - 4x^2)^2$ h) $(3 + 2x + x^2)^2$

5 a) Expand $(1 - x)^{10}$ in ascending powers of x up to and including the term in x^4.
  b) Hence evaluate $(0.99)^{10}$ correct to six decimal places.

6 a) Expand $(1 + 2x)^{14}$ in ascending powers of x up to and including the term in x^3.
  b) Hence evaluate $(1.02)^{14}$ correct to three decimal places.

7 a) Write down the first three terms in the binomial expansion of $(1 - 3x)^8$.
  b) Hence evaluate $(0.997)^8$ correct to five decimal places.

8 a) Write down the first three terms in the binomial expansion of $(2 + 5x)^9$.
  b) Hence evaluate $(2.005)^9$ correct to two decimal places.

9 By first expanding $(3 - 2x)^{12}$ in ascending powers of x up to and including the term in x^2, work out the value of $(2.998)^{12}$ correct to the nearest whole number.

10 By first expanding $(5 - 4x)^5$ in ascending powers of x up to and including the term in x^3, work out the value of $(4.96)^5$ correct to the nearest whole number.

11 Given that $(1 + ax)^n \equiv 1 + 30x + 375x^2 + \ldots$, find the values of the constants a and n.

12 Given that $(1 + bx)^n \equiv 1 - 15x + 90x^2 + \ldots$, find the values of the constants b and n.

13 When $(1 + cx)^n$ is expanded as a series in ascending powers of x, the first three terms are given by $1 + 20x + 150x^2$.
Calculate the values of the constants c and n.

14 When $(1 + ax)^n$ is expanded as a series in ascending powers of x, the first three terms are given by $1 - 8x + 30x^2$.
Calculate the values of a and n.

15 a) Show that $(x + y)^6 + (x - y)^6 \equiv 2x^6 + 30x^4y^2 + 30x^2y^4 + 2y^6$.
b) Hence deduce that $(\sqrt{3} + \sqrt{2})^6 + (\sqrt{3} - \sqrt{2})^6 = 970$.

16 Without using a calculator, simplify each of these.
a) $(\sqrt{5} + \sqrt{2})^4 + (\sqrt{5} - \sqrt{2})^4$ b) $(\sqrt{2} + 1)^5 - (\sqrt{2} - 1)^5$
c) $(\sqrt{7} + \sqrt{3})^6 + (\sqrt{7} - \sqrt{3})^6$ d) $(3 + \sqrt{3})^3 - (3 - \sqrt{3})^3$

Summary

You should know how to ...

▶ Solve problems involving arithmetic series.
 ▷ The nth term $u_n = u_1 + (n - 1)d$
 ▷ The sum of the first n terms $S_n = \dfrac{n}{2}[2u_1 + (n - 1)d]$

▶ Solve problems involving geometric series.
 ▷ The nth term $u_n = u_1 \times r^{n-1}$
 ▷ The sum of the first n terms $S_n = \dfrac{u_1(1 - r^n)}{1 - r}$
 $\qquad\qquad = \dfrac{u_1(r^n - 1)}{r - 1}$
 ▷ The sum to infinity $S_\infty = \dfrac{u_1}{1 - r}$, if $-1 < r < 1$.

▶ Write terms of a binomial expansion.
 ▷ $(a + b)^n = a^n + \binom{n}{1}a^{n-1}b^1 + \binom{n}{2}b^{n-2}b^2 + \ldots + b^n$, where $n \in \mathbb{N}$
 ▷ Pascal's Triangle is:

$$\begin{array}{ccccccccc} & & & & 1 & & & & \\ & & & 1 & & 1 & & & \\ & & 1 & & 2 & & 1 & & \\ & 1 & & 3 & & 3 & & 1 & \\ 1 & & 4 & & 6 & & 4 & & 1 \end{array}$$ and so on.

Revision exercise 6

1 The first three terms of an arithmetic sequence are 7, 9.5, 12.

a) What is the 41st term of the sequence?

b) What is the sum of the first 101 terms of the sequence?   © IBO [2001]

2 In an arithmetic sequence, the first term is -2, the fourth term is 16 and the nth term is 11 998.

a) Find the common difference d.

b) Find the value of n.   © IBO [2002]

3 Find

a) $\displaystyle\sum_{r=1}^{10} 6r$ b) $\displaystyle\sum_{r=10}^{20} 3 \times 2^r$

4 Let $S_n = 1 + 2 + 4 + 8 + \ldots + 2^{n-1}$.

a) Find S_{10}.

b) Find the least value of n for which $S_n > 10^6$.   © IBO [1997]

5 A geometric series has first term 10 and common ratio 0.02.

a) Find the sum to infinity.

b) How many terms are needed so that their sum differs from the sum to infinity by less than 0.000 01?

6 Consider the expansion of $\left(3x^2 - \dfrac{1}{x}\right)^9$.

a) How many terms are there in this expansion?

b) Find the constant term in this expansion.   © IBO [2002]

7 Each year for the past five years the population of a certain country has increased at a steady rate of 2.7% per annum. The present population is 15.2 million.

a) What was the population one year ago?

b) What was the population five years ago?   © IBO [2001]

8 Use the binomial theorem to complete this expansion.

$$(3x + 2y)^4 = 81x^4 + 216x^3y + \ldots$$ © IBO [2001]

9 Consider the binomial expansion

$$(1 + x)^4 = 1 + \binom{4}{1}x + \binom{4}{2}x^2 + \binom{4}{3}x^3 + x^4$$

a) By substituting $x = 1$ into both sides, or otherwise, evaluate

$$\binom{4}{1} + \binom{4}{2} + \binom{4}{3}$$

b) Evaluate $\binom{9}{1} + \binom{9}{2} + \binom{9}{3} + \binom{9}{4} + \binom{9}{5} + \binom{9}{6} + \binom{9}{7} + \binom{9}{8}$ © *IBO* [2001]

10 Michele invested 1500 francs at an annual rate of interest of 5.25 per cent.

a) Find the value of Michele's investment after 3 years. Give your answer to the nearest franc.

b) How many complete years will it take for Michele's investment to double in value?

c) What should the interest rate be if Michele's initial investment were to double in value in 10 years?   © *IBO* [2001]

11 Ashley and Billie are swimmers training for a competition.

a) Ashley trains for 12 hours in the first week. She decides to increase the amount of time she spends training by 2 hours each week. Find the total number of hours she spends training during the first 15 weeks.

b) Billie also trains for 12 hours in the first week. She decides to train for 10% longer each week than the previous week.

i) Show that in the third week she trains for 14.52 hours.

ii) Find the total number of hours she spends training during the first 15 weeks.

iii) In which week will the time Billie spends training first exceed 50 hours?   © *IBO* [2002]

12 The nth term of a sequence is given by $t_n = 3n + 1$, $n = 1, 2, 3, \ldots$

a) Find the value of t_{21}.

b) Write down an expression for $t_{n+1} - t_n$ and simplify it. Hence explain why the sequence is arithmetic.

c) Let S_n be the sum of the first n terms of the sequence. Write down an expression for S_n and simplify it.

d) Evaluate $t_{16} + t_{17} + t_{18} + \ldots + t_{26}$.   © *IBO* [1998]

13 Portable telephones are first sold in the country Cellmania in 1990. During 1990, the number of units sold is 160. In 1991, the number of units sold is 240 and in 1992 it is 360.

In 1993 it was noticed that the annual sales formed a geometric sequence with first term 160, the 2nd and 3rd terms being 240 and 360 respectively.

a) What is the common ratio of this sequence?

Assume that this sales trend continues.

b) How many units will be sold during 2002?

c) In what year does the number of units sold first exceed 5000?

d) What is the total number of units sold between 1990 and 2002?

During this period, the total population of Cellmania remains approximately 80 000.

e) Use this information to suggest a reason why the geometric growth in sales would not continue.

© IBO [2001]

7 Differentiation 2

7.2 The chain rule for composite functions. The product and quotient rules.

7.7 Graphical behaviour of functions: tangents and normals, behaviour for large |x|, horizontal and vertical asymptotes.

7.1 Function of a function

If $y = (x + 2)^4$, then y can be said to be a function of x.
If you let $u = x + 2$ then:

$$y = u^4 \quad \text{where } u = x + 2$$

y is now a function of u, and u is a function of x. The new variable, u, is the link between the two expressions.

To differentiate the expression $y = (x + 2)^4$ with respect to x, you would first need to expand the bracket. In this particular case, it would certainly be feasible. However, an expression such as $y = (x + 2)^{12}$ would involve more work. The **chain rule** allows you to differentiate such expressions more easily.

The chain rule

If y is a function of u, and u is a function of x,

$$\frac{dy}{dx} = \frac{dy}{du} \cdot \frac{du}{dx}$$

> The dot between $\dfrac{dy}{du}$ and $\dfrac{du}{dx}$ means 'multiply'.

Using the example of $y = (x + 2)^4$,

$$y = u^4 \quad \text{where } u = x + 2$$

Differentiating y with respect to u gives

$$\frac{dy}{du} = 4u^3$$

Differentiating u with respect to x gives

$$\frac{du}{dx} = 1$$

By the chain rule,

$$\frac{dy}{dx} = \frac{dy}{du} \cdot \frac{du}{dx} = (4u^3)(1)$$

$$\therefore \frac{dy}{dx} = 4u^3$$

Substituting $u = x + 2$ gives

$$\frac{dy}{dx} = 4(x + 2)^3$$

which is the required derivative.

Example 1

If $y = (3x - 1)^7$ find $\frac{dy}{dx}$.

....

Let $u = 3x - 1$, then $y = u^7$

Differentiating each expression gives

$$\frac{dy}{du} = 7u^6 \quad \text{and} \quad \frac{du}{dx} = 3$$

By the chain rule,

$$\frac{dy}{dx} = \frac{dy}{du} \cdot \frac{du}{dx}$$

$$= (7u^6)(3) = 21u^6$$

$$\therefore \frac{dy}{dx} = 21(3x - 1)^6$$

> You usually let u be the function in brackets.

Example 2

Find $\frac{dy}{dx}$ for each of these functions.

a) $y = 2(1 - x)^5$ b) $y = (x^2 + 3)^4$

c) $y = \dfrac{1}{(3 - 7x)}$ d) $y = \sqrt{6x + 1}$

....

a) $y = 2(1 - x)^5$

   Let $u = (1 - x)$, then $y = 2u^5$

   Differentiating each expression gives

$$\frac{dy}{du} = 10u^4 \quad \text{and} \quad \frac{du}{dx} = -1$$

   By the chain rule,

$$\frac{dy}{dx} = \frac{dy}{du} \cdot \frac{du}{dx}$$

$$= 10u^4 \times (-1)$$

$$= -10u^4$$

$$\therefore \frac{dy}{dx} = -10(1 - x)^4$$

b) $y = (x^2 + 3)^4$

Let $u = (x^2 + 3)$, then $y = u^4$

Differentiating each expression gives

$$\frac{dy}{du} = 4u^3 \quad \text{and} \quad \frac{du}{dx} = 2x$$

By the chain rule,

$$\frac{dy}{dx} = \frac{dy}{du} \cdot \frac{du}{dx}$$

$$= 4u^3 \times 2x$$

$$= 8u^3x$$

$$\therefore \frac{dy}{dx} = 8x(x^2 + 3)^3$$

c) $y = \dfrac{1}{(3 - 7x)}$

Let $u = (3 - 7x)$, then $y = \dfrac{1}{u} = u^{-1}$

Differentiating each expression gives

$$\frac{dy}{du} = -u^{-2} = -\frac{1}{u^2} \quad \text{and} \quad \frac{du}{dx} = -7$$

By the chain rule,

$$\frac{dy}{dx} = \frac{dy}{du} \cdot \frac{du}{dx}$$

$$= -\frac{1}{u^2} \times (-7)$$

$$= \frac{7}{u^2}$$

$$\therefore \frac{dy}{dx} = \frac{7}{(3 - 7x)^2}$$

> Remember to give your final answer in terms of x, not u.

d) $y = \sqrt{6x + 1}$

Let $u = (6x + 1)$, then $y = \sqrt{u} = u^{\frac{1}{2}}$

Differentiating each expression gives

$$\frac{dy}{du} = \frac{1}{2}u^{-\frac{1}{2}} = \frac{1}{2\sqrt{u}} \quad \text{and} \quad \frac{du}{dx} = 6$$

By the chain rule,

$$\frac{dy}{dx} = \frac{dy}{du} \cdot \frac{du}{dx}$$

$$= \frac{1}{2\sqrt{u}} \times 6$$

$$= \frac{3}{\sqrt{u}}$$

$$\therefore \frac{dy}{dx} = \frac{3}{\sqrt{6x + 1}}$$

Exercise 7A

1 Find $\dfrac{dy}{dx}$ for each of these.

a) $y = (2x - 1)^3$

b) $y = (3x + 4)^2$

c) $y = (5x - 3)^4$

d) $y = (3 - x)^5$

e) $y = (4 - 3x)^6$

f) $y = (x^2 + 1)^4$

g) $y = (x^3 - 6)^2$

h) $y = (1 - 2x^2)^3$

i) $y = (4 - x^4)^2$

j) $y = (7 - 5x^3)^6$

k) $y = (6x^2 - 5)^4$

l) $y = (9 - 7x^2)^3$

2 Differentiate each of these expressions with respect to x.

a) $(2x - 5)^{-3}$

b) $(3x + 2)^{-1}$

c) $(x^2 + 3)^{-2}$

d) $(5 - 2x^3)^{-1}$

e) $\dfrac{1}{3 + 4x}$

f) $\dfrac{1}{4 - x^2}$

g) $\dfrac{5}{3 - 2x}$

h) $\dfrac{3}{(x + 1)^2}$

i) $\dfrac{7}{(2 - x^2)^5}$

j) $-\dfrac{1}{(3x^2 + 8)}$

k) $(5x^3 - 4)^{-4}$

l) $\dfrac{1}{2(5 - 3x^4)^2}$

3 Find $f'(x)$ for each of these.

a) $f(x) = (2x - 1)^{\frac{1}{2}}$

b) $f(x) = (6 - x)^{\frac{1}{3}}$

c) $f(x) = (x^3 - 2)^{\frac{2}{3}}$

d) $f(x) = (4 - x^5)^{-\frac{1}{5}}$

e) $f(x) = \sqrt{4x - 5}$

f) $f(x) = \sqrt[3]{x^2 + 3}$

g) $f(x) = \dfrac{1}{\sqrt{5 - 2x}}$

h) $f(x) = \dfrac{6}{\sqrt[3]{x^2 + 5}}$

4 Find $\dfrac{dy}{dx}$ for each of these.

a) $y = (x^2 + x - 1)^4$

b) $y = \sqrt{x^3 - 6x}$

c) $y = \dfrac{1}{x^2 - 3x + 5}$

d) $y = \left(\dfrac{1}{\sqrt{x}} - 1\right)^4$

e) $y = (x^4 - 2x^2 + 3)^5$

f) $y = \left(1 + \dfrac{3}{x}\right)^2$

g) $y = (2\sqrt{x} - x)^4$

h) $y = \sqrt[4]{6x - x^3}$

Integrating a function of a function

You can also integrate a function of a function.

Consider the function $f(x) = (x^2 + 1)^4$.

$$f'(x) = 4(x^2 + 1)^3(x^2 + 1)'$$
$$= 4(x^2 + 1)^3(2x)$$
$$\therefore f'(x) = 8x(x^2 + 1)^3$$

Use the chain rule, with $y = u^4$ and $u = x^2 + 1$.

$(x^2 + 1)'$ means 'the derivative of $x^2 + 1$'.

So you can write

$$\int 8x(x^2 + 1)^3 \, dx = (x^2 + 1)^4 + c$$

In this case, the integrand is of the form

$$f'(x)[f(x)]^n$$

where $f(x) = x^2 + 1$, $f'(x) = 2x$ and $n = 3$.

Consider the integral $\int x(3x^2 - 2)^5 \, dx$. Notice that the derivative of $(3x^2 - 2)$ is $6x$ and that there is an x term outside the main function, $(3x^2 - 2)^5$, of the integrand. So consider $(3x^2 - 2)^6$, which when differentiated gives $36x(3x^2 - 2)^5$. Thereforc,

$$\int x(3x^2 - 2)^5 \, dx = \frac{1}{36}(3x^2 - 2)^6 + c$$

$$\frac{dy}{dx} = \frac{dy}{du} \cdot \frac{du}{dx}$$
$$= 6(3x^2 - 2)^5(6x)$$
$$= 36x(3x^2 - 2)^5$$

In general:

$$\int f'(x)[f(x)]^n \, dx = \frac{[f(x)]^{n+1}}{n + 1} + c, \quad n \neq -1$$

Example 3

Find each of these integrals.

a) $\int (x - 2)^2 \, dx$ b) $\int x(3x^2 + 6)^4 \, dx$

c) $\int 4x^2(x^3 - 3)^5 \, dx$ d) $\int (x + 2)(x^2 + 4x - 1)^3 \, dx$

e) $\int \dfrac{x}{\sqrt{x^2 + 3}} \, dx$

. . . .

a) $\int (x - 2)^2 \, dx$

If you differentiate $(x - 2)^3$, you get $3(x - 2)^2$. Therefore,

$$\int (x - 2)^2 \, dx = \frac{(x - 2)^3}{3} + c$$

b) $\int x(3x^2 + 6)^4 \, dx$

Notice that the derivative of $3x^2 + 6$ is $6x$, and that there is an x term outside the main function in the integrand. If you differentiate $(3x^2 + 6)^5$, you get $30x(3x^2 + 6)^4$. Therefore,

$$\int x(3x^2 + 6)^4 \, dx = \frac{(3x^2 + 6)^5}{30} + c$$

c) $\int 4x^2(x^3 - 3)^5 \, dx$

Notice that the derivative of $(x^3 - 3)$ is $3x^2$, and that there is an x^2 term outside the main function in the integrand.
If you differentiate $(x^3 - 3)^6$ you get $18x^2(x^3 - 3)^5$.
Therefore,

$$\int 4x^2(x^3 - 3)^5 \, dx = \tfrac{4}{18}(x^3 - 3)^6 + c$$

$$= \tfrac{2}{9}(x^3 - 3)^6 + c$$

d) $\int (x + 2)(x^2 + 4x - 1)^3 \, dx$

Notice that the derivative of $(x^2 + 4x - 1)$ is
$2x + 4 = 2(x + 2)$, and that there is a term $(x + 2)$ outside the main function in the integrand.
If you differentiate $(x^2 + 4x - 1)^4$, you get

$$4(2x + 4)(x^2 + 4x - 1)^3 = 8(x + 2)(x^2 + 4x - 1)^3$$

Therefore,

$$\int (x + 2)(x^2 + 4x - 1)^3 \, dx = \tfrac{1}{8}(x^2 + 4x - 1)^4 + c$$

e) $\int \dfrac{x}{\sqrt{x^2 + 3}} \, dx = \int x(x^2 + 3)^{-\frac{1}{2}} \, dx$

Notice that the derivative of $(x^2 + 3)$ is $2x$ and that there is an x term outside the main function in the integrand.
If you differentiate $(x^2 + 3)^{\frac{1}{2}}$, you get

$$2x \times \tfrac{1}{2}(x^2 + 3)^{-\frac{1}{2}} = x(x^2 + 3)^{-\frac{1}{2}}$$

Therefore,

$$\int \dfrac{x}{\sqrt{x^2 + 3}} \, dx = (x^2 + 3)^{\frac{1}{2}} + c$$

$$= \sqrt{x^2 + 3} + c$$

Exercise 7B

1 Integrate each of these with respect to x.

a) $(2x - 3)^4$ b) $(5x + 8)^2$ c) $(3x - 4)^5$

d) $3(x - 7)^2$ e) $(4 - x)^5$ f) $-(6 - 7x)^3$

g) $(3x - 4)^{-3}$ h) $6(5 - 9x)^{-2}$ i) $\dfrac{1}{(2x - 1)^7}$

j) $\dfrac{3}{(1 - x)^2}$ k) $\sqrt{2x - 3}$ l) $\dfrac{12}{\sqrt[3]{x - 4}}$

2 Find each of these integrals.

a) $\int (2x - 7)^5 \, dx$ b) $\int \dfrac{1}{\sqrt{2x - 1}} \, dx$ c) $\int x(x^2 + 2)^3 \, dx$

d) $\int 2x(4 - 3x^2)^5 \, dx$ e) $\int x^2(x^3 - 4)^2 \, dx$ f) $\int \dfrac{4x}{(3 - x^2)^2} \, dx$

g) $\int x^3 \sqrt{x^4 - 1} \, dx$ h) $\int 4x\sqrt[3]{2 - 3x^2} \, dx$ i) $\int x^{\frac{1}{3}}(x^{\frac{4}{3}} - 2)^2 \, dx$

j) $\int \dfrac{25x^4}{(3 - x^5)^2} \, dx$ k) $\int x(x^2 + 1)^2 \, dx$ l) $\int (x^2 + 1)^2 \, dx$

3 Given that $y = \sqrt{\sqrt{(1 + \sqrt{x})} - 1}$, show that

$$\frac{dy}{dx} = \frac{1}{8\sqrt{x(1 + \sqrt{x})(\sqrt{(1 + \sqrt{x})} - 1)}}$$

7.2 The product and quotient rules

The product rule

The **product rule** of differentiation says that:

> If $y = uv$, where u and v are both functions of x, then
>
> $$\frac{dy}{dx} = u\frac{dv}{dx} + v\frac{du}{dx}$$

You can prove this rule as follows.

Let δx be a small increment or increase in x and let δu, δv and δy be the resulting small increments in u, v and y, respectively. Then

> You will not be expected to prove this in the examination.

$$y + \delta y = (u + \delta u)(v + \delta v)$$

$$\therefore \ \delta y = uv + u\delta v + v\delta u + \delta u \delta v - y$$

But $y = uv$, therefore

$$\delta y = uv + u\delta v + v\delta u + \delta u\delta v - uv$$

$$= u\delta v + v\delta u + \delta u\delta v$$

Dividing throughout by δx gives

$$\frac{\delta y}{\delta x} = u\frac{\delta v}{\delta x} + v\frac{\delta u}{\delta x} + \frac{\delta u\delta v}{\delta x}$$

As $\delta x \to 0$, $\delta u \to 0$, $\delta v \to 0$ and $\delta v \to 0$, and

$$\frac{\delta y}{\delta x} \to \frac{dy}{dx} \quad \frac{\delta u}{\delta x} \to \frac{du}{dx} \quad \frac{\delta v}{\delta x} \to \frac{dv}{dx} \quad \frac{\delta u\delta v}{\delta x} \to 0$$

Therefore,

$$\frac{dy}{dx} = u\frac{dv}{dx} + v\frac{du}{dx}$$

> You can also write this as:
> $(uv)' = uv' + vu'$

as required.

Example 1

If $y = x^2(x + 2)^3$ find $\dfrac{dy}{dx}$.

In this example, y is the product of two functions u and v, where

$$u = x^2 \quad \text{and} \quad v = (x + 2)^3$$

Differentiating each with respect to x gives

$$\frac{du}{dx} = 2x \quad \text{and} \quad \frac{dv}{dx} = 3(x + 2)^2 \cdot (x + 2)'$$

$$= 3(x + 2)^2$$

Using the product rule gives

$$\frac{dy}{dx} = u\frac{dv}{dx} + v\frac{du}{dx}$$

$$= x^2[3(x + 2)^2] + (x + 2)^3(2x)$$

$$\therefore \quad \frac{dy}{dx} = 3x^2(x + 2)^2 + 2x(x + 2)^3$$

Factorising gives

$$\frac{dy}{dx} = x(x + 2)^2[3x + 2(x + 2)]$$

$$\therefore \quad \frac{dy}{dx} = x(x + 2)^2(5x + 4)$$

> Use the chain rule to find $\dfrac{dv}{dx}$.

In the examples that follow, the derivative is fully factorised.

Example 2

Find $\dfrac{dy}{dx}$ for each of these functions.

a) $y = x^2\sqrt{x - 1}$ $\qquad\qquad$ b) $y = (x + 1)^5\sqrt{x}$

a) $y = x^2\sqrt{x - 1} = x^2(x - 1)^{\frac{1}{2}}$

Using the product rule gives

$$\frac{dy}{dx} = x^2 \cdot \tfrac{1}{2}(x - 1)^{-\frac{1}{2}} + (x - 1)^{\frac{1}{2}} \cdot 2x$$

$$\therefore \quad \frac{dy}{dx} = \frac{x^2}{2}(x - 1)^{-\frac{1}{2}} + 2x(x - 1)^{\frac{1}{2}}$$

> $u = x^2$
> $v = (x - 1)^{\frac{1}{2}}$

You can simplify this further if needed by factorising the RHS:

$$\frac{dy}{dx} = x(x-1)^{-\frac{1}{2}}\left(\frac{x}{2} + 2(x-1)\right)$$

$$= x(x-1)^{-\frac{1}{2}}\left(\frac{5x-4}{2}\right)$$

$$\therefore \frac{dy}{dx} = \frac{x(5x-4)}{2\sqrt{x-1}}$$

b) $y = (x+1)^5\sqrt{x} = (x+1)^5 x^{\frac{1}{2}}$

Using the product rule gives

$$\frac{dy}{dx} = (x+1)^5 \cdot \frac{1}{2}x^{-\frac{1}{2}} + x^{\frac{1}{2}} \cdot 5(x+1)^4$$

$$\therefore \frac{dy}{dx} = \frac{x^{-\frac{1}{2}}}{2}(x+1)^5 + 5x^{\frac{1}{2}}(x+1)^4$$

$$u = (x+1)^5$$
$$v = x^{\frac{1}{2}}$$

You can simplify this further if needed by factorising the RHS:

$$\frac{dy}{dx} = x^{-\frac{1}{2}}(x+1)^4\left(\tfrac{1}{2}(x+1) + 5x\right)$$

$$= x^{-\frac{1}{2}}(x+1)^4\left(\frac{11x+1}{2}\right)$$

$$\therefore \frac{dy}{dx} = \frac{(x+1)^4(11x+1)}{2\sqrt{x}}$$

The quotient rule

The quotient rule of differentiation says that:

If $y = \dfrac{u}{v}$, where u and v are functions of x, then

$$\frac{dy}{dx} = \frac{v\dfrac{du}{dx} - u\dfrac{dv}{dx}}{v^2}$$

You can prove this rule as follows.

Let δx be a small increment in x and let δu, δv and δy be the resulting increments in u, v and y respectively. Then

$$y + \delta y = \frac{u + \delta u}{v + \delta v}$$

You will not be expected to prove this in the examination.

$$\therefore \quad \delta y = \frac{u + \delta u}{v + \delta v} - \frac{u}{v}$$

$$= \frac{v(u + \delta u) - u(v + \delta v)}{v(v + \delta v)}$$

$$= \frac{uv + v\delta u - uv - u\delta v}{v(v + \delta v)}$$

$$\therefore \quad \delta y = \frac{v\delta u - u\delta v}{v^2 + v\delta v}$$

Dividing throughout by δx gives

$$\frac{\delta y}{\delta x} = \frac{v\delta u - u\delta v}{v^2 + v\delta v} \times \frac{1}{\delta x}$$

$$\therefore \quad \frac{\delta y}{\delta x} = \frac{v\dfrac{\delta u}{\delta x} - u\dfrac{\delta v}{\delta x}}{v^2 + v\delta v}$$

> Dividing by δx is the same as multiplying by $\dfrac{1}{\delta x}$.

As $\delta x \to 0$, $\delta u \to 0$, $\delta v \to 0$ and $\delta y \to 0$, and

$$\frac{\delta y}{\delta x} \to \frac{dy}{dx} \qquad \frac{\delta u}{\delta x} \to \frac{du}{dx} \qquad \frac{\delta v}{\delta x} \to \frac{dv}{dx}$$

Therefore,

$$\frac{dy}{dx} = \frac{v\dfrac{du}{dx} - u\dfrac{dv}{dx}}{v^2}$$

as required.

You can also write this as $\left(\dfrac{u}{v}\right)' = \dfrac{vu' - uv'}{v^2}$

Example 3

If $y = \dfrac{3x + 2}{2x + 1}$ find $\dfrac{dy}{dx}$.

....

In this example, y is the quotient of two functions u and v, where

$$u = 3x + 2 \quad \text{and} \quad v = 2x + 1$$

Differentiating each with respect to x gives

$$u' = 3 \quad \text{and} \quad v' = 2$$

Using the quotient rule,

$$\frac{dy}{dx} = \frac{3 \times (2x + 1) - (3x + 2) \times 2}{(2x + 1)^2}$$

$$\therefore \quad \frac{dy}{dx} = -\frac{1}{(2x + 1)^2}$$

Example 4

Find $\dfrac{dy}{dx}$ for each of these functions.

a) $y = \dfrac{x}{(x^2 + 4)^3}$ b) $y = \sqrt{\dfrac{x^3}{x^2 - 1}}$

a) When $y = \dfrac{x}{(x^2 + 4)^3}$, using the quotient rule gives

$\dfrac{dy}{dx} = \dfrac{(x^2 + 4)^3(x)' - x[(x^2 + 4)^3]'}{[(x^2 + 4)^3]^2}$

$= \dfrac{(x^2 + 4)^3(1) - x[3(x^2 + 4)^2(2x)]}{(x^2 + 4)^6}$

$\therefore \dfrac{dy}{dx} = \dfrac{(x^2 + 4)^3 - 6x^2(x^2 + 4)^2}{(x^2 + 4)^6}$

$u = x$
$v = (x^2 + 4)^3$

Factorising the numerator gives

$\dfrac{dy}{dx} = \dfrac{(x^2 + 4)^2[(x^2 + 4) - 6x^2]}{(x^2 + 4)^6}$

$\therefore \dfrac{dy}{dx} = \dfrac{(4 - 5x^2)}{(x^2 + 4)^4}$

b) $y = \sqrt{\dfrac{x^3}{x^2 - 1}} = \dfrac{x^{\frac{3}{2}}}{(x^2 - 1)^{\frac{1}{2}}}$

Using the quotient rule:

$\dfrac{dy}{dx} = \dfrac{(x^2 - 1)^{\frac{1}{2}}(x^{\frac{3}{2}})' - x^{\frac{3}{2}}[(x^2 - 1)^{\frac{1}{2}}]'}{[(x^2 - 1)^{\frac{1}{2}}]^2}$

$u = x^{\frac{3}{2}}$
$v = (x^2 - 1)^{\frac{1}{2}}$

$= \dfrac{(x^2 - 1)^{\frac{1}{2}}(\frac{3}{2}x^{\frac{1}{2}}) - x^{\frac{3}{2}}[\frac{1}{2}(x^2 - 1)^{-\frac{1}{2}}(2x)]}{(x^2 - 1)}$

$\therefore \dfrac{dy}{dx} = \dfrac{\frac{3}{2}x^{\frac{1}{2}}(x^2 - 1)^{\frac{1}{2}} - x^{\frac{5}{2}}(x^2 - 1)^{-\frac{1}{2}}}{(x^2 - 1)}$

Factorising the numerator:

$\dfrac{dy}{dx} = \dfrac{\frac{1}{2}x^{\frac{1}{2}}(x^2 - 1)^{-\frac{1}{2}}[3(x^2 - 1) - 2x^2]}{(x^2 - 1)}$

$= \dfrac{\frac{1}{2}x^{\frac{1}{2}}(x^2 - 1)^{-\frac{1}{2}}(x^2 - 3)}{(x^2 - 1)}$

$\therefore \dfrac{dy}{dx} = \dfrac{\sqrt{x}(x^2 - 3)}{2\sqrt{(x^2 - 1)^3}}$

Exercise 7C
....

In each part of questions **1** to **4**, use the product rule to differentiate the given function with respect to x.

1 a) $(x + 3)(x - 4)$ b) $(2x + 5)(x - 7)$
  c) $(3x - 4)(2x + 5)$ d) $(6 + x)(5 - x)$
  e) $(x + 4)(x^2 - 2)$ f) $(x^2 - 5)(4x - 1)$
  g) $(x + 6)(x^2 - 3x + 3)$ h) $(3x - 5)(x^2 - 2x + 7)$
  i) $(x^2 + 1)(x^3 - 5)$

2 a) $x^2(x + 3)^4$ b) $x^3(2 + x)^2$
  c) $x^4(3x - 1)^3$ d) $3x^2(2x + 5)^2$
  e) $x^3(4x^2 - 1)^3$ f) $5x^2(2 - x^3)^2$
  g) $3x^2(5x^2 + 1)^4$ h) $x^7(2 - 5x^3)^4$
  i) $x^2(x^2 + x - 1)^3$

3 a) $(x + 2)^2(x - 5)^3$ b) $(2x - 1)^3(x + 4)^2$
  c) $(5x + 2)^4(4x - 3)^3$ d) $(2 - x)^6(5 + 2x)^4$
  e) $(3 + 5x)^2(4 - 7x)^7$ f) $(x^2 + 1)^2(2x - 3)^4$
  g) $(5x + 9)^3(x^2 - 2)^3$ h) $(2x^2 - 3)^5(4x - 7)^6$
  i) $(x^3 - 1)^3(4x^2 + 5)^2$

4 a) $x\sqrt{x + 1}$ b) $2x\sqrt{3 - x}$
  c) $3x\sqrt{5 + 2x}$ d) $x^2\sqrt{x + 3}$
  e) $x^2\sqrt{3 - 4x}$ f) $(2x - 1)\sqrt{5x + 3}$
  g) $(1 - 3x)\sqrt{2x + 5}$ h) $\sqrt{x}(5x - 4)^3$
  i) $(3x + 5)^2\sqrt{x - 2}$

In each part of questions **5** and **6**, use the quotient rule to differentiate the given function with respect to x.

5 a) $\dfrac{x}{x - 2}$ b) $\dfrac{x + 3}{x - 1}$ c) $\dfrac{3 - x}{4 + x}$
  d) $\dfrac{2x - 5}{x + 4}$ e) $\dfrac{5x}{x + 2}$ f) $\dfrac{1 + 3x}{2 - 5x}$
  g) $\dfrac{x^2}{x + 3}$ h) $\dfrac{x^2}{4 - x}$ i) $\dfrac{x^3}{2x - 3}$

6 a) $\dfrac{(3x - 2)^2}{\sqrt{x}}$ b) $\dfrac{(5x + 1)^3}{\sqrt{x}}$ c) $\dfrac{(x^2 - 4)^5}{\sqrt{x}}$
  d) $\dfrac{\sqrt{x}}{2x - 1}$ e) $\dfrac{3 - \sqrt{x}}{(2 + x)^2}$ f) $\dfrac{5 + 2\sqrt{x}}{(5 - 4x)^3}$
  g) $\sqrt{\dfrac{x - 2}{x + 1}}$ h) $\sqrt{\dfrac{x - 3}{2x + 5}}$

7 Find $\dfrac{dy}{dx}$ for each of these.

a) $y = x^3(3-x)^2$

b) $y = \dfrac{x}{2x-1}$

c) $y = \sqrt{x}(5x-1)^2$

d) $y = \dfrac{2x}{\sqrt{x}+1}$

e) $y = (5x+3)^3(x-2)^2$

f) $y = \sqrt{\dfrac{3x-2}{x-3}}$

g) $y = x^3\sqrt{7-2x}$

h) $y = \dfrac{x^2}{2-x}$

i) $y = (3-x)^4(2+x)^5$

8 Differentiate each of these functions with respect to x.

a) $\dfrac{x(x-1)^3}{x-3}$

b) $\dfrac{x\sqrt{5-x^2}}{6-x}$

c) $\dfrac{x^3\sqrt{4-x^2}}{5-\sqrt{x}}$

7.3 Applications of differentiation

You can apply these rules of differentiation to various problems, such as finding the equation of a tangent to a curve.

Example 1

Find the equation of the tangent to the curve $y = x^2(x+1)^4$ at the point P(1, 16).

To find $\dfrac{dy}{dx}$ use the product rule together with the chain rule:

$$\frac{dy}{dx} = (x+1)^4(x^2)' + x^2[(x+1)^4]'$$
$$= (x+1)^4(2x) + x^2[4(x+1)^3]$$
$$= 2x(x+1)^4 + 4x^2(x+1)^3$$
$$= 2x(x+1)^3[2x + (x+1)]$$
$$\therefore \frac{dy}{dx} = 2x(x+1)^3(3x+1)$$

At P(1, 16),

$$\left.\frac{dy}{dx}\right|_{x=1} = 2(8)(4) = 64$$

Therefore, the equation of the tangent line is of the form

$$y = 64x + c$$

Since the tangent line passes through P(1, 16),

$$16 = 64(1) + c$$
$$\therefore c = -48$$

The equation of the tangent is $y = 64x - 48$.

> **Remember:**
> To find the equation, you must first find the gradient by differentiating (see page 119).

> You can use the same technique to find the equation of the normal to a curve at any point (see page 121).

Differentiation is used to find stationary points on curves.

Example 2

Given that $y = x(x + 1)^5$,

a) find the x-coordinates of the stationary points and determine their nature.

b) The curve has two points of inflexion: find the x-coordinates of both points.

c) Sketch the curve.

....

a) At a stationary point $\dfrac{dy}{dx} = 0$. Using the product rule gives

$$\frac{dy}{dx} = x.5(x + 1)^4 + (x + 1)^5$$
$$= 5x(x + 1)^4 + (x + 1)^5$$
$$\therefore \frac{dy}{dx} = (x + 1)^4(6x + 1)$$

When $\dfrac{dy}{dx} = 0$:

$$(x + 1)^4(6x + 1) = 0$$
$$\therefore x = -1 \quad \text{or} \quad x = -\tfrac{1}{6}$$

The x-coordinates of the stationary points on the curve are $x = -1$ and $x = -\tfrac{1}{6}$.

To determine their nature, use the second derivative.

Since $\dfrac{dy}{dx} = (x + 1)^4(6x + 1)$, use the product rule to find $\dfrac{d^2y}{dx^2}$:

$$\frac{d^2y}{dx^2} = (x + 1)^4.6 + (6x + 1).4(x + 1)^3$$
$$= 6(x + 1)^4 + 4(6x + 1)(x + 1)^3$$
$$= (x + 1)^3[6(x + 1) + 4(6x + 1)]$$
$$\therefore \frac{d^2y}{dx^2} = (x + 1)^3(30x + 10)$$

When $x = -1 : \dfrac{d^2y}{dx^2} = 0$. This stationary point could be a maximum, minimum or point of inflexion. To decide which it is, you must examine the gradient either side of the point.

Note that a second stationary point occurs at $x = -\tfrac{1}{6}$ and therefore the point chosen must have an x-value greater than -1 but less than $-\tfrac{1}{6}$. This gives

x	-2	-1	-0.9
$\dfrac{dy}{dx}$	-11	0	$-0.000\,44$

The gradient either side of the point $x = -1$ is negative, therefore the point $x = -1$ is a point of inflexion.

When $x = -\frac{1}{6}: \dfrac{d^2y}{dx^2} = (-\frac{1}{6} + 1)^3(30(-\frac{1}{6}) + 10) = \frac{625}{216} > 0$.

Therefore the point $x = -\frac{1}{6}$ is a minimum.

b) To investigate for points of inflexion, look at $\dfrac{d^2y}{dx^2} = 0$.

$$(x + 1)^3(30x + 10) = 0$$
$$\therefore \ x = -1 \quad \text{or} \quad x = -\frac{1}{3}$$

You know that $x = -1$ is a point of inflexion from part a).

You need to investigate the gradient either side of $x = -\frac{1}{3}$. Be careful to choose points on either side that don't go beyond $x = -1$ to the left and $x = +1$ to the right.

x	-0.9	$-\frac{1}{3}$	0
$\dfrac{dy}{dx}$	$-0.000\,44$	0	$+1$

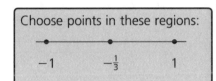

Choose points in these regions:

The gradient changes from negative, through zero to positive, therefore the point $x = -\frac{1}{3}$ is a minimum.

c) Sketching the curve gives the diagram shown.

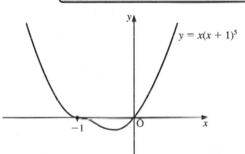

$y = x(x + 1)^5$

Example 3

Find the coordinates of the stationary points on the curve
$$f(x) = \frac{x}{x^2 + 4}. \text{ Show that}$$
$$f''(x) = \frac{2x(x^2 - 12)}{(x^2 + 4)^3}$$
and hence determine the nature of the stationary points.

· ·

At a stationary point, $f'(x) = 0$. Using the quotient rule gives
$$f'(x) = \frac{(x^2 + 4)(1) - x(2x)}{(x^2 + 4)^2}$$
$$= \frac{x^2 + 4 - 2x^2}{(x^2 + 4)^2}$$
$$\therefore \ f'(x) = \frac{4 - x^2}{(x^2 + 4)^2}$$

When $f'(x) = 0$: $\dfrac{4 - x^2}{(x^2 + 4)^2} = 0$

$$4 - x^2 = 0$$

$$\therefore \; x = \pm 2$$

When $x = 2$: $f(2) = \dfrac{2}{2^2 + 4} = \dfrac{1}{4}$

When $x = -2$: $f(-2) = \dfrac{-2}{(-2)^2 + 4} = -\dfrac{1}{4}$

The coordinates of the stationary points on the curve are $(2, \frac{1}{4})$ and $(-2, -\frac{1}{4})$.

Since $f'(x) = \dfrac{4 - x^2}{(x^2 + 4)^2}$,

$$f''(x) = \dfrac{(x^2 + 4)^2(-2x) - (4 - x^2)[2(x^2 + 4)(2x)]}{(x^2 + 4)^4}$$

$$= \dfrac{-2x(x^2 + 4)^2 - 4x(x^2 + 4)(4 - x^2)}{(x^2 + 4)^4}$$

$$= \dfrac{2x(x^2 + 4)[-(x^2 + 4) - 2(4 - x^2)]}{(x^2 + 4)^4}$$

$$= \dfrac{2x(x^2 + 4)(x^2 - 12)}{(x^2 + 4)^4}$$

$$\therefore \; f''(x) = \dfrac{2x(x^2 - 12)}{(x^2 + 4)^3}$$

as required.

> Use the quotient rule.

When $x = 2$: $f''(2) = \dfrac{-32}{512} = -\dfrac{1}{16} < 0$

Since $f''(2) < 0$, the stationary point $(2, \frac{1}{4})$ is a maximum.

When $x = -2$: $f''(-2) = \dfrac{32}{512} = \dfrac{1}{16} > 0$

Since $f''(-2) > 0$, the stationary point $(-2, -\frac{1}{4})$ is a minimum.

Curve sketching

You will often need to draw a rough sketch of a curve without plotting a large number of points. Here are five steps to follow when sketching a curve.

❖ **Zeros** Find the value of y when $x = 0$ and find (if possible) the value(s) of x when $y = 0$. This gives the points where the curve crosses the axes.

❖ **Infinities** Find the values of x for which y is not defined. In most cases these will be values of x which make the denominator of a rational function zero.

❖ **Sign** There are two places where a curve might change sign:
   i) at $y = 0$ (where the curve crosses the x-axis)
   ii) at $y = \infty$ (at infinities).

> A curve changes sign when y changes from positive to negative or vice versa.

❖ **Turning points** Calculate $\dfrac{dy}{dx}$.

When $\dfrac{dy}{dx} > 0$, the curve slopes upwards from left to right.

When $\dfrac{dy}{dx} < 0$, the curve slopes downwards from left to right.

When $\dfrac{dy}{dx} = 0$, the curve has a turning point.

> You can remember this checklist by remembering the word ZISTA – Zeros, Infinities, Sign, Turning points, Asymptotes.

❖ **Asymptotes** Examine the behaviour of the function as $x \to +\infty$ and as $x \to -\infty$.

Example 4

Sketch the curve $y = \dfrac{x + 1}{2x - 3}$.

. . . .

❖ Zeros: When $x = 0$, $y = -\frac{1}{3}$. When $y = 0$, $x = -1$.
   The curve cuts the x-axis at $(-1, 0)$ and the y-axis at $(0, -\frac{1}{3})$.

❖ Infinities: y is not defined when $2x - 3 = 0$.
   That is, when $x = \frac{3}{2}$.

❖ Sign change: There are sign changes at $x = -1$ and $x = \frac{3}{2}$.

❖ Turning points:

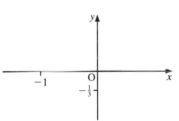

The zeros are at $x = -1$ and $y = -\frac{1}{3}$

$$\frac{dy}{dx} = \frac{(2x - 3)(1) - (x + 1)(2)}{(2x - 3)^2}$$

$$\therefore \frac{dy}{dx} = -\frac{5}{(2x - 3)^2}$$

> Use the quotient rule.

$\dfrac{dy}{dx}$ is always negative and never zero.
Therefore, the curve slopes downwards from left to right and there are no turning points.

❖ Asymptotes: As $x \to +\infty$, $y \to \frac{1}{2}$ from the positive direction.

As $x \to -\infty$, $y \to \frac{1}{2}$ from the negative direction.

The curve is as shown.

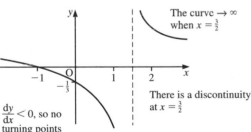

The curve $\to \infty$ when $x = \frac{3}{2}$

There is a discontinuity at $x = \frac{3}{2}$

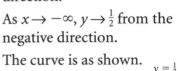

$\dfrac{dy}{dx} < 0$, so no turning points

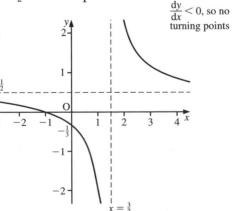

Example 5

Sketch the curve $y = \dfrac{x^2}{x+1}$.

....

❖ Zeros: When $x = 0$, $y = 0$. The curve passes through the origin.
❖ Infinities: y is not defined when $x + 1 = 0$.
That is, when $x = -1$.
❖ Sign: There is a change of sign at $x = -1$.
❖ Turning points:

$$\frac{dy}{dx} = \frac{(x+1)(2x) - x^2(1)}{(x+1)^2}$$

$$= \frac{x^2 + 2x}{(x+1)^2}$$

$$\therefore \frac{dy}{dx} = \frac{x(x+2)}{(x+1)^2}$$

> Use the quotient rule.

When $\dfrac{dy}{dx} = 0$,

$x(x+2) = 0$

$\therefore x = 0$ or $x = -2$

When $x = 0$, $y = 0$. When $x = -2$, $y = -4$.

Therefore, there are two stationary points on the curve, $(0, 0)$ and $(-2, -4)$.

Examine the gradient either side of each of the stationary points:

> In algebra, an improper fraction is one where the numerator (here x^2) is of higher degree than the denominator (here, $x + 1$).

x	$-\frac{1}{2}$	0	1
$\dfrac{dy}{dx}$	-3	0	$\frac{3}{4}$

\ — /

x	-3	-2	$-\frac{8}{9}$
$\dfrac{dy}{dx}$	$\frac{3}{4}$	0	-80

/ — \

Therefore, $(0, 0)$ is a minimum turning point and $(-2, -4)$ is a maximum turning point.

❖ Asymptotes: $y = \dfrac{x^2}{x+1}$ is an improper fraction, so you can write it in the form

$$y = x - 1 + \frac{1}{x+1}$$

Now, as $x \to +\infty$, $\dfrac{1}{x+1} \to 0$ from the positive direction and $y = x - 1$ is an asymptote.

As $x \to -\infty$, $\dfrac{1}{x+1} \to 0$ from the negative direction.

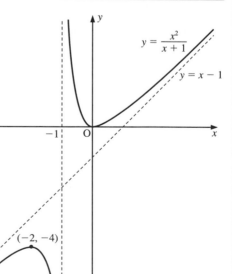

Example 6

The diagram shows a sketch of $y = \dfrac{x}{2x + a}$ $(a \neq 0)$, where $a > 0$.

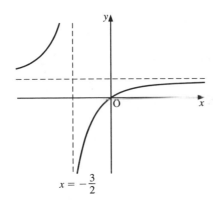

$x = -\dfrac{3}{2}$

a) Find the equation of the horizontal asymptote.

b) Find the value of the constant a.

c) Sketch the graph of $y = \dfrac{x}{2x + a}$ $(a \neq 0)$, where $a < 0$.

. . . .

a) As $x \to +\infty$, $y - \dfrac{x}{2x + a} \to \dfrac{1}{2}$.

> As x gets bigger and bigger ($\to\infty$), the value of a becomes negligible, so the function is approximately $\dfrac{x}{2x} = \dfrac{1}{2}$.

  Similarly as $x \to -\infty$, $y = \dfrac{x}{2x + a} \to \dfrac{1}{2}$.

  The equation of the horizontal asymptote is $y = \frac{1}{2}$.

b) $y = \dfrac{x}{2x + a}$ is not defined when

> You cannot divide anything by zero.

  $2x + a = 0$

  $\therefore x = -\dfrac{a}{2}$

  From the sketch graph above you know that y is not defined when $x = -\frac{3}{2}$. Therefore

  $-\dfrac{a}{2} = -\dfrac{3}{2}$

  $\therefore a = 3$

c) When $a < 0$ the horizontal asymptote is still $y = \frac{1}{2}$ but the vertical asymptote is positive. This is shown in the diagram on the right.

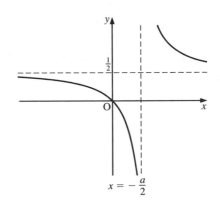

$x = -\dfrac{a}{2}$

Example 7

The diagram shows a sketch of the graph of $y = \dfrac{ax}{x^2 + 1}$ $(a \neq 0)$.

a) Show that the x-coordinates of the turning points P and Q are $x = 1$ and $x = -1$ respectively.

b) By examining the gradient either side of the turning points P and Q, show that $a > 0$.

c) Sketch the graph of $y = \dfrac{ax}{x^2 + 1}$ $(a \neq 0)$ for $a < 0$.

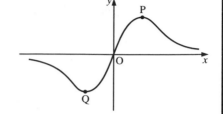

....

a) Turning points occur when $\dfrac{dy}{dx} = 0$.

$$\frac{dy}{dx} = \frac{(x^2 + 1)(a) - ax(2x)}{(x^2 + 1)^2}$$

> Use the quotient rule.

$$\therefore \frac{dy}{dx} = \frac{-ax^2 + a}{(x^2 + 1)^2}$$

When $\dfrac{dy}{dx} = 0$,

$$-ax^2 + a = 0$$
$$a(1 - x^2) = 0$$
$$\therefore x = \pm 1$$

The x-coordinates of P and Q are $x = 1$ and $x = -1$ respectively.

b) Examine the gradient either side of each of the points P and Q:

P

x	0	1	2
$\dfrac{dy}{dx}$	a	0	$-\dfrac{3a}{25}$

Q

x	-2	-1	0
$\dfrac{dy}{dx}$	$-\dfrac{3a}{25}$	0	a

You know that P is a maximum (from the sketch) so the gradient is positive to the left of $x = 1$ and negative to the right of $x = 1$. From the first table you can see that this is only true if $a > 0$. With $a > 0$ the second table shows a minimum turning point at Q which agrees with the sketch graph.

c) When $a < 0$ the only change to the sketch shown in the question is that the maximum occurs when $x = -1$ and the minimum occurs when $x = 1$. This is shown here.

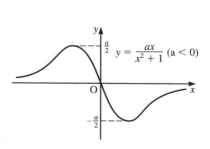

Exercise 7D

. . . .

1 Find the equation of the tangent and the normal to the curve
$y = x(4 - x)^2$ at the point $(2, 8)$.

2 Find the equation of the tangent and the normal to the curve
$y = \dfrac{2x}{x - 1}$ at the point $(3, 3)$.

3 The tangent to the curve $y = x^3(x - 2)^2$ at the point
$(-1, -9)$, meets the normal to the same curve at the point
$(1, 1)$, at the point P. Find the coordinates of P.

4 The tangent to the curve $y = 3x\sqrt{1 + 2x}$ at the point $(4, 36)$,
meets the x-axis at P, and the y-axis at Q.
Calculate the area of the triangle OPQ, where O is the origin.

5 Find the coordinates of the two points on the curve $y = \dfrac{x}{1 + x}$
where the gradient is $\frac{1}{9}$.

6 Show there is just one point, P, on the curve $y = x(x - 1)^3$,
where the gradient is 7. Find the coordinates of P.

7 Given that $y = \dfrac{x^2}{2 - x}$, show that

$$\frac{dy}{dx} = \frac{x(4 - x)}{(2 - x)^2}$$

Hence find the coordinates of the two points on the curve

$y = \dfrac{x^2}{2 - x}$ where the gradient of the curve is zero.

8 Given that $y = x\sqrt{3 + 2x}$, show that

$$\frac{dy}{dx} = \frac{3(1 + x)}{\sqrt{3 + 2x}}$$

Hence find the point on the curve $y = x\sqrt{3 + 2x}$ where the
gradient is zero.

9 Find the coordinates of the stationary points on the curve
$y = (x + 3)^2(2 - x)$, and determine their nature.

10 Show that the curve $y = \dfrac{x + 3}{(x + 4)^2}$ has a single stationary
point. Find the coordinates of that stationary point, and
determine its nature.

11 Find the coordinates of the stationary point on the curve
$y = \dfrac{x}{\sqrt{x - 5}}$, and determine its nature.

12 Find and classify all the stationary values on the curve
$y = (4x - 1)(x^2 - 4)^2$.

> Use your GDC to sketch the
> curves in questions 9–12.

13 Use ZISTA to sketch each of these curves.
In each case $a > 0$ and $b > 0$.

a) $y = \dfrac{1}{x + a}$ b) $y = \dfrac{1}{a - x}$

c) $y = \dfrac{x}{x + a}$ d) $y = \dfrac{x + a}{x - b}$

e) $y = \dfrac{x - a}{x - b}, \quad a > b$ f) $y = \dfrac{x^2}{x + a}$

g) $y = \dfrac{x}{(x + a)^2}$ h) $y = \dfrac{x - b}{x^2 - a^2}, \quad a > b$

i) $y = \dfrac{1}{x^2 + a^2}$ j) $y = \dfrac{x}{a^2 + x^2}$

k) $y = \dfrac{1}{x^2 - a^2}$ l) $y = \dfrac{x}{a^2 - x^2}$

14 The total profit, y thousand dollars, generated from the production and sale of x items of a particular product is given by the formula

$$y = \frac{300\sqrt{x}}{100 + x}$$

Calculate the value of x which gives a maximum profit, and determine that maximum profit.

15 A rectangle is drawn inside a semicircle of radius 5 cm, in such a way that one of its sides lies along the diameter of the semicircle, as in the diagram.

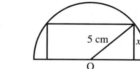

Given that the width of the rectangle is x cm, show that the area of the rectangle is $[2x\sqrt{25 - x^2}]$ cm².
Calculate the maximum value of this area.

16 a) Given that $y = \dfrac{x^2}{ax + b}, (a \neq 0, b \neq 0)$, show that

$$\frac{d^2y}{dx^2} = \frac{2b^2}{(ax + b)^3}.$$

b) Given further that the curve $y = \dfrac{x^2}{ax + b}$ has a stationary point at $(3, 3)$, find the values of the constants a and b, and show that the point $(3, 3)$ is a minimum.

17 The tangent to the curve $y = x^2 - 4$ at the point where $x = a$ meets the x- and y-axes at the points P and Q respectively.

a) Show that the area of the triangle OPQ, where O is the origin, is given by $\dfrac{(a^2 + 4)^2}{4a}$.

b) Hence find the minimum area of the triangle OPQ.

Summary

You should know how to ...

▶ Use the chain rule for differentiation.

▷ $\dfrac{dy}{dx} = \dfrac{dy}{du} \times \dfrac{du}{dx}$, where $y = f(u)$ and $u = f(x)$

▶ Integrate functions of the form $f'(x)[f(x)]^n$.

▷ $\displaystyle\int f'(x)[f(x)]^n \, dx = \dfrac{[f(x)]^{n+1}}{n+1} + c$

▶ Use the product rule for differentiation.

▷ $\dfrac{dy}{dx} = u\dfrac{dv}{dx} + v\dfrac{du}{dx}$ where $y = uv$

▶ Use the quotient rule for differentiation.

▷ $\dfrac{dy}{dx} = \dfrac{v\dfrac{du}{dx} - u\dfrac{dv}{dx}}{v^2}$ where $y = \dfrac{u}{v}$

▶ Sketch curves.

Look for:

▷ **Zeros** – the value of y when $x = 0$, and the value of x when $y = 0$

▷ **Infinities** – values of x for which y is not defined (usually when the denominator equals zero)

▷ **Sign** – if a function changes sign, it is usually at $y = 0$ or $y = \infty$

▷ **Turning points** – values for which $\dfrac{dy}{dx} = 0$

▷ **Asymptotes** – limiting values of y as $x \to \pm\infty$ (horizontal asymptotes) or limiting values of x as $y \to \pm\infty$ (vertical asymptotes)

Revision exercise 7

1 Differentiate the following with respect to x.

a) $(1 + x^2)^4$

b) $\sqrt{(3 - 4x)}$

c) $\dfrac{2}{(1 + \sqrt{x})^3}$

d) $x^3(5x + 2)^2$

e) $(x + 1)\sqrt{(3 - 2x)}$

f) $\dfrac{x}{x^2 - 1}$

g) $\dfrac{5x^2}{1 + \sqrt{x}}$

2 The function $f(x)$ is defined by $f(x) = \dfrac{x - 2}{x^2 + 5}$

a) Find $f'(x)$.

b) Hence show that there is a maximum at $x = 5$ and a minimum at $x = -1$.

3 An open box is made from a rectangular sheet of card that measures 40 cm by 30 cm, by removing a square of side x cm from each corner of the sheet as shown in the diagram.

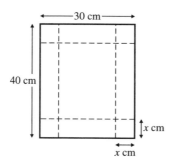

a) Find an expression in terms of x for the volume of the box.

b) Differentiate this expression with respect to x.

c) Hence find the maximum possible volume of the box.

4 A piece of wire 24 cm long is cut into two parts. One piece is bent into a circle with circumference $3x$ cm. The remainder is bent into a rectangle. The length of the rectangle is $2a$ cm and its width is a cm.

a) Find an expression for a in terms of x.

b) Find an expression for the combined area of the two shapes in terms of x.

c) Show that the largest possible area is obtained when
$$x = \frac{16\pi}{(9 + 2\pi)}.$$

5 A curve has equation $y = x^3(x - 2)^2$.

  a) Find $\dfrac{dy}{dx}$ and $\dfrac{d^2y}{dx^2}$.

  b) Show that $\dfrac{dy}{dx} = 0$ when $x = 2$.

  c) Use the second derivative test to show that the curve has a minimum point at $x = 2$.

6 The function f is defined by $f(x) = \dfrac{x^2}{1 - x}, x \neq 1$.

  a) Find $f'(x)$ and $f''(x)$.

  b) Find the values of x for which $f'(x) = 0$.

  c) Use the second derivative to find which of these values of x minimises $f(x)$.

7 Consider the function $y = \dfrac{3x - 2}{2x + 5}$. The graph of this function has a vertical and a horizontal asymptote.

  a) Write down the equation of

   i) the vertical asymptote   ii) the horizontal asymptote.

  b) Find $\dfrac{dy}{dx}$, simplifying your answer as much as possible.   © *IBO*[*2001*]

8 The function f is given by $f(x) = 1 - \dfrac{2x}{1 + x^2}$.

  a) i) To display the graph of $y = f(x)$ for $-10 \leqslant x \leqslant 10$, a suitable interval for y, $a \leqslant y \leqslant b$ must be chosen.
     Suggest appropriate values for a and b.

   ii) Give the equation of the asymptote of the graph.

  b) Show that $f'(x) = \dfrac{2x^2 - 2}{(1 + x^2)^2}$.

  c) Use your answer to part b) to find the coordinates of the maximum point of the graph.   © *IBO*[*Spec.*]

8 Further trigonometry

3.2 Definition of $\cos \theta$ and $\sin \theta$ in terms of the unit circle.

Definition of $\tan \theta$ as $\dfrac{\sin \theta}{\cos \theta}$.

The identity $\cos^2 \theta + \sin^2 \theta = 1$.

3.3 Double angle formulae: $\sin 2\theta = 2 \sin \theta \cos \theta$; $\cos 2\theta = \cos^2 \theta - \sin^2 \theta$.

3.4 The circular functions $\sin x$, $\cos x$ and $\tan x$, their domains and ranges; their periodic nature; and their graphs.
Composite functions of the form
$f(x) = a \sin[b(x + c)] + d$.

3.5 Solution of trigonometric equations in a finite interval.
Equations of the type $a \sin(b(x + c)) + k$.
Equations leading to quadratic equations in $\sin x$, etc.
Graphical interpretation of the above.

Trigonometry literally means 'triangle measure'. However, its uses in real life extend far beyond triangles.

You studied the trigonometric ratios sine, cosine and tangent in Chapter 2.

8.1 The functions $\sin x$, $\cos x$ and $\tan x$

You can plot graphs of the functions $y = \sin x$, $y = \cos x$ and $y = \tan x$.

There is no restriction on the size of angle x, but plotting the graphs in the range $-2\pi < x < 2\pi$ shows the main features.

Properties of the sine function

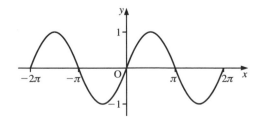

$y = \sin x$, x measured in radians

> **Remember:**
> 2π radians $= 360°$.

❖ The function $f(x) = \sin x$ is periodic, of period 2π rad. That is:
$$\sin(x + 2\pi) = \sin x$$

❖ The graph of $f(x) = \sin x$ has rotational symmetry about the origin of order 2.

❖ The maximum value of $f(x)$ is 1 and its minimum value is -1. In other words $-1 \leqslant f(x) \leqslant 1$.

Properties of the cosine function

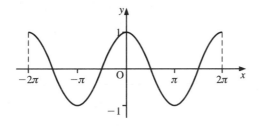

$y = \cos x,$ x measured in radians

❖ The function $f(x) = \cos x$ is periodic, of period 2π rad. That is:
$$\cos(x + 2\pi) = \cos x$$

❖ The graph of $f(x) = \cos x$ is symmetrical about the y-axis.

❖ The maximum value of $f(x)$ is 1 and its minimum value is -1. In other words $-1 \leqslant f(x) \leqslant 1$.

Properties of the tangent function

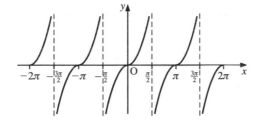

> Try plotting accurate graphs of $\sin x$, $\cos x$ and $\tan x$ for $-360° \leqslant x \leqslant 360°$. Use a calculator to obtain the values, or plot them on your GDC.

$y = \tan x,$ x measured in radians

❖ The function $f(x) = \tan x$ is periodic, of period π rad. That is:
$$\tan(x + \pi) = \tan x$$

❖ The graph of $f(x) = \tan x$ has rotational symmetry about the origin of order 2.

❖ The function $f(x) = \tan x$ is not defined when
$$x = \pm\frac{\pi}{2}, \pm\frac{3\pi}{2}, \dots$$

Example 1

Express each of these as the trigonometric ratio of an acute angle, using the same trigonometric function as given in the question.

a) sin 120°   b) cos 150°   c) tan 300°

....

a) From the graph of $y = \sin x$ you can see that
$\sin 120° = \sin 60°$.

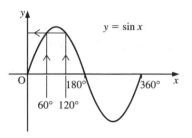

b) From the graph of $y = \cos x$ you can see that
$\cos 150° = -\cos 30°$.

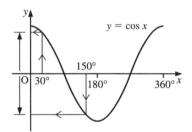

c) From the graph of $y = \tan x$ you can see that
$\tan 300° = -\tan 60°$.

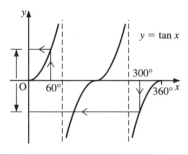

Curve sketching

 In Chapter 3 you studied the transformation of graphs.
You can use a GDC to investigate transformations of the graphs of trigonometric functions.

Example 2

Use a GDC to plot the graphs of

a) $y = \sin x$ b) $y = 2 \sin x$ c) $y = \sin 2x$ d) $y = 2 + \sin x$

for $-2\pi \leqslant x \leqslant 2\pi$.

For b), c) and d), use the GDC to determine:
i) the maximum and minimum values of the function
ii) the x-intercepts
iii) the period of the function.

Remember:

Set the GDC to read angles in radians.

Plotting the graphs gives:

a)

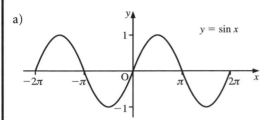

$y = \sin x$

b)

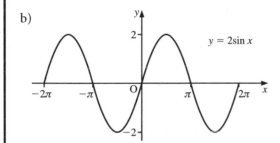

$y = 2\sin x$

c)

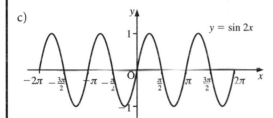

$y = \sin 2x$

d)

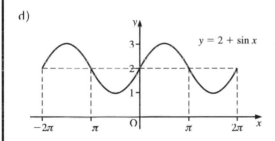

$y = 2 + \sin x$

b) i) For $y = 2\sin x$, the maximum value is 2.
The minimum value is -2.

ii) The x-intercepts are -2π, $-\pi$, 0, π and 2π.

iii) The period is 2π.

> $y = 2\sin x$ is $y = \sin x$ stretched parallel to the y-axis by a scale factor of 2.

c) i) For $y = \sin 2x$, the maximum value is 1.
The minimum value is -1.

ii) The x-intercepts are -2π, $-\dfrac{3\pi}{2}$, $-\pi$, $-\dfrac{\pi}{2}$, 0, $\dfrac{\pi}{2}$, π, $\dfrac{3\pi}{2}$, 2π.

iii) The period is π.

> $y = \sin 2x$ is $y = \sin x$ stretched parallel to the x-axis by a scale factor of $\frac{1}{2}$.

d) i) For $y = 2 + \sin x$, the maximum value is 3.
The minimum value is 1.

ii) There are no x-intercepts.

iii) The period is 2π.

> $y = 2 + \sin x$ is a translation of $y = \sin x$ by 2 units parallel to the y-axis.

Example 3

Using a GDC, plot the graph of $y = \cos\left(x + \dfrac{\pi}{2}\right)$ for $0 \leqslant x \leqslant 2\pi$.

Use the GDC to determine:

a) the maximum and minimum values of $y = \cos\left(x + \dfrac{\pi}{2}\right)$

b) the x-intercepts

c) the period of the function.

....

Plotting the graph gives:

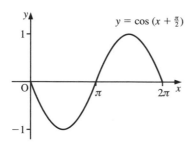

$y = \cos\left(x + \tfrac{\pi}{2}\right)$

Using the GDC functions gives:

a) the maximum value of $y = \cos\left(x + \dfrac{\pi}{2}\right)$ as 1 and the

  minimum value as -1

b) the x-intercepts as 0, π and 2π

c) the period as 2π.

> Notice that the graph of $y = \cos\left(x + \dfrac{\pi}{2}\right)$ is obtained from the graph of $y = \cos x$ by a translation of $-\dfrac{\pi}{2}$ units parallel to the x-axis.

Exercise 8A

....

1 Using your GDC, plot the graphs of each of these functions in the range $0 \leqslant x \leqslant 2\pi$.

In each case determine i) the maximum and minimum values of the function, ii) the x-intercepts and iii) the period of the function.

a) $y = \cos x$ b) $y = \cos 2x$

c) $y = \sin 4x$ d) $y = \tan 2x$

e) $y = 3 \cos x$ f) $y = -\sin x$

g) $y = \sin\left(x + \dfrac{\pi}{4}\right)$ h) $y = \tan\left(x - \dfrac{\pi}{4}\right)$

2 Write down the greatest and least values of each of these expressions, and state the smallest non-negative value of x, measured in degrees, for which they occur.

a) $\sin x$ b) $3 + \cos x$

c) $5 - 3 \sin x$ d) $\sin(x + 20°)$

e) $3 - 4 \cos(x - 40°)$ f) $3 + 7 \sin(60° - x)$

g) $\dfrac{1}{3 + \sin x}$ h) $\dfrac{6}{2 - \cos 2x}$

3 Express each of these as the trigonometric ratio of an **acute angle**, using the same trigonometric function as in the question.

a) $\sin 200°$

b) $\cos 240°$

c) $\tan 160°$

d) $\cos 310°$

e) $\tan 220°$

f) $\cos 490°$

g) $\sin(-20°)$

h) $\cos(-280°)$

4 Express each of these as the trigonometric ratio of an acute angle.

a) $\sin\left(\dfrac{3\pi}{4}\right)$

b) $\cos\left(\dfrac{2\pi}{3}\right)$

c) $\tan\left(\dfrac{7\pi}{6}\right)$

d) $\sin\left(\dfrac{5\pi}{3}\right)$

e) $\tan\left(\dfrac{7\pi}{4}\right)$

f) $\cos\left(\dfrac{11\pi}{6}\right)$

g) $\sin\left(-\dfrac{3\pi}{4}\right)$

h) $\tan\left(\dfrac{5\pi}{3}\right)$

Portfolio work: Investigating the graphs of trigonometric functions

Examples of graphs to support your findings should be included with your work.

Part 1

1 Set your graphical display calculator (GDC) to work in radian mode.

a) On your GDC draw the graph of $y = \sin x$ for $-2\pi \leqslant x \leqslant 2\pi$.

b) Describe the curve in terms of its amplitude and period.

c) Draw, on the same set of axes, the graphs of $y = 2 \sin x$, $y = \frac{1}{2} \sin x$.

d) Investigate other graphs of the form $y = A \sin x$.

e) Make a conjecture about the relationship between the value of A and the shape of the graph.

f) Consider for what range of values of A your conjecture is valid.

g) Explain why your conjecture is always true for these values of A.

2 Investigate graphs of the form $y = \sin Bx$ in a similar way.

3 Investigate the family of curves $y = \sin(x + C)$.

4 Use your answers to questions **1**, **2** and **3** to explain how you can predict the shape and position of the graph $y = A \sin(B(x + C))$ from a knowledge of the values of A, B and C.

Part 2

This part of the work investigates graphs of the form $y = P\sin x + Q\cos x$. For this part of the work you should work in degrees using the domain $-180° \leqslant x \leqslant 360°$.

1 Draw the graph of $y = \sin x + \cos x$.
Use your GDC to answer the following questions.

a) Write down the value of the maximum
(try to give the exact value, not the decimal value).

b) Write down the value of the minimum
(try to give the exact value, not the decimal value).

c) The graph appears to have been translated from the graph of $y = A\sin x$.
Write down the translation.

d) Express the equation of the function $y = \sin x + \cos x$ in the form $y = A\sin(x + C)$.

2 Repeat question **1** for the following functions:

a) $y = \sin x - \cos x$

b) $y = \sqrt{3}\sin x + \cos x$

c) $y = \sin x + \sqrt{3}\cos x$.

3 Make a conjecture relating the forms

$$y = P\sin x + Q\cos x \text{ and } y = A\sin(x + c)$$

4 For what range of values of P and Q is your conjecture valid?

5 Justify your conjecture graphically.

8.2 Trigonometric equations

You have met trigonometric equations many times before. For example, $\cos x = \frac{1}{2}$ is a trigonometric equation in which the unknown is x. In this particular case, the acute angle x which would satisfy this equation is $\frac{\pi}{3}$. However, because the cosine graph is periodic, you know that there are other solutions to this equation.

> You have met this type of equation when solving triangles.

Drawing the graph of $y = \cos x$ and $y = \frac{1}{2}$ on the same set of axes gives:

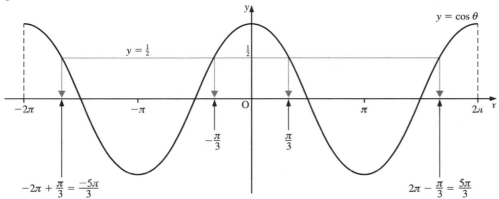

This shows that in the range $-2\pi \leq x \leq 2\pi$ there are actually *four*

solutions to the equation $\cos x = \frac{1}{2}$. These are $x = \pm\dfrac{\pi}{3}, \pm\dfrac{5\pi}{3}$.

If no range for x is stated, there is an infinite number of solutions to this equation. For this reason, trigonometric equations are usually accompanied by a range for x.

Example 1

Solve the following equations for x, where $-360° \leq x \leq 360°$.

a) $\sin x = \dfrac{\sqrt{3}}{2}$ b) $\cos x = \dfrac{1}{3}$ c) $\tan x = -\dfrac{1}{4}$

. . . .

a) Drawing the graphs of $y = \sin x$ and $y = \dfrac{\sqrt{3}}{2}$ on the same set of axes for $-360° \leq x \leq 360°$ gives:

> You know that $\sin 60° = \dfrac{\sqrt{3}}{2}$ (see page 52).

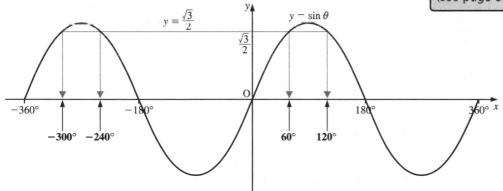

One solution is $x = 60°$. The other solutions in this range are

$$x = 180° - 60° = 120°$$
$$x = -180° - 60° = -240°$$
$$x = -360° + 60° = -300°$$

The solutions are $x = 60°$, $120°$, $-240°$ and $-300°$.

b) Drawing the graphs of $y = \cos x$ and $y = \frac{1}{3}$ on the same set of axes for $-360° \leqslant x \leqslant 360°$ gives:

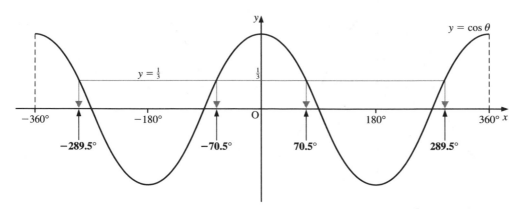

The calculator gives the value of x as $\cos^{-1}(\frac{1}{3}) = 70.5°$, to one decimal place. The other positive solution in this range is $x = 360° - 70.5° = 289.5°$. Since the cosine graph is symmetrical about the y-axis the other solutions in this range are $x = -70.5°$ and $x = -289.5°$.

The solutions are $x = \pm70.5°, \pm289.5°$.

c) Drawing the graphs of $y = \tan x$ and $y = -\frac{1}{4}$ on the same set of axes for $-360° \leqslant x \leqslant 360°$ gives:

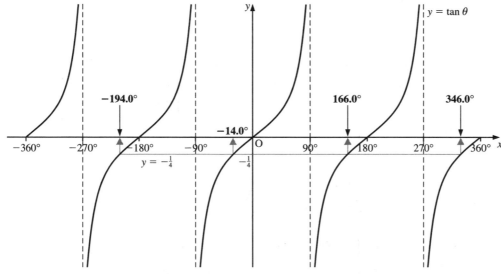

The calculator gives the value of x as $\tan^{-1}(-\frac{1}{4}) = -14.0°$, to one decimal place. The other solutions in this range are

$$x = 180° - 14.0° = 166.0°$$
$$x = 360° - 14.0° = 346.0°$$
$$x = -180° - 14.0° = -194.0°$$

Therefore, the solutions are $x = 166.0°, 346.0°, -14.0°$ and $-194.0°$.

To solve an equation involving a compound angle such as $(x + a°)$ you must change the range.

Example 2

Solve each of these equations for x, for the given range.

a) $\sin(x + 30°) = 0.2$, $0 \leqslant x \leqslant 360°$

b) $\cos\left(x - \dfrac{\pi}{3}\right) = 0.5$, $0 \leqslant x \leqslant 2\pi$

....

a) Changing the range to $x + 30°$ gives $30° \leqslant x + 30° \leqslant 390°$.

$$\sin(x + 30°) = 0.2$$
$$\therefore x + 30° = 11.5° \text{ to one decimal place.}$$

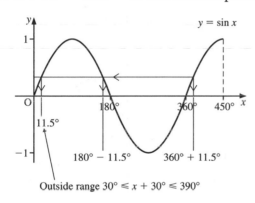

Outside range $30° \leqslant x + 30° \leqslant 390°$

The solutions in the range are

$$180° - 11.5° = 168.5° \text{and} 360° + 11.5° = 371.5°$$

So:

$$x + 30° = 168.5°, 371.5°$$
$$\therefore x = 138.5°, 341.5°$$

b) Changing the range to $x - \dfrac{\pi}{3}$ gives $-\dfrac{\pi}{3} \leqslant x - \dfrac{\pi}{3} \leqslant \dfrac{5\pi}{3}$.

$$\cos\left(x - \dfrac{\pi}{3}\right) = 0.5$$
$$\therefore x - \dfrac{\pi}{3} = \dfrac{\pi}{3}$$

The other solution is $2\pi - \dfrac{\pi}{3} = \dfrac{5\pi}{3}$.

So:

$$x - \dfrac{\pi}{3} = \dfrac{\pi}{3}, \dfrac{5\pi}{3}$$
$$\therefore x = \dfrac{2\pi}{3}, 2\pi$$

> **Remember:**
> $\cos\dfrac{\pi}{3} = \cos 60°$ (page 59)
> $\cos 60° = \frac{1}{2}$ (page 52)

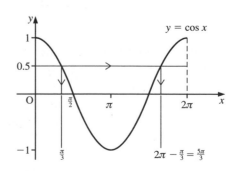

An equation may involve powers of trigonometric functions.

Example 3

Solve each of these equations for x, where $0 \leqslant x \leqslant 2\pi$.

a) $\sin^2 x = \frac{1}{4}$ b) $\tan^2 x = 3$

....

a) When $\sin^2 x = \frac{1}{4}$: $\sin x = \pm\sqrt{\frac{1}{4}} = \pm\frac{1}{2}$.

When $\sin x = \frac{1}{2}$, one solution is $x = \dfrac{\pi}{6}$.

In the range $0 \leqslant x \leqslant 2\pi$, the other solution is

$$x = \pi - \frac{\pi}{6} = \frac{5\pi}{6}.$$

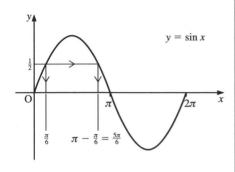

When $\sin x = -\frac{1}{2}$, one solution is $x = -\dfrac{\pi}{6}$.

The other solution is $x = -\dfrac{5\pi}{6}$.

However, in the range $0 \leqslant x \leqslant 2\pi$, the solutions are $\dfrac{7\pi}{6}$ and $\dfrac{11\pi}{6}$, as the diagram shows.

The solutions are $x = \dfrac{\pi}{6}, \dfrac{5\pi}{6}, \dfrac{7\pi}{6}, \dfrac{11\pi}{6}$.

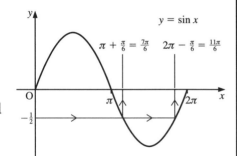

b) When $\tan^2 x = 3$: $\tan x = \pm\sqrt{3}$.

When $\tan x = \sqrt{3}$, one solution is $x = \dfrac{\pi}{3}$.

In the range $0 \leqslant x \leqslant 2\pi$, the other solution is

$$x = \pi + \frac{\pi}{3} = \frac{4\pi}{3}.$$

When $\tan x = -\sqrt{3}$, one solution is $x = -\dfrac{\pi}{3} + \pi = \dfrac{2\pi}{3}$.

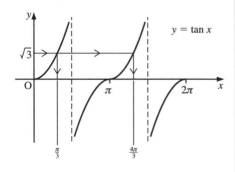

In the range $0 \leqslant x \leqslant 2\pi$, the other solution is

$$x = 2\pi - \frac{\pi}{3} = \frac{5\pi}{3}.$$

The solutions are $x = \dfrac{\pi}{3}, \dfrac{4\pi}{3}, \dfrac{2\pi}{3}, \dfrac{5\pi}{3}$.

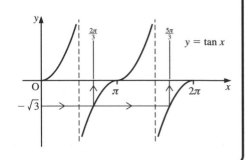

You can often solve trigonometric equations quickly using a GDC.

Example 4

Using the GDC, plot on the same axes the graphs of

a) $y = \sin x$ for $0 \leqslant x \leqslant 2\pi$ b) $y = \dfrac{x}{6}$ for $0 \leqslant x \leqslant 2\pi$.

Using the GDC, find solutions to the equation

$$\sin x = \frac{x}{6}, \quad 0 \leqslant x \leqslant 2\pi$$

....

Plotting the graphs gives:

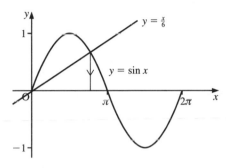

The graphs show that the equation $\sin x = \dfrac{x}{6}$ has two solutions in the given range. One is $x = 0$ and the other is the x-coordinate of the point of intersection of the two graphs.

Using the GDC, the second solution is approximately 2.67 radians.

Example 5

Using the GDC, plot on the same axes the graphs of

a) $y = \sin x$ for $0 \leqslant x \leqslant 2\pi$ b) $y = \cos x$ for $0 \leqslant x \leqslant 2\pi$.

Using the GDC, solve the equation

$$\sin x = \cos x, \quad 0 \leqslant x \leqslant 2\pi$$

....

Plotting the graphs gives:

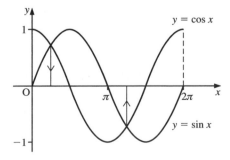

The graphs show that the equation $\sin x = \cos x$ has two solutions in the given range, which are the x-coordinates of the points of intersection of the two graphs.

Using the GDC these solutions are $\dfrac{\pi}{4}$ and $\dfrac{5\pi}{4}$ exactly.

Note: If you use the GDC to plot the graph of $y = \dfrac{\sin x}{\cos x}$ for $0 \leqslant x \leqslant 2\pi$, you get the graph of $y = \tan x$.

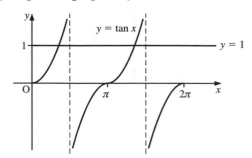

Rearranging the original equation gives:

$$\sin x = \cos x$$

$$\frac{\sin x}{\cos x} = 1$$

That is: $\tan x = 1$

From the graph of $y = \tan x$ you can see that the line $y = 1$ intersects the curve at $\dfrac{\pi}{4}$ and $\dfrac{5\pi}{4}$ in the range $0 \leqslant x \leqslant 2\pi$, which agrees with the original solutions.

Many real-life problems can be modelled by trigonometric equations.

Example 6

The depth of water at a particular point in a harbour is y metres at time t hours after low tide, where y is given by

$$y = 8 - 3\cos(0.5t)$$

Find the depth of water in the harbour

a) at low tide   b) at high tide.

Find the length of time after low tide when the water is at a depth of 9 metres.

....

a) The cosine function has a maximum value of 1 and a minimum value of -1. Therefore

$$y_{\min} = 8 - 3(1) = 5$$

The depth of water at low tide is 5 metres.

b) The maximum depth of water is given by

$$y_{\max} = 8 - 3(-1) = 11$$

The depth of water at high tide is 11 metres.

When the depth of water is 9 metres:

$$8 - 3\cos(0.5t) = 9$$

$$3\cos(0.5t) = -1$$

$$\therefore \ \cos(0.5t) = -\tfrac{1}{3}$$

The sketch shows the graph of $y = \cos(0.5t)$.

Solving the equation gives

$$0.5t = 1.91$$

$$\therefore \ t = 3.82$$

The water is 9 metres deep 3.82 hours after low tide.

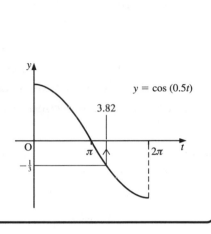

Exercise 8B

1 Solve each of these equations for $0° \leqslant x \leqslant 360°$, giving your answers correct to one decimal place.

a) $\sin x = 0.3$ b) $\cos x = 0.7$

c) $\tan x = 2$ d) $\cos x = -0.5$

e) $\sin x = -0.35$ f) $\tan x = -7$

g) $\cos x = 0.4$ h) $\sin x = -1$

2 Solve each of these equations for $0 \leqslant x \leqslant 2\pi$, giving your answers correct to two decimal places.

a) $\sin x = 0.8$ b) $\cos x = 0.2$

c) $\tan x = 3$ d) $\cos x = -0.6$

e) $\sin x = -0.75$ f) $\tan x = 6$

g) $\cos x = 0.9$ h) $\sin x = 0.3$

3 Solve each of these equations for $0° \leqslant x \leqslant 360°$, giving your answers correct to one decimal place.

a) $\sin(x - 40°) = 0.8$ b) $\cos(x + 20°) = 0.2$

c) $\tan(x - 50°) = 4$ d) $\sin(x - 30°) = -0.7$

e) $\tan(x + 23°) = -8$ f) $\cos(x + 46°) = 0.25$

g) $\sin(x + 15°) = -0.9$ h) $\tan(x - 76°) = 0.4$

4 The depth of water, d metres, at the entrance to a harbour is given by the formula $d = 12 - 7\sin\left(\dfrac{\pi t}{6}\right)$

where t is the time in hours after midnight one day.

a) Sketch the graph of d against t for $0 \leqslant t \leqslant 12$.

b) Find the value of t when the depth is least, and the value of d at that time.

c) Find the value of t when the depth is greatest, and the value of d at that time.

5 A weight hangs on the end of an elastic string. The weight is pulled down and released from rest at time $t = 0$. The length, x centimetres, of the string at time t seconds, is given by the formula $x = 7 - 2 \sin(\pi t)$.

a) Sketch the graph of x against t for $0 \leqslant t \leqslant 4$.

b) Find the two times at which the length of the string is least, and the value of x at those times.

c) After how many seconds does the weight return to its starting point for the first time?

6 On a fair ride a passenger capsule moves on a vertical circle such that its height, h metres above the ground, at a time t seconds after the ride has started, is given by the formula

$$h = 10.5 - 10 \cos\left(\frac{\pi t}{7}\right)$$

a) Sketch the graph of h against t for $0 \leqslant t \leqslant 14$.

b) How high above the ground is the capsule when the ride starts?

c) Find the value of t when the height of the capsule is greatest, and the value of h at that time.

d) How long does it take the capsule to make one complete cycle?

More complicated trigonometric equations

More complicated trigonometric equations can be treated like algebraic equations.

Example 7

Solve each of these equations for x, where $-180° \leqslant x \leqslant 180°$.

a) $\tan^2 x - \tan x = 0$ b) $2 \cos^2 x - \cos x - 1 = 0$

··

a) Notice that $\tan^2 x - \tan x = 0$ is a quadratic equation in $\tan x$. Factorising gives:

$$\tan x(\tan x - 1) = 0$$

$$\therefore \ \tan x = 0 \text{ or } \tan x = 1$$

When $\tan x = 0$, $x = -180°$, $0°$ and $180°$ in the required range.

When $\tan x = 1$, one solution is $x = 45°$.
The other solution is $x = -180° + 45° = -135°$.

The solutions are $x = -180°, -135°, 0°, 45°$ and $180°$.

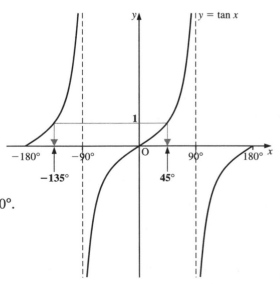

b) Notice that $2\cos^2 x - \cos x - 1 = 0$ is a quadratic equation in $\cos x$. Factorising gives:

$$(2\cos x + 1)(\cos x - 1) = 0$$

Solving gives $\cos x = -\frac{1}{2}$ or $\cos x = 1$

When $\cos x = -\frac{1}{2}$, one solution is $x = -120°$.

The other solution in the required range is $x = 120°$.

When $\cos x = 1$, the only solution in the required range is $0°$.

The solutions are $x = -120°, 0°$ and $120°$.

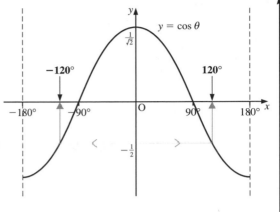

Multiple angles

All the trigonometric equations you have looked at so far have involved solving $\sin\theta = k$, $\cos\theta = k$ or $\tan\theta = k$. Now let's look at equations that involve $2\theta, 3\theta, \ldots$, etc.

Consider the equation $\sin 2\theta = \frac{1}{2}$, where $-180° \leqslant \theta \leqslant 180°$.

If you solve $\sin x = \frac{1}{2}$ in the range $-180° \leqslant x \leqslant 180°$, you get $x = 30°$ and $x = 150°$.

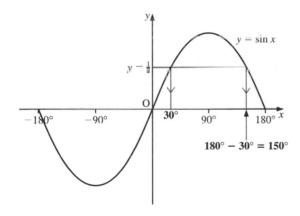

Since $x = 2\theta$, $\theta = 15°$ and $\theta = 75°$.

But you have, in fact, lost two other solutions to the equation $\sin 2\theta = \frac{1}{2}$: $\theta = -105°$ and $\theta = -165°$. These two solutions have been missed because the range in which you have been working is for θ and not 2θ.

To completely solve this equation, you must change the range to match the multiple angle. That is:

$$\sin 2\theta = \frac{1}{2} \quad -360° \leqslant 2\theta \leqslant 360°$$

Now you have:

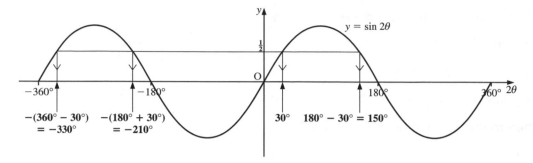

So:

$$2\theta = -330°, -210°, 30° \text{ and } 150°$$
$$\therefore \quad \theta = -165°, -105°, 15° \text{ and } 75°$$

Example 8

Solve each of these equations in the stated range.

a) $\cos^2 2x - 1 = 0$, $\quad -180° \leqslant x \leqslant 180°$

b) $\tan(2x + 45°) = \sqrt{3}$, $\quad -90° \leqslant x \leqslant 90°$

....

a) Changing the range to $2x$ gives $-360° \leqslant 2x \leqslant 360°$. We have

$$\cos^2 2x - 1 = 0$$

$$\cos^2 2x = 1$$

$$\therefore \quad \cos 2x = \pm 1$$

When $\cos 2x = 1$: $\quad 2x = -360°, 0° \text{ and } 360°$

$$\therefore \quad x = -180°, 0° \text{ and } 180°$$

> To see this, look at the graph of $\cos x$ on page 64.

When $\cos 2x = -1$: $\quad 2x = -180° \text{ and } 180°$

$$\therefore \quad x = -90° \text{ and } 90°$$

The solutions are $x = -180°, -90°, 0°, 90° \text{ and } 180°$.

b) Changing the range to $(2x + 45°)$ gives

$$-180° \leqslant 2x \leqslant 180°$$

$$-180° + 45° \leqslant 2x + 45° \leqslant 180° + 45°$$

$$\therefore \quad -135° \leqslant 2x + 45° \leqslant 225°$$

> To see this, look at the graph of $\tan x$ on page 64.

One solution of $\tan(2x + 45°) = \sqrt{3}$ is

$$2x + 45° = 60° \quad \text{so } 2x = 15° \quad \text{giving } x = 7.5°$$

In the required range, the other solution is

$$2x + 45° = -120° \quad \text{so } 2x = -165° \quad \text{giving } x = -82.5°$$

The solutions are $x = 7.5° \text{ and } -82.5°$.

Exercise 8C

1 Solve each of these equations for $0 \leqslant x \leqslant 2\pi$, giving your answers correct to two decimal places.

a) $4 \cos^2 x = 1$

b) $\tan^2 x = 9$

c) $2 \sin^2 x - \sin x = 0$

d) $3 \cos^2 x = \cos x$

e) $\tan^2 x + 4 \tan x = 0$

f) $6 \sin^2 x - 5 \sin x + 1 = 0$

g) $8 \cos^2 x - 6 \cos x + 1 = 0$

h) $(\tan x + 1)^2 = 9$

2 Solve each of these equations for $0° \leqslant x \leqslant 180°$, giving your answers correct to two decimal places.

a) $\sin 2x - 0.3$

b) $4 \tan 3x = 1$

c) $4 \sin 4x = 2$

d) $\cos 2x = -0.4$

e) $2 + 3 \sin 2x = 0$

f) $5 \cos 5x = 2$

g) $1 + 8 \cos 4x = 0$

h) $1 - \sin 5x = 0$

3 Solve each of these equations for $-\pi \leqslant x \leqslant \pi$, leaving each of your answers as a multiple of π.

a) $\sin x = \dfrac{\sqrt{3}}{2}$

b) $\sin x = -\dfrac{1}{2}$

c) $\tan^2 x = 3$

d) $4 \cos^2 x - 1$

e) $\tan 2x = -1$

f) $1 + 2 \cos 4x = 0$

g) $\tan\left(x + \dfrac{\pi}{6}\right) = \sqrt{3}$

h) $\sin\left(2x - \dfrac{\pi}{4}\right) - 1$

Standard trigonometric identities

Here are two trigonometric identities that you should learn:

For any angle θ,

i) $\tan \theta \equiv \dfrac{\sin \theta}{\cos \theta}$

ii) $\sin^2 \theta + \cos^2 \theta \equiv 1$

> These two identities are very useful in solving trigonometric equations. You should memorise them.

To prove these two results for an acute angle θ, consider the right-angled triangle shown.

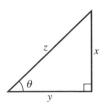

Proof of i)

You know that $\sin \theta = \dfrac{x}{z}$ and $\cos \theta = \dfrac{y}{z}$. Therefore,

$$\frac{\sin \theta}{\cos \theta} = \frac{\left(\dfrac{x}{z}\right)}{\left(\dfrac{y}{z}\right)} = \frac{x}{y} = \tan \theta$$

as required.

b) Replace $\cos^2 x$ by $1 - \sin^2 x$:

$$5(1 - \sin^2 x) = 3(1 + \sin x)$$

$$5 - 5\sin^2 x = 3 + 3\sin x$$

$$5\sin^2 x + 3\sin x - 2 = 0$$

$$\therefore \ (5\sin x - 2)(\sin x + 1) = 0$$

Solving gives $\sin x = \frac{2}{5}$ or $\sin x = -1$.

When $\sin x = \frac{2}{5}$, one solution is $x = 0.41$ radians. In the range $0 \leqslant x \leqslant \pi$, the other solution is $\pi - 0.41 = 2.73$ radians.

When $\sin x = -1$, one solution is $x = -\dfrac{\pi}{2}$, but this is not in the given range. In the range $0 \leqslant x \leqslant \pi$ there are no solutions to $\sin x = -1$.

Use your calculator in RAD mode.

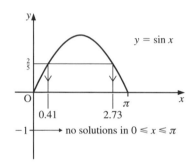

The solutions are 0.41 radians and 2.73 radians.

Exercise 8D

1 Given that x is acute and that $\sin x = \frac{3}{5}$, express each of these in surd form.

a) $\cos x$ b) $\tan x$

2 Given that x is acute and that $\cos x = \frac{12}{13}$, express each of these in surd form.

a) $\sin x$ b) $\tan x$

3 Given that x is acute and that $\tan = 3$, express each of these in surd form.

a) $\sin x$ b) $\cos x$

4 Given that x is acute and that $\sin x = \frac{2}{3}$, express each of these in surd form.

a) $\cos x$ b) $\tan x$

5 Given that x is acute and that $\cos x = \frac{1}{4}$, express each of these in surd form.

a) $\sin x$ b) $\tan x$

6 Given that x is acute and that $\tan x = \frac{7}{24}$, express each of these in surd form.

a) $\sin x$ b) $\cos x$

7 Solve each of these equations for $0° \leqslant x \leqslant 360°$, giving your answers correct to one decimal place.

a) $\sin x = 3\cos x$ b) $5\cos x = 3\sin x$

c) $\sin x + \cos x = 0$ d) $2\cos x - 3\sin x = 0$

e) $\sin x = 2\cos x$ f) $3\cos x + 5\sin x = 0$

g) $\sin x - 5\cos x = 0$ h) $3\cos x = 7\sin x$

8 Solve each of these equations for $-\pi \leqslant x \leqslant \pi$, giving your answers correct to two decimal places.

a) $6\cos^2 x - \sin x - 5 = 0$ b) $2\sin^2 x + 3\cos x - 3 = 0$

c) $6\sin^2 x - 5\cos x + 7$ d) $4\cos^2 x = 4\sin x + 5$

e) $2\cos^2 x + 3\sin x = 3$ f) $3\sin^2 x - 5\cos x - 1 = 0$

g) $8\sin^2 x = 11 - 10\cos x$ h) $\sin^2 x - 2 = 2\cos^2 x - 4\sin x$

9 a) Use your GDC to solve the equation $6y^3 + 5y^2 - 2y - 1 = 0$.

b) Hence solve the equation

$6\sin^3 x + 5\sin^2 x - 2\sin x - 1 = 0$, for $-180° \leqslant x \leqslant 180°$.

8.3 Double angles

Trigonometric equations often contain double angles, such as $2x$. You should memorise these useful formulae for the sine and cosine of double angles.

❖ For any angle x,

$$\sin 2x \equiv 2\sin x\cos x$$

❖ For any angle x, these three identities are true:

$$\cos 2x \equiv \cos^2 x - \sin^2 x$$
$$\equiv 2\cos^2 x - 1$$
$$\equiv 1 - 2\sin^2 x$$

From the diagram, you can see that:

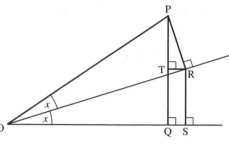

$$\sin(x + x) = \frac{PQ}{OP}$$

$$= \frac{PT + TQ}{OP}$$

$$= \frac{PT + RS}{OP}$$

$$= \frac{PT}{OP} + \frac{RS}{OP}$$

$$= \frac{PT}{PR} \times \frac{PR}{OP} + \frac{RS}{OR} \times \frac{OR}{OP}$$

$$= \cos x \sin x + \sin x \cos x$$

$$= 2 \sin x \cos x, \quad \text{as required.}$$

> Can you see from the geometry why $T\hat{P}R = x$?

From the same diagram,

$$\cos(x + x) = \frac{OQ}{OP}$$

$$= \frac{OS - TR}{OP}$$

$$= \frac{OS}{OP} - \frac{TR}{OP}$$

$$= \frac{OS}{OR} \times \frac{OR}{OP} - \frac{TR}{PR} \times \frac{PR}{OP}$$

$$= \cos x \cos x - \sin x \sin x$$

$$= \cos^2 x - \sin^2 x, \quad \text{as required.}$$

You can use your GDC to illustrate these results.

❖ To illustrate

$$\sin 2x \equiv 2 \sin x \cos x$$

plot the graph of $y = \sin 2x$ for $-2\pi \leqslant x \leqslant 2\pi$.
Now plot the graph of $y = 2 \sin x \cos x$ and you should see that both are the same.

❖ To illustrate

$$\cos 2x \equiv \cos^2 x - \sin^2 x \equiv 2 \cos^2 x - 1 \equiv 1 - 2 \sin^2 x$$

plot the graph of $y = \cos 2x$ for $-2\pi \leqslant x \leqslant 2\pi$.
Now plot the graphs of:

i) $y = \cos^2 x - \sin^2 x$
ii) $y = 2 \cos^2 x - 1$
iii) $y = 1 - 2 \sin^2 x$

Again, it should be clear that the graphs are the same.

Example 1

Given that θ is acute and that $\tan \theta = \frac{1}{2}$, evaluate each of the following:

a) $\sin 2\theta$ b) $\cos 2\theta$

....

When $\tan \theta = \frac{1}{2}$, from this right-angled triangle

$$x^2 = 2^2 + 1^2$$
$$\therefore \ x = \sqrt{5}$$

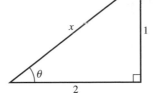

So $\sin \theta = \dfrac{1}{\sqrt{5}}$ and $\cos \theta = \dfrac{2}{\sqrt{5}}$.

a) $\sin 2\theta = 2 \sin \theta \cos \theta$

$$= 2\left(\frac{1}{\sqrt{5}}\right)\left(\frac{2}{\sqrt{5}}\right)$$

$$\therefore \ \sin 2\theta = \frac{4}{5}$$

b) $\cos 2\theta = 2 \cos^2 \theta - 1$

$$= 2\left(\frac{2}{\sqrt{5}}\right)^2 - 1 = \frac{8}{5} - 1$$

$$\therefore \ \cos 2\theta = \frac{3}{5}$$

Example 2

Solve the equation $4 \cos 2x - 2 \cos x + 3 = 0$, for $0° \leqslant x \leqslant 360°$.

....

If you substitute $\cos 2x = 2 \cos^2 x - 1$ into the equation you will get a quadratic in $\cos x$:

$$4(2 \cos^2 x - 1) - 2 \cos x + 3 = 0$$

$$8 \cos^2 x - 2 \cos x - 1 = 0$$

$$\therefore \ (4 \cos x + 1)(2 \cos x - 1) = 0$$

Solving gives $\cos x = -\frac{1}{4}$ or $\cos x = \frac{1}{2}$.

When $\cos x = -\frac{1}{4}$, one solution is $x = 104.5°$.
In the range $0° \leqslant x \leqslant 360°$, the other solution is
$x = 360° - 104.5° = 255.5°$.

When $\cos x = \frac{1}{2}$, one solution is $x = 60°$.
In the range $0° \leqslant x \leqslant 360°$, the other solution is
$x = 360° - 60° = 300°$.

The solutions are $x = 60°, 104.5°, 255.5°$ and $300°$.

Exercise 8E

1 Given that x is an acute angle and that $\sin x = \frac{4}{5}$, find the value of each of these.

   a) $\sin 2x$ b) $\cos 2x$

2 Given that x is an acute angle and that $\cos x = \frac{5}{13}$, find the value of each of these.

   a) $\cos 2x$ b) $\sin 2x$

3 Given that x is an acute angle and that $\tan x = 2$, find the value of each of these.

   a) $\sin 2x$ b) $\cos 2x$

4 Given that x is an acute angle and that $\sin x = \frac{1}{2}$, find the value of each of these.

   a) $\sin 2x$ b) $\cos 2x$

5 Given that x is an acute angle and that $\cos x = \frac{7}{25}$, find the value of each of these.

   a) $\cos 2x$ b) $\sin 2x$

6 Given that x is an acute angle and that $\tan x = \frac{3}{2}$, find the value of each of these.

   a) $\sin 2x$ b) $\cos 2x$

7 Solve each of these equations for $0° \leqslant x \leqslant 360°$, giving your answers correct to one decimal place.

   a) $3 \sin 2x = \sin x$ b) $4 \cos x = 3 \sin 2x$

   c) $\sin 2x + \cos x = 0$ d) $3 \cos 2x - \cos x + 2 = 0$

   e) $6 \cos 2x - 7 \sin x + 6 = 0$ f) $2 \cos 2x = 1 - 3 \sin x$

   g) $2 \cos 2x = 2 + 15 \cos x$ h) $\cos 2x \sin 2x = 0$

8 Solve each of these equations for $0 \leqslant x \leqslant 2\pi$, giving your answers correct to a multiple of π.

   a) $\sin 2x = \sin x$ b) $\sqrt{3} \cos x = \sin 2x$

   c) $\sin 2x + \sin x = 0$ d) $\cos 2x + \sin x = 0$

   e) $\cos 2x + 9 \cos x - 3 = 0$ f) $2 + \cos 2x = 3 \cos x$

   g) $\cos 2x = \cos x$ h) $\sin x \sin 2x = \cos x$

In questions **9** to **15** prove each of the given identities.

9 $2\cos^2 x - \cos 2x \equiv 1$

10 $2\cos^3 x + \sin 2x \sin x \equiv 2 \cos x$

11 $\dfrac{\sin x}{\cos x} + \dfrac{\cos x}{\sin x} \equiv \dfrac{2}{\sin 2x}$

12 $\cos^4 x - \sin^4 x \equiv \cos 2x$

13 $\dfrac{1 - \cos 2x}{1 + \cos 2x} \equiv \tan^2 x$

14 $\dfrac{\cos 2x}{\cos x + \sin x} \equiv \cos x - \sin x$

<u>15</u> $\dfrac{1}{\cos x + \sin x} + \dfrac{1}{\cos x - \sin x} \equiv \dfrac{2 \cos x}{\cos 2x}$

Summary

You should know how to …

► Recognise the graphs of sin x, cos x and tan x.

▷ $y = \sin x$ ▷ $y = \cos x$ ▷ $y = \tan x$

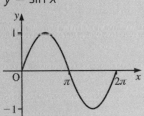 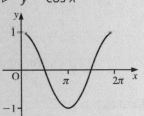 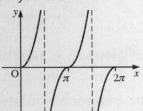

► Solve trigonometric equations in a given range.

► Use trigonometric identities.

▷ $\dfrac{\sin x}{\cos x} \equiv \tan x$

▷ $\sin^2 x + \cos^2 x \equiv 1$

▷ $\sin 2x \equiv 2 \sin x \cos x$

    $\cos 2x \equiv \cos^2 x - \sin^2 x$

         $\equiv 2 \cos^2 x - 1$

         $\equiv 1 - 2 \sin^2 x$

Revision exercise 8

1 Given that $f(x) = x^2$ and $g(x) = \cos \pi x$, evaluate

    a) $(f \circ g)(1)$;   b) $(g \circ f)(1)$ © IBO[1997]

2 Given the functions $f(x) = x^2$ and $g(x) = \tan x$, find

    a) an expression for $(g \circ f)(x)$

    b) the exact value of $(f \circ g)\left(\dfrac{2\pi}{3}\right)$.   © IBO[1998]

3 Two functions f and g are defined as follows:

$$f(x) = \cos x, \qquad 0 \leqslant x \leqslant 2\pi$$
$$g(x) = 2x + 1, \qquad x \in \mathbb{R}.$$

    Solve the equation $(g \circ f)(x) = 0$.   © IBO[1999]

4 a) Write the expression $3 \sin^2 x + 4 \cos x$ in the form
       $a \cos^2 x + b \cos x + c$.

    b) Hence or otherwise, solve the equation
       $3 \sin^2 x + 4 \cos x - 4 = 0, \quad 0° \leqslant x \leqslant 90°$ © IBO[2001]

5 Consider the trigonometric equation $2 \sin^2 x = 1 + \cos x$.

    a) Write this equation in the form $f(x) = 0$, where
       $f(x) = a \cos^2 x + b \cos x + c$ and $a, b, c \in \mathbb{Z}$.

    b) Factorise $f(x)$.

    c) Solve $f(x) = 0$ for $0° \leqslant x \leqslant 360°$.   © IBO[2002]

6 a) Factorise the expression $3 \sin^2 x - 11 \sin x + 6$.

    b) Consider the equation $3 \sin^2 x - 11 \sin x + 6 = 0$.

       i) Find the two values of $\sin x$ which satisfy this equation.

       ii) Solve the equation, for $0° \leqslant x \leqslant 180°$.   © IBO[2003]

7 Let $f(x) = \sin 2x$ and $g(x) = \sin(0.5x)$.

    a) Write down

       i) the minimum value of the function f

       ii) the period of the function g.

    b) Consider the equation $f(x) = g(x)$.

       Find the number of solutions to this equation, for $0 \leqslant x \leqslant \dfrac{3\pi}{2}$.   © IBO[2002]

8 The depth, y metres, of sea water in a bay t hours after midnight may be represented by the function

$$y = a + b\cos\left(\frac{2\pi}{k}t\right), \text{ where } a, b \text{ and } k \text{ are constants.}$$

The water is at a maximum depth of 14.3 m at midnight and noon, and is at a minimum depth of 10.3 m at 06:00 and at 18:00.

Write down the value of

a) a b) b c) k.   © *IBO*[2001]

9 Given that $\sin\theta = \frac{1}{2}$, $\cos\theta = \frac{\sqrt{3}}{2}$ and $0° \leqslant \theta \leqslant 360°$,

a) find the values of θ

b) write down the exact value of $\tan\theta$.   © *IBO*[2000]

10 If $\hat{A}$ is an obtuse angle in a triangle and $\sin A = \frac{5}{13}$, calculate the exact value of $\sin 2A$.   © *IBO*[2000]

11 The diagram shows the graph of $y = a\cos b\theta$, $-\frac{\pi}{2} \leqslant \theta \leqslant \pi$.

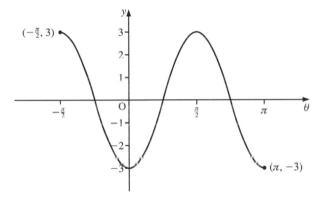

Find the values of a and b.   © *IBO*[1997]

12 The diagram below shows the graph of $y = x\sin\left(\frac{x}{3}\right)$,

for $0 < x < m$, and $0 < y < n$, where x is in radians and m and n are integers.

Find the value of

a) m b) n.

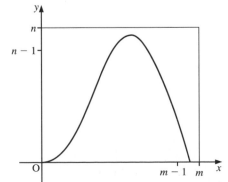

© *IBO*[2001]

13 Three of the following diagrams I, II, III, IV represent the graphs of

a) $y = 3 + \cos 2x$

b) $y = 3 \cos(x + 2)$

c) $y = 2 \cos x + 3$.

Identify which diagram represents which graph.

I

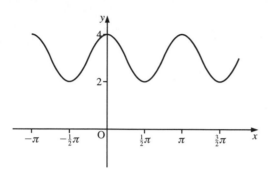

II

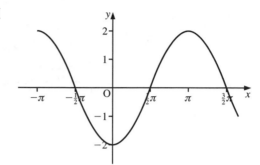

III

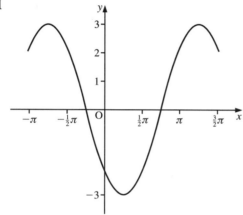

IV

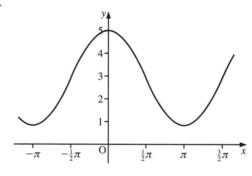

© IBO [1999]

9 Calculus with trigonometry

7.1 Derivatives of $\sin x$, $\cos x$, $\tan x$.

7.4 Indefinite integral of $\sin x$ and $\cos x$.
The composites of any of these with the linear function $ax + b$.

You can apply the techniques of calculus, differentiation and
integration, to trigonometric functions as well as to polynomials.

9.1 $\sin x$ and $\cos x$

Differentiating $\sin x$ and $\cos x$

❖ If $y = \sin x$, then $\dfrac{dy}{dx} = \cos x$

❖ If $y = \cos x$, then $\dfrac{dy}{dx} = -\sin x$

> You need to memorise these results.

where x is measured in radians.

Differentiating of $\sin nx$ and $\cos nx$

To differentiate functions of the form $\sin nx$ and $\cos nx$, use the
chain rule. For example, if $y = \sin 4x$, let $u = 4x$. Then:

$$y = \sin u \quad \text{and} \quad u = 4x$$

$$\therefore \frac{dy}{du} = \cos u \quad \text{and} \quad \frac{du}{dx} = 4$$

By the chain rule,

$$\frac{dy}{dx} = \frac{dy}{du} \cdot \frac{du}{dx} = \cos u \times 4$$

$$\therefore \frac{dy}{dx} = 4\cos 4x$$

In practice, it is quicker to write:

$$\frac{dy}{dx} = \cos 4x \times (4x)' = 4\cos 4x$$

where $(4x)'$ denotes the first derivative with respect to x.

Example 1

Find $\dfrac{dy}{dx}$ for each of these functions.

a) $y = \cos 3x$ b) $y = \sin(x^2 + 2)$ c) $y = \cos \sqrt{x}$

. . . .

a) When $y = \cos 3x$: $\dfrac{dy}{dx} = -\sin 3x \times (3x)' = -3\sin 3x$

b) When $y = \sin(x^2 + 2)$: $\dfrac{dy}{dx} = \cos(x^2 + 2) \times (x^2 + 2)'$

$$= 2x\cos(x^2 + 2)$$

c) When $y = \cos \sqrt{x}$: $\dfrac{dy}{dx} = -\sin \sqrt{x} \times (\sqrt{x})'$

$$= -\frac{1}{2}x^{-\frac{1}{2}}\sin \sqrt{x}$$

$$= -\frac{1}{2\sqrt{x}}\sin \sqrt{x}$$

Integrating sin x and cos x

Since $\dfrac{d(\sin x)}{dx} = \cos x$ and $\dfrac{d(\cos x)}{dx} = -\sin x$

you can deduce these integrals:

❖ $\displaystyle\int \sin x\, dx = -\cos x + c$

❖ $\displaystyle\int \cos x\, dx = \sin x + c$

You can use these results to integrate functions of the form

$$g'(x)f(g(x))$$

where f is a trigonometric function.

Example 2

Find each of these integrals.

a) $\displaystyle\int \sin 5x\, dx$ b) $\displaystyle\int x^2 \cos(x^3 - 2)\, dx$

. . . .

a) To find $\displaystyle\int \sin 5x\, dx$, notice that the derivative of $5x$ is 5 and therefore

$$\int \sin 5x\, dx = -\tfrac{1}{5}\cos 5x + c$$

You can check your answers by differentiation.

b) To find $\int x^2 \cos(x^3 - 2) \, dx$, notice that the derivative of $x^3 - 2$ is $3x^2$ and that there is an x^2 term outside the main integrand. Therefore,

$$\int x^2 \cos(x^3 - 2) \, dx = \tfrac{1}{3} \sin(x^3 - 2) + c$$

Differentiating $\sin^n x$ and $\cos^n x$

You can also use the chain rule to differentiate functions of the form $\sin^n x$ and $\cos^n x$. For example, if $y = \cos^2 x$, you can write this as $y = (\cos x)^2$. Letting $u = \cos x$ gives

$$y = u^2 \quad \text{and} \quad u = \cos x$$

$$\therefore \ \frac{dy}{du} = 2u \quad \text{and} \quad \frac{du}{dx} = -\sin x$$

By the chain rule,

$$\frac{dy}{dx} = \frac{dy}{du} \cdot \frac{du}{dx} = 2u(-\sin x)$$

$$\therefore \ \frac{dy}{dx} = -2 \cos x \sin x$$

In practice, it is quicker to write:

$$\frac{dy}{dx} = 2 \cos x \times (\cos x)' = -2 \cos x \sin x$$

Example 3

Find $\dfrac{dy}{dx}$ for each of these functions.

a) $y = \sin^4 x$ b) $y = (\sin x + \cos x)^5$ c) $y = \cos^3 2x$

....

a) $y = \sin^4 x$:

$$\frac{dy}{dx} = 4 \sin^3 x \times (\sin x)'$$
$$= 4 \sin^3 x \cos x$$

b) $y = (\sin x + \cos x)^5$:

$$\frac{dy}{dx} = 5(\sin x + \cos x)^4 \times (\sin x + \cos x)'$$
$$= 5(\sin x + \cos x)^4 \times (\cos x - \sin x)$$

c) $y = \cos^3 2x$
$= (\cos 2x)^3$:

$$\frac{dy}{dx} = 3(\cos 2x)^2 \times (\cos 2x)'$$
$$= 3(\cos 2x)^2 \times (-2 \sin 2x)$$
$$= -6 \cos^2 2x \sin 2x$$

You may need to use the product rule (see page 215) or the quotient rule (see page 217) to differentiate some functions.

Example 4

Find $\dfrac{dy}{dx}$ for each of these functions.

a) $y = x \sin x$ b) $y = \sin^2 \cos 2x$ c) $y = \dfrac{1 + \sin x}{1 - \cos x}$

....

a) $y = x \sin x$

Using the product rule,

$$\frac{dy}{dx} = x(\sin x)' + \sin x(x)'$$

$$= x \cos x + \sin x$$

b) $y = \sin^2 x \cos 2x$

Using the product rule,

$$\frac{dy}{dx} = \sin^2 x(\cos 2x)' + \cos 2x(\sin^2 x)'$$

$$= \sin^2 x(-2 \sin 2x) + \cos 2x(2 \sin x \cos x)$$

$$= -2 \sin^2 x \sin 2x + \cos 2x \sin 2x$$

$$\therefore \frac{dy}{dx} = \sin 2x(\cos 2x - 2 \sin^2 x)$$

> **Remember:**
>
> $\sin 2x = 2 \sin x \cos x$ (page 255)

c) $y = \dfrac{1 + \sin x}{1 - \cos x} = \dfrac{u}{v}$, where $u = 1 + \sin x$ and $v = 1 - \cos x$.

Using the quotient rule,

$$\frac{dy}{dx} = \frac{u'v - uv'}{v^2}$$

$$= \frac{(1 + \sin x)'(1 - \cos x) - (1 + \sin x)(1 - \cos x)'}{(1 - \cos x)^2}$$

$$= \frac{\cos x(1 - \cos x) - (1 + \sin x)\sin x}{(1 - \cos x)^2}$$

$$= \frac{\cos x - \cos^2 x - \sin x - \sin^2 x}{(1 - \cos x)^2}$$

> **Remember:**
>
> $\sin^2 x + \cos^2 x = 1$

$$\therefore \frac{dy}{dx} = \frac{\cos x - \sin x - 1}{(1 - \cos x)^2}$$

Integrating $\sin^n x$ and $\cos^n x$

In complex expressions, you need to look for something that is the derivative of the main part of the expression.

Example 5

Find each of these integrals.

a) $\int \cos x \sin^2 x \, dx$ b) $\int \dfrac{\cos x}{\sqrt{2 + \sin x}} \, dx$

....

a) To find $\int \cos x \sin^2 x \, dx = \int \cos x (\sin x)^2 \, dx$, notice that the derivative of $\sin x$ is $\cos x$, and that the function $\cos x$ is outside the main function of the integrand. Therefore,

$$\int \cos x \sin^2 x \, dx = \frac{(\sin x)^3}{3} + c$$

$$= \frac{\sin^3 x}{3} + c$$

b) To find

$$\int \frac{\cos x}{\sqrt{2 + \sin x}} \, dx = \int \cos x (2 + \sin x)^{-\frac{1}{2}} \, dx$$

notice that the derivative of $2 + \sin x$ is $\cos x$ and that the function $\cos x$ is outside the main function of the integrand. Therefore,

$$\int \frac{\cos x}{\sqrt{2 + \sin x}} \, dx = \int \cos x (2 + \sin x)^{-\frac{1}{2}} \, dx$$

$$= \frac{(2 + \sin x)^{\frac{1}{2}}}{\left(\frac{1}{2}\right)} + c$$

$$= 2\sqrt{2 + \sin x} + c$$

Exercise 9A

....

1 Find $\dfrac{dy}{dx}$ for each of these functions.

a) $y = \sin 3x$ b) $y = \cos 2x$ c) $y = \sin 5x$

d) $y = -\sin 6x$ e) $y = 2 \cos 7x$ f) $y = -6 \cos 5x$

g) $y = 8 \sin \frac{1}{2}x$ h) $y = \cos(x + 3)$ i) $y = \sin(x - 4)$

2 Differentiate each of these with respect to x.

a) $\sin(x^2)$ b) $\cos(x^3)$ c) $2 \cos(x^2 - 1)$

d) $3 \sin(2x^3 + 3)$ e) $-4 \sin(1 - x^2)$ f) $6 \cos(4 - 3x^4)$

g) $-\cos(x^2 - 2x)$ h) $\sin(x^3 - 3x^2)$ i) $6 \sin \sqrt{x}$

3 Find each of these integrals.

a) $\displaystyle\int 2\cos 2x\,dx$ b) $\displaystyle\int \sin 4x\,dx$

c) $\displaystyle\int 3\sin 5x\,dx$ d) $\displaystyle\int \cos(2x-1)\,dx$

e) $\displaystyle\int -6\sin(3x+2)\,dx$ f) $\displaystyle\int \sin\left(\frac{5x-\pi}{4}\right)dx$

g) $\displaystyle\int x\cos(x^2)\,dx$ h) $\displaystyle\int 8x^3\sin(x^4)\,dx$

i) $\displaystyle\int 3x\cos(x^2-7)\,dx$

4 Find $f'(x)$ for each of these functions.

a) $f(x)=\sin^2 x$ b) $f(x)=\cos^3 x$ c) $f(x)=\sqrt{\cos x}$

d) $f(x)=\dfrac{1}{\cos^2 x}$ e) $f(x)=2\sin^7 x$ f) $f(x)=-3\cos^6 x$

g) $f(x)=\sin^4 5x$ h) $f(x)=\cos^6 \frac{1}{2}x$ i) $f(x)=2\sqrt{\cos 4x}$

5 Find $\dfrac{dy}{dx}$ for each of these functions.

a) $y=(1+\sin x)^2$ b) $y=(3-\cos x)^4$

c) $y=(5+3\cos x)^6$ d) $y=(\sin x+\cos 2x)^3$

e) $y=\dfrac{1}{1+\cos x}$ f) $y=\sqrt{1-6\sin x}$

g) $y=-\dfrac{3}{1+\cos 3x}$ h) $y=\dfrac{4}{\sqrt{1-\sin 6x}}$

i) $y=(1+\sin^2 x)^3$

6 Find each of these integrals.

a) $\displaystyle\int 4\cos x\sin^3 x\,dx$ b) $\displaystyle\int \sin x\cos^2 x\,dx$

c) $\displaystyle\int \sin x(4-\cos x)^5\,dx$ d) $\displaystyle\int 2\cos x(3+\sin x)^3\,dx$

e) $\displaystyle\int \frac{\sin x}{(1+\cos x)^2}\,dx$ f) $\displaystyle\int -\frac{\cos x}{\sqrt{4-\sin x}}\,dx$

g) $\displaystyle\int 6\cos 3x\sin^5 3x\,dx$ h) $\displaystyle\int (1-\cos x)(x-\sin x)^2\,dx$

i) $\displaystyle\int \sin x\cos x\cos 2x\,dx$

7 Differentiate each of these expressions with respect to x.

a) $x \sin x$ b) $x^2 \cos x$ c) $x \cos 3x$

d) $x^3 \sin 6x$ e) $x \sin^5 x$ f) $3x^2 \cos^4 2x$

g) $\dfrac{x}{\sin x}$ h) $\dfrac{\cos 2x}{x + 1}$ i) $\dfrac{1}{1 + \sin x}$

j) $\dfrac{1 + \sin 2x}{\cos 2x}$ k) $\dfrac{x}{1 + \cos^2 x}$ l) $\dfrac{1 + \sin x}{1 + \cos x}$

8 Show that $\dfrac{d}{dx}\left(\dfrac{\cos x + \sin x}{\cos x - \sin x}\right) \equiv \dfrac{2}{1 - \sin 2x}$

9 Given that $y = A \sin 3x + B \cos 3x$, where A and B are

constants, show that $\dfrac{d^2 y}{dx^2} + 9y \equiv 0$.

10 Given that $y = \sin x + 3 \cos x$, show that

$\dfrac{d^2 y}{dx^2} - 3\dfrac{dy}{dx} + 2y = 10 \sin x$.

··

9.2   Differentiating $\tan x$

> If $y = \tan x$, then $\dfrac{dy}{dx} = \dfrac{1}{\cos^2 x} = \sec^2 x$

> The **secant** of x, $\sec x$, is the reciprocal of $\cos x$.
>
> $\sec x = \dfrac{1}{\cos x}$
>
> Note that $\sec x$ is not on the syllabus for this module and is only included here for completeness.

You can easily derive this from the quotient rule.

$$y = \tan x = \frac{\sin x}{\cos x} = \frac{u}{v}, \text{ where } u = \sin x \text{ and } v = \cos x.$$

Using the quotient rule:

$$\frac{dy}{dx} = \frac{u'v - uv'}{v^2}$$

$$= \frac{(\sin x)'\cos x - \sin x(\cos x)'}{\cos^2 x}$$

$$= \frac{\cos^2 x + \sin^2 x}{\cos^2 x}$$

$$= \frac{1}{\cos^2 x}$$

$$= \sec^2 x$$

Example 1

Find $\dfrac{dy}{dx}$ for each of these functions.

a) $y = \tan 3x$ b) $y = \tan(x^2 - 4)$ c) $y = \tan^4 x$

· ·

a) When $y = \tan 3x$: $\dfrac{dy}{dx} = \sec^2 3x \times (3x)' = 3 \sec^2 3x$

b) When $y = \tan(x^2 - 4)$: $\dfrac{dy}{dx} = \sec^2(x^2 - 4) \times (x^2 - 4)'$

$= 2x \sec^2(x^2 - 4)$

c) When $y = \tan^4 x$: $\dfrac{dy}{dx} = 4 \tan^3 x \times (\tan x)'$

$= 4 \sec^2 x \tan^3 x$

> **Remember:** $\sec x$ is not on the syllabus for Mathematics Standard Level.

You can use your knowledge of the derivative of $\tan x$ to integrate expressions involving $\sec x$.

Example 2

Find $\dfrac{dy}{dx}$ for each of these functions.

a) $y = x^3 \tan x$ b) $y = \dfrac{x^2}{\tan 4x}$

· ·

a) $y = x^3 \tan x = uv$ where $u = x^3$ and $v = \tan x$.

$\dfrac{dy}{dx} = u'v + uv'$

$= \tan x(x^3)' + x^3(\tan x)$

$= 3x^2 \tan x + x^3 \sec^2 x$

$\therefore \dfrac{dy}{dx} = 3x^2 \tan x + x^3 \sec^2 x$

> Use the product rule.

b) $y = \dfrac{x^2}{\tan 4x} = \dfrac{u}{v}$ where $u = x^2$ and $v = \tan 4x$.

$\dfrac{dy}{dx} = \dfrac{u'v - uv'}{v^2}$

$= \dfrac{(x^2)' \tan 4x - x^2(\tan 4x)'}{\tan^2 4x}$

$= \dfrac{2x(\tan 4x) - x^2(4 \sec^2 4x)}{\tan^2 4x}$

$\therefore \dfrac{dy}{dx} = \dfrac{2x \tan 4x - 4x^2 \sec^2 4x}{\tan^2 4x}$

> Use the quotient rule.

Example 3

Find each of the following integrals.

a) $\displaystyle\int \sec^2 5x \, dx$
b) $\displaystyle\int x \sec^2(x^2 + 3) \, dx$

....

a) To find $\displaystyle\int \sec^2 5x \, dx$, notice that the derivative of $5x$ is 5 and therefore

$$\int \sec^2 5x \, dx = \frac{1}{5} \tan 5x + c$$

b) To find $\displaystyle\int x \sec^2(x^2 + 3) \, dx$, notice that the derivative of $x^2 + 3$ is $2x$ and there is an x term outside the main integrand. Therefore

$$\int x \sec^2(x^2 + 3) \, dx = \frac{1}{2} \tan(x^2 + 3) + c$$

$\dfrac{d}{dx}(\tan x) = \sec^2 x$

So $\displaystyle\int \sec^2 x \, dx = \tan x + c$

Exercise 9B

....

1 Find $\dfrac{dy}{dx}$ for each of these.

a) $y = \tan 2x$
b) $y = \tan 5x$
c) $y = \tan(3x - 2)$

d) $y = \tan(4x + 1)$
e) $y = \tan(x^3)$
f) $y = \tan(x^5)$

2 Differentiate each of these with respect to x.

a) $\tan^3 x$
b) $\tan^7 x$
c) $2\tan^3 x - 3\tan x$

d) $(1 + \tan x)^2$
e) $2\sqrt{1 + \tan x}$
f) $\dfrac{5}{4 - 3\tan x}$

3 Find each of these integrals.

a) $\displaystyle\int \sec^2 3x \, dx$
b) $\displaystyle\int \sec^2 6x \, dx$

c) $\displaystyle\int \sec^2(5x - 2) \, dx$
d) $\displaystyle\int x \sec^2(x^2) \, dx$

Remember: $\sec x$ is not on the syllabus for Mathematics Standard Level.

4 Differentiate each of these with respect to x.

a) $x \tan x$
b) $x^3 \tan x$
c) $x^5 \tan 3x$

d) $x^2 \tan 5x$
e) $\dfrac{x}{\tan x}$
f) $\dfrac{\tan 2x}{x^2}$

g) $\dfrac{\tan 4x}{\tan 6x}$
h) $\dfrac{1 + \tan x}{1 - \tan x}$

5 Given that $y = \dfrac{1}{\cos x} + 2\tan x$, show that

$$\cos x \frac{dy}{dx} + 3\tan x \equiv 2y.$$

....

9.3 Applications

Example 1

Find the equation of the tangent to the curve $y = x + \tan x$ at the point where $x = \dfrac{\pi}{4}$.

· ·

To find the gradient of the tangent, you need $\dfrac{dy}{dx}$ when $x = \dfrac{\pi}{4}$.

Since $y = x + \tan x$:

$$\frac{dy}{dx} = 1 + \frac{1}{\cos^2 x}$$

When $x = \dfrac{\pi}{4}$: $\left.\dfrac{dy}{dx}\right|_{x = \frac{\pi}{4}} = 1 + \dfrac{1}{\cos^2\left(\frac{\pi}{4}\right)} = 1 + \dfrac{1}{\left(\frac{1}{2}\right)} = 3$

The gradient of the tangent line is 3. Therefore, the equation of the tangent is of the form $y = 3x + c$.

When $x = \dfrac{\pi}{4}$, $y = \dfrac{\pi}{4} + 1$.

Therefore, the tangent passes through the point $\left(\dfrac{\pi}{4}, \dfrac{\pi}{4} + 1\right)$. So,

$$\frac{\pi}{4} + 1 = 3\left(\frac{\pi}{4}\right) + c$$

$$\therefore\ c = 1 - \frac{\pi}{2}$$

The equation of the tangent is

$$y = 3x + 1 - \frac{\pi}{2} \quad \text{or} \quad 2y - 6x = 2 - \pi$$

Use your GDC to plot the graph $y = x + \tan x$ for $-\dfrac{\pi}{2} < x < \dfrac{\pi}{2}$.

On the same set of axes plot the graph of $y = 3x + 1 - \dfrac{\pi}{2}$ to see that this is a tangent to the curve at the point where $x = \dfrac{\pi}{4}$.

Using the zoom function, find the approximate coordinates of the point of intersection of the curve and the line.

Example 2

Find the area enclosed between the curve $y = \cos 2x$, the x-axis and the y-axis.

· ·

The graph shows the required area. The area A is:

$$A = \int_0^{\frac{\pi}{4}} \cos 2x \, dx$$

$$= \left[\frac{\sin 2x}{2}\right]_0^{\frac{\pi}{4}}$$

$$= \frac{\sin 2\left(\frac{\pi}{4}\right)}{2} - \frac{\sin 2(0)}{2} = \frac{1}{2}$$

The required area is $\frac{1}{2}$.

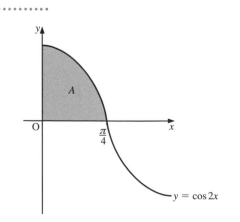

Example 3

Find and classify the stationary points on the curve $y = x + \sin x$ in the range $0 \leqslant x \leqslant 2\pi$.

··

If $y = x + \sin x$ then $\dfrac{dy}{dx} = 1 + \cos x$.

Stationary points occur when $\dfrac{dy}{dx} = 0$, giving

$$1 + \cos x = 0$$
$$\cos x = -1$$
$$\therefore \qquad x = \pi \quad \text{in the range } 0 \leqslant x \leqslant 2\pi$$

When $x = \pi$: $y = \pi + \sin \pi = \pi$.

Therefore a stationary point in the given range is (π, π).

To determine the nature of the stationary point find $\dfrac{d^2y}{dx^2}$, giving

$$\dfrac{d^2y}{dx^2} = -\sin x$$

When $x = \pi$: $\left. \dfrac{dy}{dx} \right|_{x=\pi} = -\sin \pi - 0.$

The second derivative being zero implies a point of inflexion but the sign of the gradient either side of $x = \pi$ must be checked.

x	$\dfrac{\pi}{2}$	π	$\dfrac{3\pi}{2}$
$\dfrac{dy}{dx}$	$+1$	0	$+1$
	/	−	/

The gradient either side of $x = \pi$ is positive and therefore the stationary point is a point of inflexion.

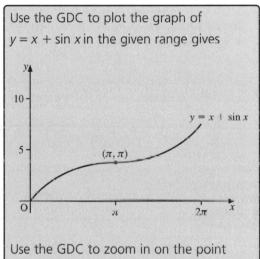

Use the GDC to plot the graph of $y = x + \sin x$ in the given range gives

Use the GDC to zoom in on the point of inflexion and check that this occurs at (π, π).

Exercise 9C

··

This exercise revises the calculus techniques which were developed in earlier chapters, and applies these techniques to trigonometric functions.

Tangents and normals

1 Find the equation of the tangent to the curve $y = x + \sin x$ at the point where $x = \dfrac{\pi}{3}$.

2 Find the equation of the tangent and normal to the curve $y = x \cos x$ at the point where $x = \pi$.

3 The normals to the curve $y = \cos 2x$ at the points $A\left(\dfrac{\pi}{4}, 0\right)$ and $B\left(\dfrac{3\pi}{4}, 0\right)$ meet at the point C. Find the coordinates of the point C, and the area of the triangle ABC.

4 Find the equation of the tangent and the normal to the curve $y = \dfrac{1}{1 + 2 \sin x}$ at the point where $x = \dfrac{\pi}{6}$.

5 Find the coordinates of the two points on the curve $y = \sin x(2 \cos x + 1)$, in the range $-\dfrac{\pi}{2} \leqslant x \leqslant \dfrac{\pi}{2}$, where the gradient is $-\frac{1}{2}$.

6 Show that there are two points on the curve $y = \dfrac{\sin x}{1 + \cos x}$, in the range $0 \leqslant x \leqslant 2\pi$, where the gradient is $\frac{2}{3}$. Find the coordinates of these points.

7 Find the coordinates of the points on the curve $y = \dfrac{2 - \cos x}{\sin x}$ in the range $0 \leqslant x \leqslant 2\pi$, where the gradient is $\frac{8}{3}$.

Stationary points

8 Given that $y = \sin x(1 - \cos x)$, show that
$$\frac{\mathrm{d}y}{\mathrm{d}x} = (1 + 2 \cos x)(1 - \cos x)$$
Hence find the coordinates of the points on the curve $y = \sin x(1 - \cos x)$, in the range $0 \leqslant x \leqslant \pi$, where the gradient is zero.

9 Given that $y = \dfrac{1}{\cos x} + \dfrac{1}{\sin x}$, show that
$$\frac{\mathrm{d}y}{\mathrm{d}x} = \frac{\sin^3 x - \cos^3 x}{\sin^2 x \cos^2 x}$$
Hence find the coordinates of the point on the curve $y = \dfrac{1}{\cos x} + \dfrac{1}{\sin x}$, in the range $0 \leqslant x \leqslant \pi$, where the gradient is zero.

10 Given that $y = \dfrac{x - \sin x}{1 + \cos x}$, show that

$$\frac{dy}{dx} = \frac{x \sin x}{(1 + \cos x)^2}$$

Hence find the coordinates of the point on the curve

$$y = \frac{x - \sin x}{1 + \cos x}$$

in the range $0 \leqslant x \leqslant 2\pi$, where the gradient is zero.

11 Find and classify the stationary values on each of the following curves in the range $0 \leqslant x \leqslant 2\pi$.
Sketch each curve using your GDC.

   a) $y = x + 2 \cos x$ b) $y = 2 \cos x - \cos 2x$

   c) $y = \dfrac{\sin x}{2 - \sin x}$ d) $y = \sin x \cos^3 x$

Areas and volumes of revolution

12 Evaluate the following definite integrals.

   a) $\displaystyle\int_0^{\frac{\pi}{2}} (1 - \cos x)\, dx$ b) $\displaystyle\int_0^{\frac{\pi}{6}} \sin 3x\, dx$

   c) $\displaystyle\int_{\frac{\pi}{6}}^{\frac{\pi}{4}} \cos x \sin^3 x\, dx$ d) $\displaystyle\int_{-\frac{\pi}{6}}^{\frac{\pi}{3}} \sec 2x \tan 2x\, dx$

13 Find the area between the curve $y = \sin x$ and the x-axis from $x = 0$ to $x = \pi$.

14 Find the area between the curve $y = 3 \cos x + 2 \sin x$ and the x axis from $x = \dfrac{\pi}{6}$ to $x = \dfrac{\pi}{3}$.

15 In the interval $0 \leqslant x < \pi$, the line $y = \frac{1}{2}$ meets the curve $y = \sin x$ at the points A and B.

   a) Find the coordinates of A and B.

   b) Calculate the area enclosed between the curve and the line between A and B.

16 In the interval $0 \leqslant x \leqslant \pi$, the curve $y = \sin x$ meets the curve $y = \sin 2x$ at the origin and at the point P.

   a) Use your GDC to find the coordinates of P.

   b) Calculate the area enclosed by the two curves between the origin and P.

17 Find the volume of the solid of revolution formed by rotating about the x-axis the area between the curve $y = \dfrac{1}{\cos 2x}$ and the x-axis, from $x = 0$ to $x = \dfrac{\pi}{12}$.

Maxima and minima

18 A farmer has three pieces of fencing, each of length 5 metres, and wishes to enclose a pen, as in the diagram.
A long wall, which is already in place, will comprise the fourth side of the pen.

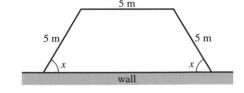

a) Show that the area of the pen is $[25 \sin x(1 + \cos x)]$ m².

b) Calculate the maximum value of this area.

19 A rectangle is drawn inside a semicircle of base radius 10 cm, as in the diagram.

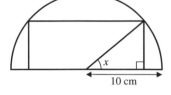

a) Show that the area of the rectangle is $200 \sin x \cos x$ cm².

b) Calculate the maximum value of this area.

20 A solid cylinder just fits under a hemispherical shell of radius *a* cm, as in the diagram.

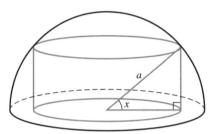

a) Show that the volume of the cylinder is $\pi a^3 \sin x \cos^2 x$.

b) Hence find an expression, in terms of *a*, for the maximum value of this volume.

Summary

You should know how to ...

▶ Differentiate trigonometric functions.

▷ $y = \sin x,\ \dfrac{dy}{dx} = \cos x$

▷ $y = \cos x,\ \dfrac{dy}{dx} = -\sin x$ ⎫ *x* is measured in radians.

▷ $y = \tan x,\ \dfrac{dy}{dx} = \dfrac{1}{\cos^2 x}$

▶ Integrate trigonometric functions.

▷ $\displaystyle\int \cos x\ dx = \sin x + c$

▷ $\displaystyle\int \sin x\ dx = -\cos x + c$ ⎫ *x* is measured in radians.

▷ $\displaystyle\int \dfrac{1}{\cos^2 x}\ x\ dx = \tan x + c$

Revision exercise 9

1 Differentiate with respect to x:

a) $\cos 2x$ b) $\sin^2 x$ c) $3 \tan(x + 2)$

2 The diagram shows part of the graph of $y = 5 \cos\left(\dfrac{x}{2}\right)$

Find the area of the shaded region.

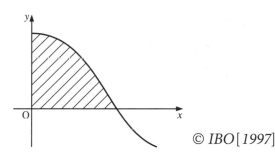

© IBO [1997]

3 Find

a) $\displaystyle\int \sin(3x + 7)\, dx$ b) $\displaystyle\int 4 \cos\left(x + \dfrac{\pi}{3}\right) dx$

4 Find

a) $\dfrac{d}{dx}(2 \sin^2 x \cos 3x)$ b) $\displaystyle\int 6 \sin(1 - 3x)\, dx$

5 If $f'(x) = \cos x$ and $f\left(\dfrac{\pi}{2}\right) = -2$, find $f(x)$. © IBO [2000]

6 The function $f(x) = a \sin x + b$ has derivative $f'(x) = 3 \cos x$.
The point $(0, 2)$ lies on the graph of the function.
Find the values of a and b.

7 The point $P(\tfrac{1}{3}, 0)$ lies on the graph of the equation $y = \sin(3x - 1)$.

a) Find $\dfrac{dy}{dx}$.

b) Hence find the gradient of the tangent to the graph at P.

c) How many tangents parallel to this one are there between $x = 0$ and $x = 2\pi$?

8 The height, h metres, of a Ferris wheel above the ground at time t seconds is given by $h = 20 + 20 \sin\left(\dfrac{t + \pi}{4}\right)$

Find the rate at which its height is changing when $t = 10$ seconds.

9 The area between the curve $y = \sin 2x$, the x-axis between 0 and 1 and the line $x = 1$ is rotated through $360°$ about the x-axis.

a) Sketch the area described.

b) Write down an integral expression for the volume of the solid formed.

c) Find this volume.

10 The diagram shows the graph of the function

$$g: x \rightarrow x \sin x \quad \text{for} \quad -0.5 \leqslant x \leqslant 3.5.$$

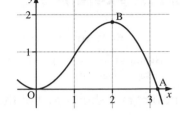

The graph intersects the x-axis at the point A with coordinates $(a, 0)$, and B is the maximum point on this part of the graph.

a) Find the value of a.

b) i) Find $g'(x)$.

ii) Show that, at the maximum point B, x satisfies the equation

$$x + \tan x = 0$$

c) Show that the second derivative $g''(x) = -x \sin x + 2 \cos x$. © *IBO* [1997]

11 A right-angled triangle has hypotenuse 15 cm and one angle x radians.

a) Find an expression for the perimeter of the triangle in terms of x.

b) Find an expression for the rate of change of the perimeter as x varies.

c) Hence find the value of x when this rate of change is greatest.

10 Exponents and logarithms

You should know some basic rules relating to exponents.
For positive integers m and n:

$$x^m \times x^n = x^{m+n}$$

$$x^m \div x^n = x^{m-n}$$

$$(x^m)^n = x^{mn}$$

For example:

$2^3 \times 2^6 = 2^9$

$3^5 \div 3^2 = 3^3$

$(5^3)^2 = 5^6$

10.1 Negative and fractional exponents

x^0

You know that $3 \div 3 = 1$. The second law of exponents tells you that:

$$3^1 \div 3^1 = 3^0$$

$$\therefore \ 3^0 = 1$$

Generally: $\quad x^m \div x^m = x^{m-m} = x^0$

and $\quad\quad\quad x^m \div x^m = \dfrac{x^m}{x^m} = 1$

So:

$$x^0 = 1 \quad (x \neq 0)$$

x^{-m}

You know that $3^0 \div 3^1 = 1 \div 3 = \frac{1}{3}$.
From the second law of exponents

$$3^0 \div 3^1 = 3^{-1}$$

$$\therefore \ 3^{-1} = \tfrac{1}{3}$$

Generally,

$$x^{-1} = \dfrac{1}{x} \quad (x \neq 0)$$

Now:
$$(x^{-1})^m = \left(\frac{1}{x}\right)^m = \frac{1}{x^m}$$

and by the third law of exponents:
$$(x^{-1})^m = x^{-m}$$

Therefore,

$$x^{-m} = \frac{1}{x^m}$$

$x^{\frac{1}{n}}$

By the first law of exponents:
$$5^{\frac{1}{2}} \times 5^{\frac{1}{2}} = 5^1$$

Therefore, $5^{\frac{1}{2}} = \sqrt{5}$.

Similarly,
$$5^{\frac{1}{3}} \times 5^{\frac{1}{3}} \times 5^{\frac{1}{3}} = 5^1$$

Therefore, $5^{\frac{1}{3}} = \sqrt[3]{5}$

Generally,

$$x^{\frac{1}{n}} = \sqrt[n]{x}$$

$x^{\frac{m}{n}}$

To interpret $x^{\frac{m}{n}}$ consider
$$x^{\frac{m}{n}} = \left(x^{\frac{1}{n}}\right)^m \quad \text{or} \quad x^{\frac{m}{n}} = (x^m)^{\frac{1}{n}}$$
$$= \left(\sqrt[n]{x}\right)^m \qquad \qquad = \sqrt[n]{x^m}$$

Generally,

$$x^{\frac{m}{n}} = \sqrt[n]{x^m}$$

Example 1

Simplify each of these quantities.

a) 5^{-2} b) $49^{\frac{1}{2}}$ c) $25^{-\frac{1}{2}}$

....

a) Rewrite this in a form that has no negative exponents:
$$5^{-2} = \frac{1}{5^2} = \frac{1}{25}$$

$$x^{-m} = \frac{1}{x^m}$$

b) You know that $x^{\frac{1}{2}} = \sqrt{x}$, therefore

$$49^{\frac{1}{2}} = \sqrt{49} = 7$$

c) Again, rewrite in a form that has no negative exponents:

$$25^{-\frac{1}{2}} = \frac{1}{25^{\frac{1}{2}}} = \frac{1}{\sqrt{25}} = \frac{1}{5}$$

Example 2

Simplify each of these quantities.

a) $4^{\frac{3}{2}}$ b) $\left(\frac{2}{3}\right)^{-3}$ c) $\left(\frac{1}{8}\right)^{-\frac{4}{3}}$ d) $\dfrac{4^{-1} \times 9^{\frac{1}{2}}}{8^{-2}}$

......

a) Rewriting $4^{\frac{3}{2}}$ gives

$$4^{\frac{3}{2}} = (4^{\frac{1}{2}})^3 = 2^3 = 8$$

$\boxed{4^{\frac{1}{2}} = \sqrt{4} = 2}$

b) Again, rewrite in a form that has no negative exponents:

$$\left(\frac{2}{3}\right)^{-3} = \frac{1}{\left(\frac{2}{3}\right)^3} = \left(\frac{3}{2}\right)^3 = \frac{27}{8}$$

c) Rewriting gives:

$$\left(\frac{1}{8}\right)^{-\frac{4}{3}} = \frac{1}{\left(\frac{1}{8}\right)^{\frac{4}{3}}} = 8^{\frac{4}{3}} = (\sqrt[3]{8})^4 = 2^4 = 16$$

d) $\dfrac{4^{-1} \times 9^{\frac{1}{2}}}{8^{-2}} = \dfrac{\frac{1}{4} \times \sqrt{9}}{\left(\frac{1}{8^2}\right)} = \frac{3}{4} \times 64 = 48$

You should be able to solve equations involving negative and fractional exponents.

Example 3

Solve each of these equations.

a) $x^{\frac{1}{5}} = 3$ b) $x^{\frac{4}{3}} = 81$ c) $2x^{\frac{3}{4}} = x^{\frac{1}{2}}$

......

a) When $x^{\frac{1}{5}} = 3$: $\left(x^{\frac{1}{5}}\right)^5 = 3^5$

$$\therefore x = 243$$

$\boxed{\left(x^{\frac{1}{5}}\right)^5 = x^{\frac{1}{5} \times 5} = x^1}$

b) When $x^{\frac{4}{3}} = 81$: $\left(x^{\frac{4}{3}}\right)^{\frac{1}{4}} = 81^{\frac{1}{4}}$

$$\therefore x^{\frac{1}{3}} = 3$$

$$(x^{\frac{1}{3}})^3 = 3^3$$

$$\therefore x = 27$$

c) When $2x^{\frac{3}{4}} = x^{\frac{1}{2}}$: $\left(2x^{\frac{3}{4}}\right)^4 = \left(x^{\frac{1}{2}}\right)^4$

$$2^4\left(x^{\frac{3}{4}}\right)^4 = x^2$$

$$16x^3 = x^2$$

$$16x^3 - x^2 = 0$$

$$\therefore \ x^2(16x - 1) = 0$$

Solving gives $x = 0$ or $16x - 1 = 0$, that is, $x = \frac{1}{16}$.

Exercise 10A

1 Simplify each of these expressions.

a) $x^5 \times x^4$ b) $p^3 \times p^{-1}$ c) $(3k^3)^2$

d) $y^{\frac{1}{2}} \times y^{\frac{1}{3}}$ e) $c^7 \div c^3$ f) $9h^2 \div 6h^{-4}$

g) $(4d^2)^2 \div (2d)^3$ h) $(6p^{-3})^4 \div (9p^{-4})^2$

2 Evaluate each of these quantities.

a) $4^{\frac{1}{2}}$ b) $27^{\frac{1}{3}}$ c) $9^{\frac{3}{2}}$

d) $8^{\frac{5}{3}}$ e) $125^{\frac{2}{3}}$ f) $49^{\frac{3}{2}}$

g) $\left(\dfrac{1}{25}\right)^{\frac{1}{2}}$ h) $\left(\dfrac{8}{27}\right)^{\frac{2}{3}}$

3 Evaluate each of these quantities.

a) 7^{-1} b) 3^{-2} c) $4^{-\frac{1}{2}}$

d) $25^{-\frac{3}{2}}$ e) $\left(\dfrac{2}{3}\right)^{-1}$ f) $27^{-\frac{2}{3}}$

g) $\left(\dfrac{9}{4}\right)^{-\frac{1}{2}}$ h) $\left(\dfrac{125}{8}\right)^{-\frac{1}{3}}$

4 Solve each of these equations for x.

a) $3x^3 = 375$ b) $98x^2 = 2$ c) $x^3 + 343 = 0$

d) $9x^{-1} = 5$ e) $x^{-3} = 8$ f) $x^{-2} = 25$

g) $x^{-6} - 64 = 0$ h) $25x^{-2} = 9$

5 Solve each of these equations for x.

a) $x^{\frac{1}{2}} = 3$ b) $x^{\frac{1}{5}} = 2$ c) $7x^{\frac{1}{2}} + 2 = 0$

d) $x^{-\frac{1}{4}} = 4$ e) $4x^{\frac{1}{2}} = x^{-\frac{3}{2}}$ f) $5x^{\frac{1}{3}} = x^{-\frac{1}{3}}$

g) $7x^{\frac{1}{6}} = x^{-\frac{5}{6}}$ h) $9x^{\frac{2}{3}} - 4x^{-\frac{4}{3}} = 0$

6 Solve each of these equations for x.

a) $x^{\frac{2}{3}} = 9$ b) $x^{\frac{3}{2}} = 64$ c) $5x^{\frac{3}{4}} + 40 = 0$

d) $x^{-\frac{2}{3}} = 81$ e) $x^{-\frac{1}{2}} = 5$ f) $x^{\frac{3}{4}} = 27$

g) $6x^{\frac{1}{3}} + 1 = 0$ h) $x^{\frac{3}{5}} + 8 = 0$

7 Solve these equations for x.

a) $x^{\frac{2}{3}} - x^{\frac{1}{3}} - 2 = 0$ b) $2x^{\frac{1}{4}} = 9 - 4x^{-\frac{1}{4}}$

10.2 Surds

You know that $\sqrt{16} = 4$ and that $\sqrt{\frac{1}{4}} = \frac{1}{2}$. These are examples of **rational numbers**.

> A rational number can be expressed as a fraction of two integers: $\frac{p}{q}$.

However, $\sqrt{2}$ cannot be expressed as a fraction of two integers. $\sqrt{2}$ is an example of an **irrational number**.

Roots such as $\sqrt{2}, \sqrt{3}, \sqrt{5}, \ldots$ are called **surds**.

The solutions to mathematical problems often contain surds. You could use your calculator to work out a decimal equivalent of the surds, but the decimal goes on and on. An answer rounded to, say, three decimal places is not accurate. It is accurate in surd form.

> For example,
> $\sqrt{2} = 1.414213562 \ldots$
> $= 1.414$ (to 3 d.p.)

Here are some properties of surds.

> ❖ $\sqrt{a} \times \sqrt{b} = \sqrt{ab}$
> ❖ $\dfrac{\sqrt{a}}{\sqrt{b}} = \sqrt{\dfrac{a}{b}}$
> ❖ $a\sqrt{c} \pm b\sqrt{c} = (a \pm b)\sqrt{c}$

> Note that a, b and c are integers.

Example 1

Simplify each of these quantities.

a) $\sqrt{48}$ b) $3\sqrt{50} + 2\sqrt{18} - \sqrt{32}$

a) To simplify $\sqrt{48}$, notice that
$$48 = \underbrace{16}_{\substack{\text{largest square} \\ \text{factor of 48}}} \times 3 = 4^2 \times 3$$

Therefore,
$$\sqrt{48} = \sqrt{4^2 \times 3} = 4\sqrt{3}$$

b) $3\sqrt{50} + 2\sqrt{18} - \sqrt{32} = 3\sqrt{25 \times 2} + 2\sqrt{9 \times 2} - \sqrt{16 \times 2}$
$$= 15\sqrt{2} + 6\sqrt{2} - 4\sqrt{2}$$
$$= 17\sqrt{2}$$

> 25, 9 and 16 are the largest square factors of 50, 18 and 32 respectively.

When surds appear in the denominator of a fraction, it is usual to eliminate them. This is called **rationalising the denominator**.

For example, to rationalise the fraction $\dfrac{1}{\sqrt{3}}$, multiply its numerator and denominator by $\sqrt{3}$:

$$\frac{1}{\sqrt{3}} \times \frac{\sqrt{3}}{\sqrt{3}} = \frac{\sqrt{3}}{3}$$

> **Remember:**
> You can obtain equivalent fractions by multiplying the numerator and denominator by the same amount.

To rationalise the fraction $\dfrac{1}{1 + \sqrt{3}}$, multiply its numerator and denominator by $1 - \sqrt{3}$:

$$\dfrac{1}{(1 + \sqrt{3})} \times \dfrac{(1 - \sqrt{3})}{(1 - \sqrt{3})} = \dfrac{1 - \sqrt{3}}{1 - \sqrt{3} + \sqrt{3} - 3}$$

$$= \dfrac{1 - \sqrt{3}}{-2}$$

$$= -\dfrac{1}{2} + \dfrac{1}{2}\sqrt{3}$$

In general, to rationalise the fraction $\dfrac{1}{a \pm \sqrt{b}}$, multiply its numerator and denominator by $a \mp \sqrt{b}$.

Example 2

Express each of these fractions in the form $a + b\sqrt{c}$.

a) $\dfrac{3}{\sqrt{5}}$

b) $\dfrac{2 + \sqrt{3}}{1 - \sqrt{3}}$

....

a) Multiplying numerator and denominator by $\sqrt{5}$ gives

$$\dfrac{3}{\sqrt{5}} \times \dfrac{\sqrt{5}}{\sqrt{5}} = \dfrac{3\sqrt{5}}{5}$$

This is in the form $a + b\sqrt{c}$, where $a = 0$, $b = \frac{3}{5}$ and $c = 5$.

b) Multiplying numerator and denominator by $1 + \sqrt{3}$ gives

$$\dfrac{(2 + \sqrt{3})}{(1 - \sqrt{3})} \times \dfrac{(1 + \sqrt{3})}{(1 + \sqrt{3})} = \dfrac{2 + 2\sqrt{3} + \sqrt{3} + 3}{1 - 3}$$

$$= \dfrac{5 + 3\sqrt{3}}{-2} = -\dfrac{5}{2} - \dfrac{3}{2}\sqrt{3}$$

This is in the form $a + b\sqrt{c}$, where $a = -\frac{5}{2}$, $b = -\frac{3}{2}$ and $c = 3$.

Exercise 10B

....

1 Simplify each of these expressions.

a) $\sqrt{12}$

b) $\sqrt{50}$

c) $\sqrt{112}$

d) $\sqrt{75} + 2\sqrt{27}$

e) $5\sqrt{20} + 2\sqrt{45}$

f) $2\sqrt{8} + \sqrt{200} - 4\sqrt{18}$

g) $\sqrt{32} + \sqrt{128} - \sqrt{200}$

h) $7\sqrt{5} + 3\sqrt{20} - \sqrt{80}$

2 Express each of these fractions in the form $\dfrac{a\sqrt{c}}{b}$, where a, b and c are integers.

a) $\dfrac{3}{\sqrt{2}}$

b) $\dfrac{5}{\sqrt{3}}$

c) $\dfrac{2}{\sqrt{6}}$

d) $\dfrac{\sqrt{7}}{\sqrt{2}}$

e) $\dfrac{10\sqrt{7}}{\sqrt{5}}$

f) $\dfrac{3\sqrt{5}}{2\sqrt{6}}$

g) $\dfrac{3\sqrt{50}}{5\sqrt{27}}$

h) $\dfrac{4\sqrt{45}}{5\sqrt{8}}$

3 Express each of these fractions in the form $\dfrac{a + b\sqrt{c}}{d}$, where a, b, c and d are integers.

a) $\dfrac{1}{2 - \sqrt{3}}$ b) $\dfrac{1}{3 + \sqrt{5}}$ c) $\dfrac{2}{5 - \sqrt{7}}$

d) $\dfrac{3}{6 + \sqrt{3}}$ e) $\dfrac{2 + \sqrt{2}}{2 - \sqrt{2}}$ f) $\dfrac{3 + \sqrt{2}}{5 + \sqrt{2}}$

g) $\dfrac{6 + \sqrt{5}}{2 - \sqrt{5}}$ h) $\dfrac{3 + \sqrt{24}}{2 + \sqrt{6}}$

4 Simplify $\dfrac{(2 + \sqrt{2})(3 + \sqrt{5})(\sqrt{5} - 2)}{(\sqrt{5} - 1)(1 + \sqrt{2})}$

10.3 Logarithms

A **logarithm** (log for short) is an index.
To see this, consider the result

$$10^2 = 100$$

You can write this using log notation as

$$\log_{10} 100 = 2$$

The number 10 is called the **base** of the logarithm. Similarly,

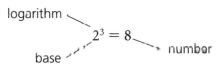

can be written as

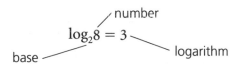

In this case the base is 2.

Generally,

| $x = m^n$ is written as $\log_m x = n$ |

A special case occurs when $n = \log_m m$, that is when $x = m$.
In this case you have to find the value of n such that

$$m^n = m$$

It is obvious that $n = 1$. Therefore $\log_m m = 1$. In other words, the log of any number to the same base equals 1.

> The words 'power', 'index', 'exponent' and 'logarithm' all mean the same thing.

Example 1

Write each of these equations using logarithm notation.

a) $5^2 = 25$
b) $6^3 = 216$

....

a) Using the general form gives

$$\log_5 25 = 2$$

b) Using the general form gives

$$\log_6 216 = 3$$

Example 2

Write $\log_8 64 = 2$ in the form $a^b = c$. Hence find $\log_2 64$.

....

Using index notation gives

$$8^2 = 64, \quad \text{which is in the required form.}$$

Writing $8 = 2^3$ gives

$$(2^3)^2 = 64$$
$$2^6 = 64$$

Rewritten in log form, this is

$$\log_2 64 = 6$$

Laws of logarithms

You should learn these three laws of logarithms:

❖ $\log ab = \log a + \log b$

❖ $\log\left(\dfrac{a}{b}\right) = \log a - \log b$

❖ $\log a^n = n \log a$

> These three results are true for any base. If you are using logs to the same base then you usually leave the base out and write simply $\log x$.

To prove these laws, let $x = \log_m a$ and $y = \log_m b$. Then

$$m^x = a \quad \text{and} \quad m^y = b \qquad\qquad [1]$$

❖ From [1]:

$$ab = m^x m^y = m^{x+y}$$

and by definition of log,

$$\log_m(ab) = x + y$$
$$= \log_m a + \log_m b$$

as required.

❖ Using [1] again,

$$\frac{a}{b} = \frac{m^x}{m^y} = m^{x-y}$$

$$\therefore \log_m\left(\frac{a}{b}\right) = x - y$$

$$= \log_m a - \log_m b$$

as required.

❖ Using [1] again,

$$a^n = (m^x)^n = m^{xn}$$

$$\therefore \log_m a^n = xn$$

$$= n \log_m a$$

as required.

Example 3

Calculate each of these.

a) $\log_3 81$ b) $\log_5\left(\frac{1}{25}\right)$

....

a) Writing $81 = 3^4$ gives

$$\log_3 81 = \log_3 3^4$$

$$= 4 \log_3 3$$ Using $\log a^n = n \log a$

$$= 4 \times 1$$

$$\therefore \log_3 81 = 4$$ Using $\log_m m = 1$

b) Writing $\frac{1}{25} = 25^{-1} = (5^2)^{-1} = 5^{-2}$ gives

$$\log_5\left(\frac{1}{25}\right) = \log_5 5^{-2}$$

$$= -2 \log_5 5$$ Using $\log a^n = n \log a$

$$= -2 \times 1$$

$$\therefore \log_5\left(\frac{1}{25}\right) = -2$$ Using $\log_m m - 1$

You can simplify logarithmic expressions using the laws of logarithms.

Example 4

Express each of these as a single logarithm.

a) $\log 3 + \log 5$ b) $\log 27 - \log 9$

c) $3 \log 2 + \log 4 - \log 8$ d) $2 \log x - 3 \log y + 2 \log xy$

....

a) $\log 3 + \log 5 = \log(3 \times 5)$

$$= \log 15$$

b) $\log 27 - \log 9 = \log\left(\frac{27}{9}\right)$

$$= \log 3$$

c) Notice that $3 \log 2 = \log 2^3 = \log 8$. Therefore,

$$3 \log 2 + \log 4 - \log 8 = \log 8 + \log 4 - \log 8$$
$$= \log 4$$

d) $2 \log x = \log x^2 \quad 3 \log y = \log y^3 \quad 2 \log xy = \log(xy)^2$

Therefore,

$$2 \log x - 3 \log y + 2 \log xy = \log x^2 - \log y^3 + \log(xy)^2$$
$$= \log\left(\frac{x^2}{y^3}\right) + \log(xy)^2$$
$$= \log\left(\frac{x^2}{y^3} \times x^2 y^2\right)$$
$$= \log\left(\frac{x^4}{y}\right)$$

Example 5

Express each of these in terms of $\log a$, $\log b$, $\log c$.

a) $\log\left(\dfrac{1}{a^2}\right)$
b) $\log\left(\dfrac{ab}{c}\right)$

....

a) $\qquad \log\left(\dfrac{1}{a^2}\right) = \log a^{-2}$
$\qquad\qquad\qquad = -2 \log a$

b) $\qquad \log\left(\dfrac{ab}{c}\right) = \log(ab) - \log c$
$\qquad\qquad\qquad = \log a + \log b - \log c$

Exercise 10C

....

1 Write each of these in terms of logarithms.

a) $2^5 = 32$
b) $3^4 = 81$
c) $4^{-2} = \frac{1}{16}$
d) $9^3 = 729$
e) $6^2 = 36$
f) $7^{-3} = \frac{1}{343}$
g) $12^0 = 1$
h) $10^6 = 1\,000\,000$
i) $2^{-9} = \frac{1}{512}$
j) $16^{\frac{1}{2}} = 4$
k) $1000^{\frac{1}{3}} = 10$
l) $(\frac{1}{2})^3 = \frac{1}{8}$

2 Evaluate each of these:

a) $\log_3 27$
b) $\log_2 32$
c) $\log_{10} 100$
d) $\log_5 125$
e) $\log_4 4$
f) $\log_7 49$
g) $\log_3(\frac{1}{9})$
h) $\log_4(\frac{1}{256})$
i) $\log_{10}(0.0001)$
j) $\log_6 1$
k) $\log_2 1024$
l) $\log_3(\frac{1}{243})$

3 Express each of these in terms of $\log a$, $\log b$.

a) $\log(ab)$ b) $\log\left(\dfrac{a}{b}\right)$ c) $\log(a^2 b)$

d) $\log(\sqrt{a})$ e) $\log\left(\dfrac{1}{a^2}\right)$ f) $\log(a\sqrt{b})$

g) $\log\left(\dfrac{a^3}{b}\right)$ h) $\log\left(\dfrac{a^2}{b^3}\right)$ i) $\log\left(\sqrt{\dfrac{a}{b}}\right)$

j) $\log\left(\dfrac{1}{ab^4}\right)$ k) $\log\left(\dfrac{1}{\sqrt{ab}}\right)$ l) $\log(\sqrt[6]{a^2 b})$

4 Express each of these as a single logarithm.

a) $\log 3 + \log 4$ b) $\log 2 + \log 7$

c) $\log 15 - \log 3$ d) $\log 24 - \log 4$

e) $\log 2 + \log 3 + \log 5$ f) $\log 6 + \log 3 - \log 9$

g) $2\log 3 + \log 4 - \log 12$ h) $3\log 2 + 2\log 5 - \log 20$

i) $\frac{1}{2}\log 80 - \frac{1}{2}\log 5$ j) $\log 15 - \frac{1}{2}\log 9$

k) $2\log a - \log b - \log c$ l) $\log a + \frac{1}{2}\log b - 3\log c$

5 Given x and y are both positive, solve the simultaneous equations

$$\log(xy) = 7 \quad \log\left(\dfrac{x}{y}\right) = 1$$

6 Given that $\log(p - q + 1) = 0$ and $\log(pq) + 1 = 0$, show that $p = q = \dfrac{1}{\sqrt{10}}$.

Change of base

Sometimes it is useful to change the base of a logarithm.

Let $y = \log_b a$

Then $a = b^y$

> b is the base and y is the power.

You can take logarithms of both sides of the equation:

$$\log_c a = \log_c b^y$$

> 'Taking logs of both sides' is a useful technique, particularly when solving equations involving powers.

Using the laws of logarithms:

$$\log_c a = y \log_c b$$

$$y = \dfrac{\log_c a}{\log_c b}$$

This leads to a general rule.

$$\log_b a = \dfrac{\log_c a}{\log_c b}$$

> So $\log_5 8 = \dfrac{\log_{10} 8}{\log_{10} 5}$
>
> $= 1.292$

Using logarithms to solve equations

You can often solve an equation involving exponents by taking logarithms of both sides of the equation.

This key point relates exponents and logarithms:

$$y = a^x \Longleftrightarrow x = \log_a y$$

Example 6

Solve these equations for x.

a) $3^x = 10$　　　　b) $5^{2x} = 8$　　　　c) $2^{-3x} = 5$

．．．

a)　　　　$\log 3^x = \log 10$

　　　　　$x \log 3 = \log 10$

　　　　　$\therefore\ x = \dfrac{\log 10}{\log 3} = 2.10$

b)　　　　$\log 5^{2x} = \log 8$

　　　　　$2x \log 5 = \log 8$

　　　　　$\therefore\ x = \dfrac{\log 8}{2 \log 5} = 0.646$

c)　　　　$\log 2^{-3x} = \log 5$

　　　　　$-3x \log 2 = \log 5$

　　　　　$x = \dfrac{\log 5}{-3 \log 2} = -0.774$

> Take logarithms of both sides.

> Use your calculator to evaluate $\dfrac{\log 10}{\log 3}$.
> Note that the log key on your calculator is probably to base 10.

If the unknown, x, appears on both sides of the equation, you must rearrange it to get x on the LHS only.

Example 7

Solve the equation $5 \times 2^x = 3 \times 7^x$ for x.

．．．

　　　$\log(5 \times 2^x) = \log(3 \times 7^x)$

　　　$\log 5 + \log 2^x = \log 3 + \log 7^x$

　　　$\log 5 + x \log 2 = \log 3 + x \log 7$

　　$x \log 2 - x \log 7 = \log 3 - \log 5$

　　$x(\log 2 - \log 7) = \log 3 - \log 5$

　　　$\therefore\ x = \dfrac{\log 3 - \log 5}{\log 2 - \log 7} = 0.408$, to three significant figures

> $\log ab = \log a + \log b$

> $\log a^n = n \log a$

> The function $y = \log_e x$ is the inverse function of $y = e^x$.

The next graph shows $y = \ln x$ on its own.

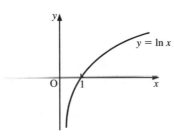

Notice that the function is only defined for $x > 0$. The curve intersects the x axis at $(1, 0)$.

Example 2

Using the GDC plot the graph of $y = \ln x$. On the same axes plot the graph of

a) $y = 2 \ln x$

b) $y = \ln 2x$

Comment on your findings.

....

a) The graphs of $y = \ln x$ and $y = 2 \ln x$ are shown below.

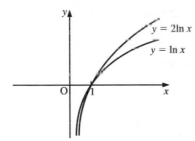

The graph of $y = 2 \ln x$ intersects the x axis at $(1, 0)$ but is steeper than $y = \ln x$.

b) The graphs of $y = \ln x$ and $y = \ln 2x$ are shown below.

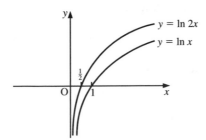

The graph of $y = \ln 2x$ intersects the x axis at $(\frac{1}{2}, 0)$ but has the same gradient as the graph of $y = \ln x$.

Exponential functions are often combined with trigonometric functions. Using the GDC to plot the graph of $y = e^{-x} \sin 4x$ gives:

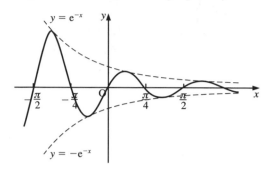

In real life situations, functions such as $y = e^{-x} \sin 4x$ represent dampened vibrations.

An example might be sound waves that gradually fade away as time progresses.

Exercise 10E

1 The value of a car depreciates in such a way that when it is t years old, its value $\$V$ is given by the formula

$$V = 12\,000e^{-0.6t}$$

a) What is the purchase price of the car?

b) What is the value of the car after 3 years?

c) The owner decides to sell the car when its value has halved. Find, to the nearest month, the age of the car when it is sold.

2 Rabbits are introduced to a deserted island and the population grows in such a way that at a time t months, the size of the population, N, is given by formula

$$N = 20e^{0.1t}$$

a) What is the initial size of the population?

b) How many rabbits are on the island after one year?

c) How many months does it take for the population to double?

3 A pan of water is heated and then allowed to cool. Its temperature, $T°C$, t minutes after cooling has begun, is given by the formula

$$T = 80e^{-0.2t}$$

a) What is the temperature of the water when its starts to cool?

b) What is the temperature after 5 minutes?

c) Find, to the nearest minute, the time taken for the water to reach a temperature of 12°C.

Summary

You should know how to ...

▶ Use the laws of exponents.

▷ $x^m \times x^n = x^{m+n}$

▷ $x^m \div x^n = x^{m-n}$

▷ $(x^m)^n = x^{mn}$

▶ Use the laws of logarithms.

▷ $\log ab = \log a + \log b$

▷ $\log\left(\dfrac{a}{b}\right) = \log a - \log b$

▷ $\log a^n = n \log a$

▶ Change the base of a logarithm.

▷ $\log_b a = \dfrac{\log_c a}{\log_c b}$

▶ Solve equations using logarithms.

▶ Recognise the graphs of exponential and logarithmic functions.

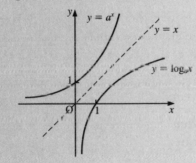

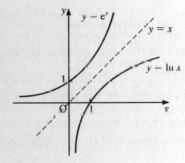

Revision exercise 10

1 Given that $f(x) = 2e^{3x}$, find the inverse function $f^{-1}(x)$. © *IBO* [*2001*]

2 Let $f(x) = \sqrt{x}$, $x \geqslant 0$; $g(x) = 2^x$.
Solve the equation $(f^{-1} \circ g)(x) = 0.25$. © *IBO* [*1999*]

3 Let $f(x) = \ln x$, $x \geqslant 0$; $g(x) = 2f(x) - f\left(\dfrac{1}{x}\right) + \dfrac{3}{f(x)}$.
Express $g(x)$ in terms of logarithms and simplify.

4 Solve the equation $\log_9 81 + \log_9(\frac{1}{9}) + \log_9 3 = \log_9 x$ © *IBO* [*2001*]

5 Solve the equation $\log_{27} x = 1 - \log_{27}(x - 0.4)$ © *IBO* [*2002*]

6 If $\log_a 2 = x$ and $\log_a 5 = y$, find in terms of x and y, expressions for

    a) $\log_2 5$ b) $\log_a 20$ *© IBO [2000]*

7 The diagram shows three graphs.

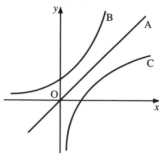

A is part of the graph of $y = x$. B is part of the graph of $y = 2^x$.

C is the reflection of graph B in line A.

Write down

a) the equation of C in the form $y = f(x)$

b) the coordinates of the point where C cuts the x-axis.   *© IBO [2000]*

8 The population of a city is 600 000 in 2004 and is increasing by 4% each year.

a) What is the growth factor?

b) What will be the total percentage increase in population after 5 years?

c) Write an expression for the percentage increase after n years.

d) Assuming the increase remains steady at 4% per year, after how many years will the population first exceed 1 million?

9 A pan of soup is heated. It is then removed from the heat and allowed to cool. Its temperature, $T°C$, after t minutes is given by $T = 100e^{-0.1t}$.

a) What is its temperature when first removed from the heat?

b) What is its temperature after 3 minutes?

c) Find the time for it to cool to 50 °C.

10 A radioactive substance decays according to the formula $m = m_0 e^{-kt}$ where m_0 is the initial mass, k is the decay constant and t is the time in years. The half-life is the time taken for half the material to decay.

a) If the half-life is 100 years find the value of k to three significant figures.

b) If $k = 0.2$ find the half-life.

11 Solve the simultaneous equations

$$y = 2\log_3 x$$
$$y + 1 = \log_3 9x$$

12 The formula $P = A(1.05)^t$ models the population, P, of a bacterial culture at time t hours.

a) What does A represent?

b) A culture starts with 100 bacteria. Find the population after
24 hours.

c) What simplifying assumptions are made in using this model?

d) Why might the model not be reliable in giving the population after one week?

11 Calculus with exponents and logarithms

7.1 Derivative of e^x and $\ln x$.

7.4 Integral of $\dfrac{1}{x}$ and e^x.

11.1 The function $y = a^x$

Functions in which the variable is in the exponent are called **exponential functions**.

> For example, $y = 2^x$ is an exponential function.

Draw a table of values for the function $y = 2^x$:

x	-3	-2	-1	0	1	2
2^x	$\frac{1}{8}$	$\frac{1}{4}$	$\frac{1}{2}$	1	2	4

You can see that as x becomes more negative, the value of 2^x decreases and approaches zero but never equals zero.
As x increases positively, the value of 2^x increases and tends to infinity. The graph of the function is shown.

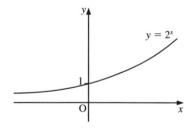

If you plot the functions $y = 3^x$ and $y = 4^x$ you get these graphs:

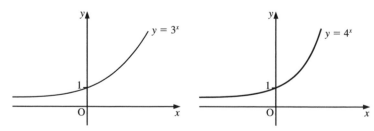

Approximate derivatives of $y = a^x$

Using your GDC, plot the graph of $y = 2^x$.

On the same set of axes plot the graphs of $y = 0.5x + 1$, $y = 0.6x + 1$, $y = 0.7x + 1$, and so on, and use trial and error to determine which of these straight lines is an approximate tangent to the curve at the point $(0, 1)$.

Trial and error shows that $y = 0.7x + 1$ is the best approximate tangent line to the curve at $(0, 1)$. This implies that the gradient of the curve $y = 2^x$ at the point $(0, 1)$ is approximately 0.7. So

$$\frac{dy}{dx} \approx 0.7$$

> Find the intersection of the curve and the straight line: it is nearest to $(0, 1)$ for $y = 0.7x + 1$.

By using approximation methods, you can draw up this table of values for the function $y = 2^x$:

x	-2	-1	0	1	2
$y = 2^x$	0.25	0.5	1	2	4
$\dfrac{dy}{dx}$	0.2	0.3	0.7	1.4	2.8

This shows that

$$\frac{d(2^x)}{dx} \approx 0.7 \times 2^x$$

Similarly, for $y = 3^x$, plot $y = 3^x$ and the graphs of $y = 0.9x + 1$, $y = x + 1$, $y = 1.1x + 1$, etc. and use trial and error to determine which of these straight lines is an approximate tangent to the curve at the point $(0, 1)$.

Trial and error shows that $y = 1.1x + 1$ is the best approximate tangent line to the curve at $(0, 1)$. This implies that the gradient of the curve $y = 3^x$ at the point $(0, 1)$ is approximately 1.1.

You can draw up this table of values for the function $y = 3^x$:

x	-2	-1	0	1	2
$y = 3^x$	$\frac{1}{9}$	$\frac{1}{3}$	1	3	9
$\dfrac{dy}{dx}$	0.1	0.4	1.1	3.3	9.9

This shows that

$$\frac{d(3^x)}{dx} \approx 1.1 \times 3^x$$

In fact, in general,

$$\frac{d(a^x)}{dx} = k \times a^x, \quad \text{where } k \text{ is a constant.}$$

The results for $y = 2^x$ and $y = 3^x$ suggest that there is a value of a between 2 and 3 for which $k = 1$.

> You have met the exponential constant e on page 293.

This number is called e, and its value is approximately 2.718.

Notice that when $a = e$:

$$\frac{d(e^x)}{dx} = e^x$$

So:

$$\int e^x \, dx = e^x + c$$

Example 1

Find $\dfrac{dy}{dx}$ for each of these functions.

a) $y = e^{2x}$
b) $y = 3e^{x^2}$
c) $y = x^2 e^x$
d) $y = \dfrac{e^{3x}}{x}$

a) Using the chain rule gives

$$\frac{dy}{dx} = e^{2x} \times (2x)'$$

$$= 2e^{2x}$$

b) Using the chain rule gives

$$\frac{dy}{dx} = 3e^{x^2} \times (x^2)'$$

$$= 3e^{x^2} \times 2x$$

$$\therefore \frac{dy}{dx} = 6xe^{x^2}$$

c) Using the product rule,

$$\frac{dy}{dx} = e^x(x^2)' + x^2(e^x) = 2xe^x + x^2 e^x$$

$$\therefore \frac{dy}{dx} = xe^x(x + 2)$$

d) Using the quotient rule,

$$\frac{dy}{dx} = \frac{(e^{3x})'x - e^{3x}(x)'}{x^2} = \frac{3xe^{3x} - e^{3x}}{x^2}$$

$$\therefore \frac{dy}{dx} = \frac{e^{3x}(3x - 1)}{x^2}$$

Example 2

Find each of the following integrals.

a) $\displaystyle\int 2e^{-x} \, dx$
b) $\displaystyle\int (1 - e^{-3x})^2 \, dx$
c) $\displaystyle\int 5xe^{x^2} \, dx$

a) To find $\displaystyle\int 2e^{-x} \, dx$, notice that the derivative of e^{-x} is $-e^{-x}$. Therefore,

$$\int 2e^{-x} \, dx = -2e^{-x} + c$$

b) First expand the bracket, obtaining

$$\int (1 - e^{-3x})^2 \, dx = \int (1 - 2e^{-3x} + e^{-6x}) \, dx$$

Therefore,

$$\int (1 - e^{-3x})^2 \, dx = x + \frac{2}{3}e^{-3x} - \frac{e^{-6x}}{6} + c$$

c) To find $\displaystyle\int 5xe^{x^2} \, dx$, notice that the derivative of e^{x^2} is $2xe^{x^2}$ and that e^{x^2} is multiplied by an x term. Therefore,

$$\int 5xe^{x^2} \, dx = \frac{5}{2}e^{x^2} + c$$

Exercise 11A

1 Find $\dfrac{dy}{dx}$ for each of these functions.

a) $y = e^{3x}$ b) $y = 2e^{5x}$ c) $y = e^{x^2}$ d) $y = 4e^{\sqrt{x}}$

e) $y = e^{\frac{5}{x}}$ f) $y = e^{2x-3}$ g) $y = e^{x^2+1}$ h) $y = \dfrac{3}{e^{x^4}}$

2 Integrate each of these expressions with respect to x.

a) e^{4x} b) e^{6x} c) e^{-2x} d) $6e^{3x}$ e) $10e^{-2x}$

f) $2xe^{x^2}$ g) $x^2e^{x^3}$ h) $2x^3e^{x^4}$ i) $\dfrac{e^{\sqrt{x}}}{\sqrt{x}}$ j) $\dfrac{x^2}{e^{x^3}}$

3 Find $f'(x)$ for each of these functions.

a) $f(x) = (1 + e^x)^2$ b) $f(x) = (1 - e^{3x})^4$

c) $f(x) = (2 + 3e^x)^5$ d) $f(x) = (1 - e^{-3x})^2$

e) $f(x) = \dfrac{1}{1 + e^x}$ f) $f(x) = \dfrac{2}{1 - e^{-4x}}$

g) $f(x) = \sqrt{1 - 2e^{4x}}$ h) $f(x) = \dfrac{1}{\sqrt{1 + e^{-2x}}}$

> Hint for a) to d): use the chain rule, with u as the function in brackets.
>
> Hint for e) and f): use the chain rule, with u as the function in the denominator.

4 Find each of these integrals.

a) $\displaystyle\int e^x(3 + e^x)^2\, dx$ b) $\displaystyle\int e^{4x}(1 + e^{4x})\, dx$

c) $\displaystyle\int 2e^x(e^x - 4)^3\, dx$ d) $\displaystyle\int e^{-2x}\sqrt{1 - e^{-2x}}\, dx$

e) $\displaystyle\int e^{3x}\sqrt{e^{3x} + 2}\, dx$ f) $\displaystyle\int \dfrac{e^{-x}}{\sqrt{1 + e^{-x}}}\, dx$

g) $\displaystyle\int \dfrac{(e^{-x} + 7)^2}{e^x}\, dx$ h) $\displaystyle\int \dfrac{4e^{-2x}}{(1 + e^{-2x})^2}\, dx$

5 Differentiate each of these expressions with respect to x.

a) xe^{2x} b) x^2e^{4x} c) $2x^3e^{-3x}$

d) $e^{2x}(1 + e^x)^2$ e) $e^{3x}(1 - 2e^{-x})^3$ f) $e^{-x}(1 + 3e^x)^4$

g) $\dfrac{e^{2x}}{x}$ h) $\dfrac{e^{3x}}{x^2}$ i) $\dfrac{e^{2x}}{1 + e^{2x}}$

j) $\dfrac{2e^{4x}}{1 - e^x}$ k) $\dfrac{e^{3x}}{1 + e^{2x}}$ l) $\dfrac{1 + e^x}{1 - e^x}$

6 Given that $y = xe^{2x}$, show that $x\dfrac{dy}{dx} \equiv (2x + 1)y$.

7 Given that $y = \dfrac{e^x}{e^x + 1}$, show that $(1 + e^x)\dfrac{dy}{dx} \equiv y$.

11.2 Natural logarithms

Logarithms to the base e are called **natural logarithms**.
The notation $\ln x$ is used as the standard abbreviation for $\log_e x$.

You have met natural logarithms on page 294.

The function $\ln x$ is the inverse function of e^x. Notice that

$$\ln e^x = x \ln e = x \log_e e = x(1)$$

So:

$$\ln(e^x) = x \quad \text{and} \quad e^{\ln x} = x$$

The graph of $y = \ln x$ is shown in the diagram.

If $y = \ln x$, then by definition $e^y = x$.

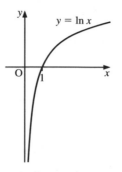

Differentiating e^y with respect to y gives

$$e^y = \frac{dx}{dy}$$

$$\frac{dy}{dx} = \frac{1}{e^y}$$

$$\therefore \frac{dy}{dx} = \frac{1}{x}$$

Remember:

$$\frac{d(e^y)}{dy} = e^y$$

So:

> If $y = \ln x$, then
> $$\frac{dy}{dx} = \frac{1}{x}$$

Conversely:

> $$\int \frac{1}{x}\,dx = \ln x + c$$

Example 1

Find $\dfrac{dy}{dx}$ for each of these.

a) $y = \ln 3x$

b) $y = \ln(x^2 - 1)$

· ·

a) Using the chain rule:

$$\frac{dy}{dx} = \frac{1}{3x} \times (3x)'$$

$$\therefore \frac{dy}{dx} = \frac{1}{3x} \times 3 = \frac{1}{x}$$

b) Using the chain rule:

$$\frac{dy}{dx} = \frac{1}{x^2 - 1} \times (x^2 - 1)'$$

$$\therefore \frac{dy}{dx} = \frac{2x}{x^2 - 1}$$

If $y = \ln(f(x))$, then

$$\frac{dy}{dx} = \frac{1}{f(x)} \times f'(x) = \frac{f'(x)}{f(x)}$$

From this you see that:

$$\int \frac{f'(x)}{f(x)}\, dx = \ln(f(x)) + c$$

Example 2

Find each of these integrals.

a) $\int \frac{1}{4x + 1}\, dx$ b) $\int \frac{x^2 + 1}{x^3 + 3x}\, dx$ c) $\int \frac{e^{2x}}{1 + e^{2x}}\, dx$

a) Rewrite the integral with the derivative of the denominator as the numerator and compensate by introducing a constant outside the integral. This gives:

$$\int \frac{1}{4x + 1}\, dx = \frac{1}{4}\int \frac{4}{4x + 1}\, dx = \frac{1}{4}\ln(4x + 1) + c$$

b) Notice that the derivative of $x^3 + 3x$ is $3x^2 + 3 = 3(x^2 + 1)$.
Therefore,

$$\int \frac{x^2 + 1}{x^3 + 3x}\, dx = \frac{1}{3}\int \frac{3x^2 + 3}{x^3 + 3x}\, dx = \frac{1}{3}\ln(x^3 + 3x) + c$$

c) Rewriting the integral gives

$$\int \frac{e^{2x}}{1 + e^{2x}}\, dx = \frac{1}{2}\int \frac{2e^{2x}}{1 + e^{2x}}\, dx = \frac{1}{2}\ln(1 + e^{2x}) + c$$

Definite integrals involving logarithms

You know that

$$\int \frac{1}{x}\, dx = \ln x + c \qquad\qquad [1]$$

However, the function $\ln x$ is only valid provided $x > 0$.

Therefore, the integral [1] is only valid when $x > 0$.

This is a problem, since $\frac{1}{x}$ exists for negative values of x.

Consider the area between the curve $y = \frac{1}{x}$, the x-axis and the ordinates $x = -1$ and $x = -2$.

Clearly, this area exists and is identical to the area between the curve, the x-axis and the ordinates $x = 1$ and $x = 2$.
Therefore, for both positive and negative values of x,

$$\int \frac{1}{x}\, dx = \ln |x| + c \quad \text{and} \quad \int \frac{f'(x)}{f(x)}\, dx = \ln |f(x)| + c$$

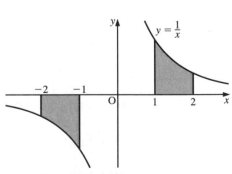

$|x|$ means the modulus of x or absolute value: $|-10| = 10$.

Since the function $y = \dfrac{1}{x}$ is not defined when $x = 0$, the definite integral

$$\int_a^b \frac{1}{x}\,dx$$

is invalid if the interval $[a, b]$ includes $x = 0$. In other words, a and b must both have the same sign for the integral to be valid.

> It is not possible to integrate across a discontinuity in the graph.

Example 3

Evaluate $\displaystyle\int_1^2 \frac{2}{1 - 3x}\,dx$

> $|-5| = 5, |-2| = 2$

$$\int_1^2 \frac{2}{1 - 3x}\,dx = -\frac{2}{3}\int_1^2 \frac{-3}{1 - 3x}\,dx$$

$$= -\frac{2}{3}\Big[\ln|1 - 3x|\Big]_1^2 = -\frac{2}{3}(\ln|-5| - \ln|-2|)$$

$$\therefore \int_1^2 \frac{2}{1 - 3x}\,dx = -\frac{2}{3}\ln\left(\frac{5}{2}\right)$$

> **Remember:** $\ln 5 - \ln 2 = \ln\frac{5}{2}$

Example 4

Find $\dfrac{dy}{dx}$ for each of these functions.

a) $y = (2 - 3\ln x)^3$

b) $y = \dfrac{1}{\sqrt{1 + \ln x}}$

c) $y = x\ln x$

d) $y = \dfrac{\ln 2x}{x^3}$

a) Using the chain rule gives

$$\frac{dy}{dx} = 3(2 - 3\ln x)^2 \times (2 - 3\ln x)'$$

$$= 3(2 - 3\ln x)^2 \times -\frac{3}{x}$$

$$\therefore \frac{dy}{dx} = -\frac{9(2 - 3\ln x)^2}{x}$$

b) You can write $y = \dfrac{1}{\sqrt{1 + \ln x}}$ in the form $y = (1 + \ln x)^{-\frac{1}{2}}$
   and then use the chain rule. This gives

$$\frac{dy}{dx} = -\frac{1}{2}(1 + \ln x)^{-\frac{3}{2}} \times (1 + \ln x)'$$

$$= -\frac{1}{2}(1 + \ln x)^{-\frac{3}{2}} \times \frac{1}{x}$$

$$\therefore \frac{dy}{dx} = -\frac{1}{2x(1 + \ln x)^{\frac{3}{2}}}$$

c) Using the product rule,

$$\frac{dy}{dx} = (x)'\ln x + x(\ln x)' = \ln x + x\left(\frac{1}{x}\right)$$

$$\therefore \frac{dy}{dx} = \ln x + 1$$

d) Using the quotient rule,

$$\frac{dy}{dx} = \frac{(\ln 2x)'x^3 - (\ln 2x)(x^3)'}{x^6}$$

$$= \frac{\left(\frac{2}{2x}\right)x^3 - (\ln 2x)3x^2}{x^6}$$

$$= \frac{x^2 - 3x^2 \ln 2x}{x^6}$$

$$\therefore \frac{dy}{dx} = \frac{1 - 3\ln 2x}{x^4}$$

Exercise 11B

1 Find $\dfrac{dy}{dx}$ for each of these functions.

a) $y = \ln(1 + 2x)$

b) $y = \ln(1 \quad 4x)$

c) $y = \ln(1 + x^2)$

d) $y = \ln(x^3 - 2)$

e) $y = \ln(x^3 \quad 3x)$

f) $y = \ln(e^x + 4)$

g) $y = \ln(1 + e^{6x})$

h) $y = \ln(\sqrt{x})$

2 Integrate each of these fractions with respect to x.

a) $\dfrac{1}{1 + x}$

b) $\dfrac{3}{2 + 3x}$

c) $\dfrac{2}{x}$

d) $\dfrac{2}{5 + x}$

e) $\dfrac{4}{2x - 1}$

f) $\dfrac{1}{4 - x}$

g) $\dfrac{3}{5 + 6x}$

h) $\dfrac{5}{2 - 3x}$

3 Integrate each of these fractions with respect to x.

a) $\dfrac{2x}{x^2 + 1}$

b) $\dfrac{3x^2}{x^3 + 4}$

c) $\dfrac{4x}{2 - x^2}$

d) $\dfrac{x^2}{3 + x^3}$

e) $\dfrac{2x - 1}{x^2 - x}$

f) $\dfrac{x - 3}{x^2 - 6x + 1}$

g) $\dfrac{e^x}{1 + e^x}$

h) $\dfrac{e^{5x}}{e^{5x} + 1}$

Exercise 11C

. . . .

This exercise revises the calculus techniques which were developed in earlier chapters, and applies the techniques to the exponential and logarithmic functions.

Tangents and normals
Use your GDC to draw the curves and the tangents and normals as appropriate.

1 Find the equation of the tangent to the curve $y = x + e^{2x}$ at the point where $x = 0$.

2 Find the equation of the tangent and the normal to the curve $y = \ln(1 + x)$ at the point where $x = 2$.

3 Find the equation of the tangent and the normal to the curve $y = xe^x$ at the point where $x = 1$.

4 The tangent to the curve $y = x \ln x$ at the point (e, e) meets the x-axis at A and the y-axis at B. Find the distance AB.

5 Find the equation of the tangent and the normal to the curve $y = e^x \ln x$ at the point where $x = 1$.

6 Find the coordinates of the points on the curve $y = x^2 + \ln x$ where the gradient is 3.

7 Show that there are two points on the curve $y = \ln(1 + x^2)$ where the gradient is $\frac{5}{13}$. Find the coordinates of these points.

8 Find the coordinates of the point on the curve $y = \ln(e^x + e^{-x})$ where the gradient is $\frac{3}{5}$.

Stationary points

9 Given that $y = x^2 e^{-x}$, show that

$$\frac{dy}{dx} = x(2 - x)e^{-x}$$

Hence find the coordinates of the two points on the curve $y = x^2 e^{-x}$ where the gradient is zero.

10 Given that $y = \dfrac{\ln x}{x}$ for $x > 0$, show that

$$\frac{dy}{dx} = \frac{1 - \ln x}{x^2}$$

Hence find the coordinates of the point on the curve

$y = \dfrac{\ln x}{x}$ where the gradient is zero.

11 Given that $y = \dfrac{e^x}{x^2 - 3}$, show that

$$\frac{dy}{dx} = \frac{e^x(x+1)(x-3)}{(x^2-3)^2}$$

Hence find the coordinates of the two points on the curve

$y = \dfrac{e^x}{x^2 - 3}$ where the gradient is zero.

12 Find and classify the stationary values on each of these curves.

a) $y = 2\ln(1 + x) - \ln x \quad (x > 0)$
b) $y = \dfrac{e^x}{x^3}$

c) $y = x(3 - \ln x) \quad (x > 0)$
d) $y = e^x(x - 1)^2$

Areas and volumes of revolution

You should check your answers to the following questions, where possible, by using the area function on your GDC.

13 Find the area between the curve $y = e^{2x}$ and the x-axis from $x = 0$ to $x = 3$.

14 Find the area between the curve $y = \dfrac{2}{x + 3}$ and the x-axis from $x = 2$ to $x = 7$.

15 Find the area between the curve $y = 4e^{2x} - 3e^x$ and the x-axis from $x = 1$ to $x = 2$.

16 The line $y = \tfrac{1}{3}$ meets the curve $y = \dfrac{1}{x + 1}$ at the point P.

a) Find the coordinates of P.

b) Calculate the area bounded by the line, the curve and the y-axis.

17 The line $y = x + 1$ meets the curve $y = \dfrac{8}{5 - x}$ at the points P and Q.

a) Find the coordinates of P and Q.

b) Use your GDC to find the area enclosed between the curve and the line between P and Q, to 3 significant figures.

c) Show that the exact value of this area is $6 - 8\ln 2$.

18 The region bounded by the curve $y = e^x + 1$, the x-axis, the line $x = 0$ and the line $x = 2$ is rotated through $360°$ about the x-axis. Calculate the volume of the solid generated.

19 The region R is bounded by the curve $y = 3 + \dfrac{2}{x + 1}$, the x-axis, the y-axis and the line $x = 4$.

a) Show that the area of R is $12 + 2\ln 5$.

R is rotated through $360°$ about the x-axis.

b) Show that the volume of the solid generated is

$$\frac{4\pi}{5}[49 + 15\ln 5].$$

Exponentials, logarithms and trigonometric functions combined

20 Show that $\dfrac{d}{dx}[\ln \cos x] = -\tan x$

21 Work out these integrals.

a) $\displaystyle\int \dfrac{\cos x}{\sin x}\,dx$

b) $\displaystyle\int \dfrac{\sin x}{\cos x}\,dx$

c) $\displaystyle\int \dfrac{\cos x}{1 + \sin x}\,dx$

d) $\displaystyle\int \dfrac{\sin 2x}{1 + \cos 2x}\,dx$

e) $\displaystyle\int \dfrac{\sin x - \cos x}{\sin x + \cos x}\,dx$

f) $\displaystyle\int \sin x\, e^{\cos x}\,dx$

g) $\displaystyle\int \cos 3x\, e^{\sin 3x}\,dx$

h) $\displaystyle\int 20 \sin 5x\, e^{\cos 5x}\,dx$

22 a) Use your GDC to sketch the curve $y = e^x \cos x$, in the range $0 \leqslant x \leqslant 2\pi$.

  b) Find and classify all the stationary values on the curve $y = e^x \cos x$, in the range $0 \leqslant x \leqslant 2\pi$.

23 a) Given that $y = e^{3x} \sin 2x$, find

  i) $\dfrac{dy}{dx}$ ii) $\dfrac{d^2y}{dx^2}$

  b) Hence show that $\dfrac{d^2y}{dx^2} - 6\dfrac{dy}{dx} + 13y \equiv 0$.

24 Show that $\dfrac{d}{dx}[\ln \tan x] = \dfrac{1}{\sin x \cos x}$

25 Find $\displaystyle\int \dfrac{2 \cos x}{\sin x + \cos x}\,dx$

> **Hint:** First show that $\dfrac{\sin x + \cos x}{\sin x + \cos x} + \dfrac{\cos x - \sin x}{\sin x + \cos x} \equiv \dfrac{2 \cos x}{\sin x + \cos x}$.

....

Summary

You should know how to …

▶ Differentiate functions involving e^x and $\ln x$.

▷ $y = e^x$, $\dfrac{dy}{dx} = e^x$

▷ $y = \ln x$, $\dfrac{dy}{dx} = \dfrac{1}{x}$

▶ Integrate functions involving e^x and $\ln x$.

▷ $\displaystyle\int e^x\,dx = e^x + c$

▷ $\displaystyle\int \dfrac{1}{x}\,dx = \ln x + c$

Revision exercise 11

1 Given that $\displaystyle\int_1^{22} \frac{1}{3x-2}\,dx = \ln b$, find the value of b. © *IBO*[1999]

2 Find, by differentiation, the gradient of the curve $y = \ln(\sin x)$ when $x = \dfrac{\pi}{4}$. © *IBO*[1997]

3 Differentiate with respect to x:

a) $e^x \sin x$

b) $\dfrac{\ln x}{x}$ © *IBO*[1998]

4 Integrate with respect to x:

a) $3e^{2-4x}$

b) $\dfrac{3}{(2-4x)}$

5 The area of the shaded region in the diagram is 6 square units.

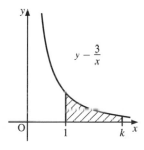

Find the *exact* value of k. © *IBO*[1998]

6 Consider the functions $f(x) = e^x(1 + x^2)$, $g(x) = e^x(1 + x)$.

a) Find $f'(x)$ and $g'(x)$.

b) Show that $g(x)$ has a minimum value at $x = -2$.

c) State the interval over which $g(x)$ is an increasing function.

d) Explain why $f(x)$ is an increasing function for all values of x.

7 A particle moves along a straight line. At time t seconds its displacement from its starting point O is given by $s = t^2 e^{-t}$ metres.

Find its maximum displacement from O and the time at which this occurs.

8 An aircraft lands on a runway. Its velocity v m s^{-1} at time
t seconds after landing is given by the equation $v = 50 + 50e^{-0.5t}$,
where $0 \leqslant t \leqslant 4$.

a) Find the velocity of the aircraft

   i) when it lands   ii)   when $t = 4$.

b) Write down an integral which represents the distance
travelled in the first four seconds.

c) Calculate the distance travelled in the first four seconds.

After four seconds, the aircraft slows down (decelerates) *at a
constant rate* and comes to rest when $t = 11$.

d) *Sketch* a graph of velocity against time for $0 \leqslant t \leqslant 11$.
Clearly label the axes and mark on the graph the point
where $t = 4$.

e) Find the constant rate at which the aircraft is slowing down
between $t = 4$ and $t = 11$.

f) Calculate the distance travelled by the aircraft between
$t = 4$ and $t = 11$.

© *IBO* [*2003*]

9 The diagram shows part of the graph of the curve with
equation $y = e^{2x} \cos x$.

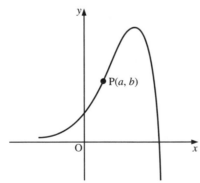

a) Show that $\dfrac{dy}{dx} = e^{2x}(2 \cos x - \sin x)$.

b) Find $\dfrac{d^2y}{dx^2}$.

At the point P (a, b), $\dfrac{d^2y}{dx^2} = 0$.

c) Use the results from parts a) and b) to prove that

   i) $\tan a = \frac{3}{4}$

   ii) the gradient of the curve at P is e^{2a}.

© *IBO* [*2000*]

10 Consider functions of the form $y = e^{-kx}$

a) Show that $\int_0^1 e^{-kx} \, dx = \frac{1}{k}(1 - e^{-k})$

b) Let $k = 0.5$.

   i) Sketch the graph of $y = e^{-0.5x}$, for $-1 \leqslant x \leqslant 3$, indicating the coordinates of the y-intercept.

   ii) Shade the region enclosed by this graph, the x-axis, the y-axis and the line $x = 1$.

   iii) Find the area of this region.

c) i) Find $\dfrac{dy}{dx}$ in terms of k, where $y = e^{-kx}$

   The point $P(1, 0.8)$ lies on the graph of the function $y = e^{-kx}$.

   ii) Find the value of k in this case.

   iii) Find the gradient of the tangent to the curve at P.   © IBO [2002]

11 The region between the curve $y = \dfrac{1}{2x - 1}$ and the lines $x = 1$ and $x = b$ (where $b > 1$) is rotated through 360° about the x-axis.

a) Find, in terms of b, the volume of the solid generated.

b) Show that, for large values of b, this volume is approximately $\dfrac{\pi}{2}$.

12 Matrices

12.1 Introduction to matrices

A matrix is a rectangular array of numbers, written as rows and columns. For example,

$$\begin{pmatrix} 2 & 3 \\ 1 & 7 \end{pmatrix}, \quad \begin{pmatrix} 2 \\ 6 \end{pmatrix}, \quad \begin{pmatrix} 3 & 0 & 3 \\ 2 & 1 & 0 \\ 1 & -2 & 1 \end{pmatrix}$$

The numbers are called the entries.

We usually represent matrices by bold capital letters. For example,

$$A = \begin{pmatrix} 2 & 6 \\ 1 & 3 \end{pmatrix}$$

The size or **order** of a matrix defines its shape. For example, the matrix

$$\begin{pmatrix} 2 \\ 3 \end{pmatrix}$$

has order 2×1 (2 by 1), which means it has 2 rows and 1 column.

$$\begin{pmatrix} 2 & 3 & 0 \\ 1 & -1 & 2 \end{pmatrix}$$

has order 2×3.

Matrices of order $n \times 1$, with just one column, are called **column matrices**. For example,

$$\begin{pmatrix} 4 \\ 1 \end{pmatrix}$$

Matrices of order $1 \times n$, with just one row, are called **row matrices**. For example,

$$(0 \quad 3 \quad 1)$$

Matrices of order $n \times n$, where the numbers of rows and columns are equal, are called **square matrices**. For example,

$$\begin{pmatrix} 5 & 1 \\ -2 & 3 \end{pmatrix}$$

The meaning of a matrix depends on its context.

Zero matrices and identity matrices

A **zero matrix** is a matrix with all entries equal to zero. For example,

$$\begin{pmatrix} 0 & 0 \\ 0 & 0 \end{pmatrix}, \quad \begin{pmatrix} 0 & 0 & 0 \\ 0 & 0 & 0 \end{pmatrix}, \quad \begin{pmatrix} 0 \\ 0 \end{pmatrix}$$

Zero matrices can have any order and are denoted by **0**.

An **identity matrix** is a square matrix with entries of 1 on the leading diagonal and entries of zero everywhere else. For example,

$$\begin{pmatrix} 1 & 0 \\ 0 & 1 \end{pmatrix}, \quad \begin{pmatrix} 1 & 0 & 0 \\ 0 & 1 & 0 \\ 0 & 0 & 1 \end{pmatrix}$$

Identity matrices are denoted by **I**.

> The leading diagonal runs from the top left to the bottom right of a matrix.

Adding and subtracting matrices

You can only add or subtract matrices if they are of the same order. To add or subtract two matrices, add or subtract corresponding entries from each matrix.

For example:

$$\begin{pmatrix} 1 & 4 \\ 3 & 7 \end{pmatrix} + \begin{pmatrix} 8 & 2 \\ 0 & 5 \end{pmatrix} = \begin{pmatrix} 1+8 & 4+2 \\ 3+0 & 7+5 \end{pmatrix} = \begin{pmatrix} 9 & 6 \\ 3 & 12 \end{pmatrix}$$

and:

$$\begin{pmatrix} 5 & 2 & 6 \\ 1 & 4 & 9 \\ 4 & 10 & 3 \end{pmatrix} - \begin{pmatrix} 1 & 0 & 3 \\ 7 & 1 & 8 \\ 0 & 5 & 2 \end{pmatrix} = \begin{pmatrix} 5-1 & 2-0 & 6-3 \\ 1-7 & 4-1 & 9-8 \\ 4-0 & 10-5 & 3-2 \end{pmatrix} = \begin{pmatrix} 4 & 2 & 3 \\ -6 & 3 & 1 \\ 4 & 5 & 1 \end{pmatrix}$$

Note that the two matrices

$$\begin{pmatrix} 3 \\ 4 \end{pmatrix}, \quad \begin{pmatrix} 0 & 5 \\ 1 & 8 \end{pmatrix}$$

cannot be added, as they are not the same size.

Multiplying a matrix by a number

To multiply a matrix by a number, multiply every entry of the matrix by the number. For example,

$$k\begin{pmatrix} a & b \\ c & d \end{pmatrix} = \begin{pmatrix} ka & kb \\ kc & kd \end{pmatrix}$$

Example 1

Given that $M = \begin{pmatrix} 4 & 7 \\ 2 & 1 \end{pmatrix}$, find $3M$.

$$3M = 3\begin{pmatrix} 4 & 7 \\ 2 & 1 \end{pmatrix} = \begin{pmatrix} 3 \times 4 & 3 \times 7 \\ 3 \times 2 & 3 \times 1 \end{pmatrix}$$

$$\therefore 3M = \begin{pmatrix} 12 & 21 \\ 6 & 3 \end{pmatrix}$$

Example 2

Given that $A = \begin{pmatrix} -1 & 4 \\ 0 & 5 \end{pmatrix}$ and $B = \begin{pmatrix} 3 & 6 \\ 2 & -2 \end{pmatrix}$, calculate

a) $A + B$ 　　　　b) $2A - B$ 　　　　c) $3A - 2B$

a) A and B are the same size, so you can add them:

$$A + B = \begin{pmatrix} -1 & 4 \\ 0 & 5 \end{pmatrix} + \begin{pmatrix} 3 & 6 \\ 2 & -2 \end{pmatrix}$$

$$= \begin{pmatrix} 2 & 10 \\ 2 & 3 \end{pmatrix}$$

b) $2A = 2\begin{pmatrix} -1 & 4 \\ 0 & 5 \end{pmatrix} = \begin{pmatrix} -2 & 8 \\ 0 & 10 \end{pmatrix}$

$2A$ and B are the same size, so you can subtract:

$$2A - B = \begin{pmatrix} -2 & 8 \\ 0 & 10 \end{pmatrix} - \begin{pmatrix} 3 & 6 \\ 2 & -2 \end{pmatrix}$$

$$= \begin{pmatrix} -5 & 2 \\ -2 & 12 \end{pmatrix}$$

c) 　　　$3A = 3\begin{pmatrix} -1 & 4 \\ 0 & 5 \end{pmatrix} = \begin{pmatrix} -3 & 12 \\ 0 & 15 \end{pmatrix}$

and 　　$2B = 2\begin{pmatrix} 3 & 6 \\ 2 & -2 \end{pmatrix} = \begin{pmatrix} 6 & 12 \\ 4 & -4 \end{pmatrix}$

$3A$ and $2B$ are the same size, so you can subtract:

$$3A - 2B = \begin{pmatrix} -3 & 12 \\ 0 & 15 \end{pmatrix} - \begin{pmatrix} 6 & 12 \\ 4 & -4 \end{pmatrix} = \begin{pmatrix} -9 & 0 \\ -4 & 19 \end{pmatrix}$$

Exercise 12A

1 Calculate:

a) $\begin{pmatrix} 2 & 3 \\ -1 & 5 \end{pmatrix} + \begin{pmatrix} 7 & 0 \\ 2 & -8 \end{pmatrix}$ 　　　　b) $\begin{pmatrix} 6 & -3 \\ 1 & 5 \end{pmatrix} + \begin{pmatrix} 0 & 3 \\ 2 & 4 \end{pmatrix}$

c) $\begin{pmatrix} 2 & -3 \\ -4 & 0 \end{pmatrix} + \begin{pmatrix} 5 & 1 \\ 2 & -3 \end{pmatrix}$ d) $\begin{pmatrix} 4 & -4 \\ 1 & 7 \end{pmatrix} + \begin{pmatrix} 2 & -6 \\ -3 & 5 \end{pmatrix}$

e) $\begin{pmatrix} 2 & 6 \\ 1 & -5 \end{pmatrix} - \begin{pmatrix} 3 & -6 \\ 2 & 1 \end{pmatrix}$ f) $\begin{pmatrix} 5 & -1 \\ 0 & 8 \end{pmatrix} - \begin{pmatrix} -4 & 1 \\ -3 & 2 \end{pmatrix}$

g) $\begin{pmatrix} 2 & -1 & 3 \\ -5 & 0 & 4 \end{pmatrix} + \begin{pmatrix} 3 & 2 & -5 \\ 2 & -3 & 0 \end{pmatrix}$

h) $\begin{pmatrix} 0 & 4 \\ -3 & -1 \\ 2 & 5 \end{pmatrix} - \begin{pmatrix} -2 & 0 \\ 3 & 5 \\ -6 & 2 \end{pmatrix}$

i) $\begin{pmatrix} 2 & 5 \\ 0 & 1 \end{pmatrix} + \begin{pmatrix} -4 & 3 \\ -1 & 6 \end{pmatrix} - \begin{pmatrix} -2 & 3 \\ 4 & -1 \end{pmatrix}$

j) $\begin{pmatrix} 5 & 6 & -2 \\ 7 & -1 & 0 \\ 3 & -8 & 3 \end{pmatrix} - \begin{pmatrix} -4 & 2 & 1 \\ 2 & -3 & 8 \\ 4 & 1 & -5 \end{pmatrix} - \begin{pmatrix} 0 & 2 & -3 \\ 5 & 1 & 0 \\ -3 & 4 & 1 \end{pmatrix}$

2 Given $A = \begin{pmatrix} 2 & 0 \\ -1 & 5 \end{pmatrix}$, $B = \begin{pmatrix} -1 & 6 \\ 5 & -2 \end{pmatrix}$ and $C = \begin{pmatrix} 2 & -3 \\ -1 & 0 \end{pmatrix}$,

calculate:

a) $A + B$ b) $2A$ c) $3A + B$

d) $2B - 3C$ e) $A - B + C$ f) $4A - 3B + 2C$

3 Given $A = \begin{pmatrix} 1 & -3 \\ -2 & 0 \\ 4 & 5 \end{pmatrix}$, $B = \begin{pmatrix} 7 & -2 \\ -1 & 4 \\ 3 & 0 \end{pmatrix}$, $C = \begin{pmatrix} 2 \\ 1 \\ -5 \end{pmatrix}$ and $D = \begin{pmatrix} 0 \\ 3 \\ 2 \end{pmatrix}$,

calculate where possible:

a) $A + B$ b) $C - D$ c) $3A$ d) $4C$

e) $D + C$ f) $B - 2A$ g) $5D - 4B$ h) $6D - 5C$

4 Given that $A = \begin{pmatrix} 1 & 3 & -1 \\ 0 & 2 & -5 \\ 3 & -6 & 4 \end{pmatrix}$, $B = \begin{pmatrix} 8 & 3 & 7 \\ 3 & 1 & 2 \\ 0 & -6 & -7 \end{pmatrix}$ and

$C = \begin{pmatrix} 2 & -1 & 3 \\ 1 & -1 & 4 \\ -2 & 3 & -5 \end{pmatrix}$, show that $2A - B + 3C = 0$.

5 Given that $\begin{pmatrix} 2 & 3 \\ y & z \end{pmatrix} + \begin{pmatrix} 5 & x \\ 3 & z \end{pmatrix} = \begin{pmatrix} w & -2 \\ 0 & 8 \end{pmatrix}$, calculate the values of

w, x, y and z.

6 Given that $\begin{pmatrix} p & 4 \\ q & p \end{pmatrix} + \begin{pmatrix} -2 & 1 \\ r & p \end{pmatrix} = \begin{pmatrix} 4 & q \\ -1 & s \end{pmatrix}$, calculate the values of

p, q, r and s.

7 Given that $\begin{pmatrix} a & 2a \\ 2c & 3c \end{pmatrix} + \begin{pmatrix} 2b & b \\ 4d & 5d \end{pmatrix} = \begin{pmatrix} 1 & 5 \\ 2 & 5 \end{pmatrix}$, use simultaneous

equations to calculate the values of a, b, c and d.

8 Given that $\begin{pmatrix} 3p & p \\ 2r & 4r \end{pmatrix} + \begin{pmatrix} 2q & -q \\ 5s & 3s \end{pmatrix} = \begin{pmatrix} 8 & 1 \\ 1 & 9 \end{pmatrix}$, use simultaneous

equations to calculate the values of p, q, r and s.

12.2 Matrix multiplication

The football results from three teams can be represented by the following matrix:

$$
\begin{array}{c}
\\
\text{Team A} \\
\text{Team B} \\
\text{Team C}
\end{array}
\begin{array}{ccc}
\text{Won} & \text{Drawn} & \text{Lost} \\
\begin{pmatrix} 3 & 2 & 3 \\ 4 & 2 & 2 \\ 2 & 1 & 5 \end{pmatrix}
\end{array}
$$

Row 1 shows that team A has won 3 games, drawn 2 games and lost 3 games.

If 3 points are awarded for a win,
   1 point is awarded for a draw and
   0 points are awarded for a loss

then   Team A is awarded $3 \times 3 + 2 \times 1 + 3 \times 0 = 11$ points
likewise Team B is awarded $4 \times 3 + 2 \times 1 + 2 \times 0 = 14$ points
and   Team C is awarded $2 \times 3 + 1 \times 1 + 5 \times 0 = 7$ points.

In matrix form, this can be represented as

$$
\begin{pmatrix} 3 & 2 & 3 \\ 4 & 2 & 2 \\ 2 & 1 & 5 \end{pmatrix} \begin{pmatrix} 3 \\ 1 \\ 0 \end{pmatrix} = \begin{pmatrix} 11 \\ 14 \\ 7 \end{pmatrix}
$$

This is an example of matrix multiplication. Notice that there are three columns in the first matrix and three rows in the second matrix.

> You can only multiply two matrices if the number of columns in the first matrix is the same as the number of rows in the second matrix.

For example,

$$
A = \begin{pmatrix} 3 \\ 2 \end{pmatrix} \text{ and } B = \begin{pmatrix} 1 & 5 \\ 0 & 4 \end{pmatrix}
$$

Matrix A has order 2×1, that is, 1 column.

Matrix B has order 2×2, that is, 2 rows.

So you cannot find the product AB.

Consider the two matrices

$$
M = \begin{pmatrix} 1 & 5 \\ 2 & 0 \end{pmatrix} \text{ and } N = \begin{pmatrix} 7 & 3 \\ 4 & 6 \end{pmatrix}
$$

Matrix M has order 2×2, that is, 2 columns.

Matrix N has order 2×2, that is, 2 rows.

So you *can* find the product **MN**.

$$\mathbf{MN} = \begin{pmatrix} 1 & 5 \\ 2 & 0 \end{pmatrix} \begin{pmatrix} 7 & 3 \\ 4 & 6 \end{pmatrix} = \begin{pmatrix} a & b \\ c & d \end{pmatrix}$$

❖ Start with the first row of matrix **M**, (1 5), and the first column of matrix **N**, $\begin{pmatrix} 7 \\ 4 \end{pmatrix}$.

Multiply the first element of the row by the first element of the column, multiply the second element of the row by the second element of the column, and add them:

$$1 \times 7 + 5 \times 4 = 27$$

This is the first row, first column entry *a*, in the matrix **MN**.

$$\mathbf{MN} = \begin{pmatrix} 27 & b \\ c & d \end{pmatrix}$$

❖ To calculate the entry *b*, stay with the first row of matrix **M**, (1 5), but now consider the second column of matrix **N**, $\begin{pmatrix} 3 \\ 6 \end{pmatrix}$.

Multiply the first element of the row by the first element of the column, multiply the second element of the row by the second element of the column, and add them:

$$1 \times 3 + 5 \times 6 = 33$$

This is the first row, first column entry *b*, in the matrix **MN**.

$$\mathbf{MN} = \begin{pmatrix} 27 & 33 \\ c & d \end{pmatrix}$$

❖ To calculate the entry *c*, start with the second row of matrix **M**, (2 0), and the first column of matrix **N**, $\begin{pmatrix} 7 \\ 4 \end{pmatrix}$.

Multiply the first element of the row by the first element of the column, multiply the second element of the row by the second element of the column, and add them:

$$2 \times 7 + 0 \times 4 = 14$$

This is the second row, first column entry *c*, in the matrix **MN**.

$$\mathbf{MN} = \begin{pmatrix} 27 & 33 \\ 14 & d \end{pmatrix}$$

❖ To calculate the entry *d*, stay with the second row of matrix **M**, (2 0), but now consider the second column of matrix **N**, $\begin{pmatrix} 3 \\ 6 \end{pmatrix}$.

Multiply the first element of the row by the first element of the column, multiply the second element of the row by the second element of the column, and add them:

$$2 \times 3 + 0 \times 6 = 6$$

This is the second row, second column entry *d*, in the matrix **MN**.

$$\mathbf{MN} = \begin{pmatrix} 27 & 33 \\ 14 & 6 \end{pmatrix}$$

To summarise:

$$MN = \begin{pmatrix} 1 & 5 \\ 2 & 0 \end{pmatrix}\begin{pmatrix} 7 & 3 \\ 4 & 6 \end{pmatrix}$$

$$= \begin{pmatrix} 1 \times 7 + 5 \times 4 & 1 \times 3 + 5 \times 6 \\ 2 \times 7 + 0 \times 4 & 2 \times 3 + 0 \times 6 \end{pmatrix}$$

$$\therefore MN = \begin{pmatrix} 27 & 33 \\ 14 & 6 \end{pmatrix}$$

> **M** is a 2 × 2 matrix.
> **N** is a 2 × 2 matrix.
> So **MN** is also a 2 × 2 matrix.

Generally, the matrix product **AB** will produce a matrix with the same number of rows as **A** and the same number of columns as **B**.

Example 1

Given that $A = \begin{pmatrix} 5 & 1 \\ 3 & 2 \end{pmatrix}$ and $B = \begin{pmatrix} 2 \\ 6 \end{pmatrix}$, find **AB**.

....

Since **A** has order 2 × 2 and **B** has order 2 × 1, the product **AB** can be found and will be a 2 × 1 matrix.

$$AB = \begin{pmatrix} 5 & 1 \\ 3 & 2 \end{pmatrix}\begin{pmatrix} 2 \\ 6 \end{pmatrix}$$

$$= \begin{pmatrix} 5 \times 2 + 1 \times 6 \\ 3 \times 2 + 2 \times 6 \end{pmatrix}$$

$$\therefore AB = \begin{pmatrix} 16 \\ 18 \end{pmatrix}$$

> These matrices can be multiplied:
> 2 × 2   2 × 1
> The product is a 2 × 1 matrix.

Example 2

Given that $A = \begin{pmatrix} 3 & 1 \\ 0 & -2 \\ 1 & 4 \end{pmatrix}$ and $B = \begin{pmatrix} 1 & 7 \\ 2 & 6 \end{pmatrix}$, find **AB** and **BA**, if possible.

....

Since **A** has order 3 × 2 and **B** has order 2 × 2, the product **AB** can be found and will be a 3 × 2 matrix.

$$AB = \begin{pmatrix} 3 & 1 \\ 0 & -2 \\ 1 & 4 \end{pmatrix}\begin{pmatrix} 1 & 7 \\ 2 & 6 \end{pmatrix}$$

$$= \begin{pmatrix} 3 \times 1 + 1 \times 2 & 3 \times 7 + 1 \times 6 \\ 0 \times 1 - 2 \times 2 & 0 \times 7 - 2 \times 6 \\ 1 \times 1 + 4 \times 2 & 1 \times 7 + 4 \times 6 \end{pmatrix}$$

$$= \begin{pmatrix} 5 & 27 \\ -4 & -12 \\ 9 & 31 \end{pmatrix}$$

> These matrices can be multiplied:
> 3 × 2   2 × 2
> The product is a 3 × 2 matrix.

Since **B** has order 2 × 2 and **A** has order 3 × 2, the product **BA** cannot be found as the number of columns of **B** is not the same as the number of rows of **A**.

> These matrices cannot be multiplied:
> 2 × 2   3 × 2

You can also use a GDC to multiply matrices.

Example 3

Given that $A = \begin{pmatrix} 3 & 1 & 5 \\ -1 & 2 & 0 \\ 4 & 6 & 3 \end{pmatrix}$ and $B = \begin{pmatrix} 0 & -2 & 3 \\ 1 & 5 & 4 \\ 7 & 2 & 1 \end{pmatrix}$, find **AB**.

Use your GDC to find **BA** and comment on your result.

Since **A** has order 3×3 and **B** has order 3×3, you can calculate both the products **AB** and **BA**. Each will be a 3×3 matrix.

$$AB = \begin{pmatrix} 3 & 1 & 5 \\ -1 & 2 & 0 \\ 4 & 6 & 3 \end{pmatrix}\begin{pmatrix} 0 & -2 & 3 \\ 1 & 5 & 4 \\ 7 & 2 & 1 \end{pmatrix}$$

$$= \begin{pmatrix} 3\times0+1\times1+5\times7 & 3\times-2+1\times5+5\times2 & 3\times3+1\times4+5\times1 \\ -1\times0+2\times1+0\times7 & -1\times-2+2\times5+0\times2 & -1\times3+2\times4+0\times1 \\ 4\times0+6\times1+3\times7 & 4\times-2+6\times5+3\times2 & 4\times3+6\times4+3\times1 \end{pmatrix}$$

$$\therefore AB = \begin{pmatrix} 36 & 9 & 18 \\ 2 & 12 & 5 \\ 27 & 28 & 39 \end{pmatrix}$$

$$BA = \begin{pmatrix} 0 & -2 & 3 \\ 1 & 5 & 4 \\ 7 & 2 & 1 \end{pmatrix}\begin{pmatrix} 3 & 1 & 5 \\ -1 & 2 & 0 \\ 4 & 6 & 3 \end{pmatrix}$$

$$= \begin{pmatrix} 14 & 14 & 9 \\ 14 & 35 & 17 \\ 23 & 17 & 38 \end{pmatrix}$$

Notice that although in Example 3 both products **AB** and **BA** exist, they are *different*. In other words $AB \neq BA$.

In general, the multiplication of two matrices is not commutative. That is, for two matrices **A** and **B**

$$AB \neq BA$$

Example 4

Given that $A = \begin{pmatrix} -2 & 1 \\ 3 & 7 \end{pmatrix}$, use your GDC to find A^2.

Show that $A^2 - 5A - 17I = 0$.

$$A^2 = \begin{pmatrix} -2 & 1 \\ 3 & 7 \end{pmatrix}\begin{pmatrix} -2 & 1 \\ 3 & 7 \end{pmatrix}$$

$$= \begin{pmatrix} 7 & 5 \\ 15 & 52 \end{pmatrix}$$

and

$$5A = 5\begin{pmatrix} -2 & 1 \\ 3 & 7 \end{pmatrix} = \begin{pmatrix} -10 & 5 \\ 15 & 35 \end{pmatrix}$$

and

$$17I = 17\begin{pmatrix} 1 & 0 \\ 0 & 1 \end{pmatrix} = \begin{pmatrix} 17 & 0 \\ 0 & 17 \end{pmatrix}$$

Therefore,

$$A^2 - 5A - 17I = \begin{pmatrix} 7 & 5 \\ 15 & 52 \end{pmatrix} - \begin{pmatrix} -10 & 5 \\ 15 & 35 \end{pmatrix} - \begin{pmatrix} 17 & 0 \\ 0 & 17 \end{pmatrix}$$

$$= \begin{pmatrix} 0 & 0 \\ 0 & 0 \end{pmatrix}$$

$$\therefore A^2 - 5A - 17I = 0, \text{ as required.}$$

Exercise 12B

1 Calculate each of these matrix products.

a) $\begin{pmatrix} 2 & 1 \\ 5 & 3 \end{pmatrix}\begin{pmatrix} 1 & -3 \\ 2 & 0 \end{pmatrix}$

b) $\begin{pmatrix} 5 & -1 \\ 2 & 4 \end{pmatrix}\begin{pmatrix} 1 & 3 \\ -3 & 8 \end{pmatrix}$

c) $\begin{pmatrix} 3 & -1 \\ 0 & 5 \end{pmatrix}\begin{pmatrix} 6 & 2 \\ -4 & 1 \end{pmatrix}$

d) $\begin{pmatrix} 1 & 1 \\ 5 & 8 \end{pmatrix}\begin{pmatrix} -2 & 9 \\ 3 & 2 \end{pmatrix}$

e) $\begin{pmatrix} 5 & 2 \\ -4 & 3 \end{pmatrix}\begin{pmatrix} 0 & 5 \\ 1 & -6 \end{pmatrix}$

f) $\begin{pmatrix} 4 & -3 \\ 0 & 5 \end{pmatrix}\begin{pmatrix} 1 & -6 \\ 3 & 2 \end{pmatrix}$

g) $\begin{pmatrix} 2 & 1 \\ -1 & 7 \end{pmatrix}\begin{pmatrix} 0 & -3 \\ -5 & 8 \end{pmatrix}$

h) $\begin{pmatrix} 4 & -3 \\ 2 & 1 \end{pmatrix}\begin{pmatrix} 6 & -7 \\ 5 & 4 \end{pmatrix}$

i) $\begin{pmatrix} 1 & 2 & 0 \\ 3 & 2 & 2 \\ 1 & -1 & 3 \end{pmatrix}\begin{pmatrix} 1 & 0 & -1 \\ 3 & 1 & 2 \\ -2 & 1 & 0 \end{pmatrix}$

j) $\begin{pmatrix} 2 & 0 & 3 \\ -1 & 1 & 2 \\ 3 & 0 & 0 \end{pmatrix}\begin{pmatrix} 4 & 1 & 2 \\ 0 & -2 & 2 \\ 3 & 1 & -1 \end{pmatrix}$

k) $\begin{pmatrix} 0 & 1 & 1 \\ 2 & -2 & 2 \\ 0 & 1 & 4 \end{pmatrix}\begin{pmatrix} -2 & 1 & 0 \\ 3 & 3 & 1 \\ 1 & 0 & -1 \end{pmatrix}$

l) $\begin{pmatrix} 1 & 1 & 1 \\ 3 & -2 & -1 \\ 0 & 0 & 3 \end{pmatrix}\begin{pmatrix} 0 & -2 & 4 \\ 1 & 0 & 1 \\ -2 & 1 & 3 \end{pmatrix}$

> You should be able to perform matrix multiplication using the row/column method and using your GDC.

2 Given $A = \begin{pmatrix} 2 & 1 \\ 4 & -1 \end{pmatrix}$, $B = \begin{pmatrix} 1 & -2 & 3 \\ 0 & 2 & 4 \end{pmatrix}$, $C = \begin{pmatrix} 5 & 1 \\ 0 & -2 \\ 1 & 3 \end{pmatrix}$, $D = \begin{pmatrix} 3 \\ -2 \\ 2 \end{pmatrix}$

and $E = \begin{pmatrix} 1 & 0 & 2 \\ -3 & 1 & 0 \\ 0 & 2 & 4 \end{pmatrix}$, calculate where possible:

a) AB b) BC c) CB d) A^2

e) AC f) BD g) C^2 h) ED

i) CD j) E^2 k) BE l) CA

3 Use your GDC to calculate these matrix products.

a) $\begin{pmatrix} 2 & 5 \\ 9 & 3 \end{pmatrix}\begin{pmatrix} 4 & -7 \\ 2 & 6 \end{pmatrix}$

b) $\begin{pmatrix} 6 & -3 \\ 5 & 2 \end{pmatrix}\begin{pmatrix} 1 & 6 & 9 \\ -4 & 3 & -2 \end{pmatrix}$

c) $\begin{pmatrix} 5 & -1 & 3 \\ 2 & 9 & -4 \\ 3 & 0 & 1 \end{pmatrix}\begin{pmatrix} 6 & -3 \\ -2 & 1 \\ 4 & 9 \end{pmatrix}$

d) $\begin{pmatrix} 6 & -1 \\ 3 & 2 \end{pmatrix}\begin{pmatrix} 4 & 5 & -3 & 0 \\ 2 & 6 & 2 & -7 \end{pmatrix}$

e) $\begin{pmatrix} 2 & 1 & 4 \\ -3 & 2 & 5 \\ 1 & 0 & 6 \end{pmatrix}\begin{pmatrix} -3 & 5 & 6 \\ 2 & -4 & 8 \\ 3 & 7 & 9 \end{pmatrix}$

f) $\begin{pmatrix} 5 & -3 & 4 \\ 4 & 0 & -7 \\ -3 & 2 & 2 \end{pmatrix}\begin{pmatrix} 6 & 1 & -4 \\ 5 & 0 & 3 \\ -7 & 1 & 2 \end{pmatrix}$

4 Given that $A = \begin{pmatrix} 2 & 1 \\ 4 & 3 \end{pmatrix}$ and $I = \begin{pmatrix} 1 & 0 \\ 0 & 1 \end{pmatrix}$,

show that $A^2 - 5A + 2I = 0$.

5 a) Given that $A = \begin{pmatrix} 1 & 2 & -4 \\ 0 & 2 & 1 \\ 3 & -1 & 0 \end{pmatrix}$, use your GDC to find A^2.

b) Given that $B = A^2 - 3A + 15I$, where $I = \begin{pmatrix} 1 & 0 & 0 \\ 0 & 1 & 0 \\ 0 & 0 & 1 \end{pmatrix}$, find B.

c) Use your GDC to find AB, and hence state the value of the constant k, such that $AB = kI$.

6 a) Given that $A = \begin{pmatrix} 2 & 1 & 0 \\ 3 & 5 & 2 \\ -2 & 0 & 3 \end{pmatrix}$, use your GDC to find A^2 and A^3.

b) By first finding $A^3 - 10A^2 + 28A$, show that
$A^3 - 10A^2 + 28A = kI$, where k is a constant.
State the value of k.

....

12.3 2 × 2 matrices

You can write a general 2 × 2 matrix as:

$\begin{pmatrix} a & b \\ c & d \end{pmatrix}$

The 2 × 2 zero matrix is

$\begin{pmatrix} 0 & 0 \\ 0 & 0 \end{pmatrix}$

and the 2 × 2 identity matrix is

$\begin{pmatrix} 1 & 0 \\ 0 & 1 \end{pmatrix}$

The determinant of a 2 × 2 matrix

The **determinant** of the 2 × 2 matrix

$$A = \begin{pmatrix} a & b \\ c & d \end{pmatrix}$$

is defined as

$$ad - bc$$

and is denoted by det $A = |A|$. In other words,

$$\det A = |A| = \begin{vmatrix} a & b \\ c & d \end{vmatrix} = ad - bc$$

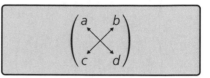

Example 1

Given that $A = \begin{pmatrix} 2 & 4 \\ 6 & 1 \end{pmatrix}$, find det A.

....

$$\det A = \begin{vmatrix} 2 & 4 \\ 6 & 1 \end{vmatrix}$$

$$= 2 \times 1 - 4 \times 6$$

$$= -22$$

Example 2

Given that $A = \begin{pmatrix} -1 & 2 \\ 10 & 1 \end{pmatrix}$ and $B = \begin{pmatrix} 0 & -3 \\ 1 & 5 \end{pmatrix}$, find:

a) det A

b) det B.

c) Use the GDC to find AB and hence show that

$$\det(AB) = \det A \times \det B$$

....

a) $\det A = \begin{vmatrix} -1 & 2 \\ 10 & 1 \end{vmatrix}$

$$= -1 \times 1 - 2 \times 10$$

$$= -21$$

b) $\det B = \begin{vmatrix} 0 & -3 \\ 1 & 5 \end{vmatrix}$

$$= 0 \times 5 - (-3 \times 1)$$

$$= 3$$

c) Using the GDC to find **AB** gives

$$AB = \begin{pmatrix} -1 & 2 \\ 10 & 1 \end{pmatrix}\begin{pmatrix} 0 & -3 \\ 1 & 5 \end{pmatrix}$$

$$= \begin{pmatrix} 2 & 13 \\ 1 & -25 \end{pmatrix}$$

Therefore,

$$\det AB = \begin{vmatrix} 2 & 13 \\ 1 & -25 \end{vmatrix}$$

$$= -63$$

From parts a) and b) you see that

$$\det A \times \det B = -21 \times 3$$

$$= -63$$

$$= \det(AB), \text{ as required.}$$

In general, the determinant of the product **AB** is equal to the product of the determinant of **A** and the determinant of **B**.

Inverse matrices

If **A** is a square matrix, its inverse, denoted by A^{-1}, is defined by

$$AA^{-1} = A^{-1}A = I$$

where **I** is the identity matrix.
The inverse of a square matrix **A** exists providing $\det A \neq 0$.

The inverse of the 2×2 matrix

$$A = \begin{pmatrix} a & b \\ c & d \end{pmatrix}$$

is given by

$$A^{-1} = \frac{1}{\det A}\begin{pmatrix} d & -b \\ -c & a \end{pmatrix}, \quad \det A \neq 0.$$

a and d change places.
b and c stay in the same place but change sign.

Example 3

Given that $P = \begin{pmatrix} 3 & 1 \\ 7 & 5 \end{pmatrix}$, find P^{-1}.

Use a GDC to show that $P^{-1}P = PP^{-1} = I$.

First calculate the determinant:

$$\det P = 3 \times 5 - 1 \times 7 = 8$$

Since $\det P \neq 0$ you know that P^{-1} exists.

Using the formula for P^{-1} gives

$$P^{-1} = \frac{1}{8}\begin{pmatrix} 5 & -1 \\ -7 & 3 \end{pmatrix} = \begin{pmatrix} \frac{5}{8} & -\frac{1}{8} \\ -\frac{7}{8} & \frac{3}{8} \end{pmatrix}$$

Using the GDC to calculate $P^{-1}P$ gives

$$P^{-1}P = \begin{pmatrix} \frac{5}{8} & -\frac{1}{8} \\ -\frac{7}{8} & \frac{3}{8} \end{pmatrix}\begin{pmatrix} 3 & 1 \\ 7 & 5 \end{pmatrix} = \begin{pmatrix} 1 & 0 \\ 0 & 1 \end{pmatrix} = I$$

Using the GDC to calculate PP^{-1} gives

$$PP^{-1} = \begin{pmatrix} 3 & 1 \\ 7 & 5 \end{pmatrix}\begin{pmatrix} \frac{5}{8} & -\frac{1}{8} \\ -\frac{7}{8} & \frac{3}{8} \end{pmatrix} = \begin{pmatrix} 1 & 0 \\ 0 & 1 \end{pmatrix} = I$$

You can use the inverse of a matrix to solve a matrix equation.

Example 4

Given that $P = \begin{pmatrix} 2 & 1 \\ 5 & 3 \end{pmatrix}$ and $Q = \begin{pmatrix} -4 & 0 \\ 1 & 1 \end{pmatrix}$, find P^{-1}.

Hence find the matrix A such that $PA = Q$.

....

Calculate the determinant of P:

$$\det P = 2 \times 3 - 1 \times 5 = 1$$

Since $\det P \neq 0$ you know that P^{-1} exists.

Using the formula for P^{-1} gives

$$P^{-1} = \frac{1}{1}\begin{pmatrix} 3 & -1 \\ -5 & 2 \end{pmatrix} = \begin{pmatrix} 3 & -1 \\ -5 & 2 \end{pmatrix}$$

You know that

$$PA = Q$$

and you need to solve this equation for A.

Multiply both sides by the inverse of P:

$$P^{-1}PA = P^{-1}Q$$
$$(P^{-1}P)A = P^{-1}Q$$

But $P^{-1}P = I$, so:

$$IA = P^{-1}Q$$
$$\therefore A = P^{-1}Q$$

Using the matrices P^{-1} and Q gives

$$A = \begin{pmatrix} 3 & -1 \\ -5 & 2 \end{pmatrix}\begin{pmatrix} -4 & 0 \\ 1 & 1 \end{pmatrix}$$

$$= \begin{pmatrix} -13 & -1 \\ 22 & 2 \end{pmatrix}$$

Solving two equations in two unknowns

Consider the simultaneous equations

$$3x + 4y = 13$$
$$2x + 3y = 9$$

Using matrices, you can rewrite these equations as

$$\begin{pmatrix} 3 & 4 \\ 2 & 3 \end{pmatrix} \begin{pmatrix} x \\ y \end{pmatrix} = \begin{pmatrix} 13 \\ 9 \end{pmatrix}$$

or

$$MX = C$$

where $M = \begin{pmatrix} 3 & 4 \\ 2 & 3 \end{pmatrix}$, $X = \begin{pmatrix} x \\ y \end{pmatrix}$ and $C = \begin{pmatrix} 13 \\ 9 \end{pmatrix}$.

If you multiply both sides of the matrix equation $MX = C$ by M^{-1} (we will assume that the inverse matrix exists), you get:

$$M^{-1}MX = M^{-1}C$$
$$(M^{-1}M)X = M^{-1}C$$

Since $M^{-1}M = I$,

$$IX = M^{-1}C$$
$$\therefore X = M^{-1}C$$

$M = \begin{pmatrix} 3 & 4 \\ 2 & 3 \end{pmatrix}$. Calculating the determinant gives

$$\det M = 3 \times 3 - 4 \times 2 = 1$$

Using the formula for 2×2 inverse matrices gives

$$M^{-1} = \frac{1}{1}\begin{pmatrix} 3 & -4 \\ -2 & 3 \end{pmatrix} = \begin{pmatrix} 3 & -4 \\ -2 & 3 \end{pmatrix}$$

So

$$X = M^{-1}C$$

becomes

$$X = \begin{pmatrix} 3 & -4 \\ -2 & 3 \end{pmatrix}\begin{pmatrix} 13 \\ 9 \end{pmatrix}$$

$$= \begin{pmatrix} 3 \\ 1 \end{pmatrix}$$

Since $X = \begin{pmatrix} x \\ y \end{pmatrix}$,

$$\begin{pmatrix} x \\ y \end{pmatrix} = \begin{pmatrix} 3 \\ 1 \end{pmatrix}$$

In other words $x = 3$ and $y = 1$, the solutions to the original simultaneous equations.

Example 5

Given the simultaneous equations

$$2x + y = 3$$
$$8x + 5y = 11$$

write them in the matrix form

$$MX = C$$

where M is a 2×2 matrix and X and C are column matrices.

By calculating M^{-1} solve the simultaneous equations.

· ·

Writing the equations in matrix form gives

$$\begin{pmatrix} 2 & 1 \\ 8 & 5 \end{pmatrix}\begin{pmatrix} x \\ y \end{pmatrix} = \begin{pmatrix} 3 \\ 11 \end{pmatrix}$$

where $M = \begin{pmatrix} 2 & 1 \\ 8 & 5 \end{pmatrix}$, $X = \begin{pmatrix} x \\ y \end{pmatrix}$ and $C = \begin{pmatrix} 3 \\ 11 \end{pmatrix}$.

You know that

$$\begin{pmatrix} x \\ y \end{pmatrix} = M^{-1}C$$

To find M^{-1}, first calculate the determinant of M:

$$\det M = 2 \times 5 - 1 \times 8 = 2$$

Using the formula for the inverse of a 2×2 matrix gives

$$M^{-1} = \frac{1}{2}\begin{pmatrix} 5 & -1 \\ -8 & 2 \end{pmatrix} = \begin{pmatrix} \frac{5}{2} & -\frac{1}{2} \\ -4 & 1 \end{pmatrix}$$

Therefore,

$$\begin{pmatrix} x \\ y \end{pmatrix} = \begin{pmatrix} \frac{5}{2} & -\frac{1}{2} \\ -4 & 1 \end{pmatrix}\begin{pmatrix} 3 \\ 11 \end{pmatrix}$$
$$= \begin{pmatrix} 2 \\ -1 \end{pmatrix}$$

The solution is $x = 2$ and $y = -1$.

Exercise 12C

· ·

1 Find the determinant of each of these matrices.

a) $\begin{pmatrix} 5 & 2 \\ 1 & 3 \end{pmatrix}$ b) $\begin{pmatrix} 2 & -1 \\ 3 & 4 \end{pmatrix}$ c) $\begin{pmatrix} 5 & 6 \\ 2 & -1 \end{pmatrix}$

d) $\begin{pmatrix} -4 & 3 \\ 0 & -2 \end{pmatrix}$ e) $\begin{pmatrix} -2 & 8 \\ 3 & 1 \end{pmatrix}$ f) $\begin{pmatrix} 6 & 6 \\ -2 & -3 \end{pmatrix}$

g) $\begin{pmatrix} 4 & 5 \\ -3 & 0 \end{pmatrix}$ h) $\begin{pmatrix} -7 & 4 \\ 5 & 3 \end{pmatrix}$

2 a) For $A = \begin{pmatrix} x & 2 \\ 5 & 3 \end{pmatrix}$, find an expression for det A in terms of x.

   b) Given det $A = 2$, find the value of x.

3 a) For $P = \begin{pmatrix} 3 & a \\ -2 & 5 \end{pmatrix}$, find an expression for det P in terms of a.

   b) Given det $P = 1$, find the value of a.

4 Given $M = \begin{pmatrix} 5 & c \\ 3 & c - 1 \end{pmatrix}$ and det $M = 12$, find the value of c.

5 Given $Q = \begin{pmatrix} 3x - 1 & 5 \\ 6 & x \end{pmatrix}$ and det $Q = 14$, find the possible values of x.

6 Given $A = \begin{pmatrix} p + 3 & p \\ p & 2 \end{pmatrix}$ and det $A = 3$, find the possible values of p.

7 Use the formula to find the inverse of each of these matrices.

   a) $\begin{pmatrix} 5 & 1 \\ 3 & 1 \end{pmatrix}$
   b) $\begin{pmatrix} 3 & 1 \\ 5 & 4 \end{pmatrix}$
   c) $\begin{pmatrix} -1 & 3 \\ -2 & 2 \end{pmatrix}$

   d) $\begin{pmatrix} 6 & 2 \\ -1 & 0 \end{pmatrix}$
   e) $\begin{pmatrix} 6 & -2 \\ -5 & 3 \end{pmatrix}$
   f) $\begin{pmatrix} -8 & 1 \\ -3 & 2 \end{pmatrix}$

   g) $\begin{pmatrix} 0 & -4 \\ 2 & 7 \end{pmatrix}$
   h) $\begin{pmatrix} 9 & 2 \\ -5 & -3 \end{pmatrix}$

8 Use your GDC to find the inverse of each of these matrices.

   a) $\begin{pmatrix} 4 & 2 \\ 3 & 2 \end{pmatrix}$
   b) $\begin{pmatrix} 0 & -4 \\ 1 & 3 \end{pmatrix}$
   c) $\begin{pmatrix} 3 & -1 \\ -2 & 1 \end{pmatrix}$

   d) $\begin{pmatrix} 6 & 1 \\ 7 & 2 \end{pmatrix}$
   e) $\begin{pmatrix} 0 & 5 \\ -2 & 6 \end{pmatrix}$
   f) $\begin{pmatrix} -4 & 2 \\ 5 & -3 \end{pmatrix}$

   g) $\begin{pmatrix} 6 & -8 \\ -4 & 7 \end{pmatrix}$
   h) $\begin{pmatrix} 1 & 3 \\ -1 & 2 \end{pmatrix}$

9 a) Given $A = \begin{pmatrix} 3 & 5 \\ 1 & 2 \end{pmatrix}$ and $B = \begin{pmatrix} 2 & -5 \\ -1 & 3 \end{pmatrix}$, calculate the product AB.

   b) Hence find the matrix C such that $BC = A$.

10 a) Given $P = \begin{pmatrix} 4 & -1 \\ 7 & -2 \end{pmatrix}$ and $Q = \begin{pmatrix} 2 & -1 \\ 7 & -4 \end{pmatrix}$, calculate the matrix product PQ.

   b) Hence find the matrix R such that $RP = Q$.

11 a) Given $A = \begin{pmatrix} 2 & 5 \\ 1 & 3 \end{pmatrix}$ and $B = \begin{pmatrix} 3 & -5 \\ 8 & 1 \end{pmatrix}$, find A^{-1}.

   b) Hence find the matrix P such that $AP = B$.

12 a) Given $C = \begin{pmatrix} 5 & 2 \\ 6 & 3 \end{pmatrix}$, find C^{-1}.

b) Hence find the values of w, x, y and z such that

$$\begin{pmatrix} 5 & 2 \\ 6 & 3 \end{pmatrix}\begin{pmatrix} w & x \\ y & z \end{pmatrix} = \begin{pmatrix} 6 & -3 \\ 2 & 5 \end{pmatrix}.$$

13 Solve each of the following equations for x and y.

a) $\begin{pmatrix} 4 & 5 \\ 2 & 1 \end{pmatrix}\begin{pmatrix} x \\ y \end{pmatrix} = \begin{pmatrix} 6 \\ 3 \end{pmatrix}$

b) $\begin{pmatrix} 4 & -1 \\ -2 & 3 \end{pmatrix}\begin{pmatrix} x \\ y \end{pmatrix} = \begin{pmatrix} 9 \\ -5 \end{pmatrix}$

c) $\begin{pmatrix} -2 & -7 \\ 1 & 5 \end{pmatrix}\begin{pmatrix} x \\ y \end{pmatrix} = \begin{pmatrix} 9 \\ 12 \end{pmatrix}$

d) $\begin{pmatrix} 2 & -5 \\ 3 & 2 \end{pmatrix}\begin{pmatrix} x \\ y \end{pmatrix} = \begin{pmatrix} 19 \\ 38 \end{pmatrix}$

e) $\begin{pmatrix} 1 & 3 \\ -2 & 4 \end{pmatrix}\begin{pmatrix} x \\ y \end{pmatrix} = \begin{pmatrix} 5 \\ -3 \end{pmatrix}$

f) $\begin{pmatrix} 3 & 2 \\ -1 & -1 \end{pmatrix}\begin{pmatrix} x \\ y \end{pmatrix} = \begin{pmatrix} -4 \\ 7 \end{pmatrix}$

g) $\begin{pmatrix} 4 & -8 \\ 2 & -3 \end{pmatrix}\begin{pmatrix} x \\ y \end{pmatrix} = \begin{pmatrix} 6 \\ -2 \end{pmatrix}$

h) $\begin{pmatrix} 9 & 5 \\ -3 & -2 \end{pmatrix}\begin{pmatrix} x \\ y \end{pmatrix} = \begin{pmatrix} 2 \\ -9 \end{pmatrix}$

14 a) Given $M = \begin{pmatrix} 5 & -8 \\ 1 & -2 \end{pmatrix}$, find M^{-1}.

b) Hence find the values of a, b, c and d such that

$$\begin{pmatrix} 5 & -8 \\ 1 & -2 \end{pmatrix}\begin{pmatrix} a & b \\ c & d \end{pmatrix}\begin{pmatrix} 5 & -8 \\ 1 & -2 \end{pmatrix} = \begin{pmatrix} 2 & 6 \\ 0 & -4 \end{pmatrix}.$$

15 Given $A = \begin{pmatrix} 4 & -1 \\ -2 & 3 \end{pmatrix}$, $B = \begin{pmatrix} 6 & 4 \\ -5 & -3 \end{pmatrix}$, $C = \begin{pmatrix} 1 & 2 \\ 3 & 4 \end{pmatrix}$,

and $APB = C$, find the matrix P.

12.4 3×3 matrices

You can write a general 3×3 matrix as:

$$\begin{pmatrix} a & b & c \\ d & e & f \\ g & h & i \end{pmatrix}.$$

The 3×3 zero matrix is

$$\begin{pmatrix} 0 & 0 & 0 \\ 0 & 0 & 0 \\ 0 & 0 & 0 \end{pmatrix}$$

and the 3×3 identity matrix is

$$\begin{pmatrix} 1 & 0 & 0 \\ 0 & 1 & 0 \\ 0 & 0 & 1 \end{pmatrix}$$

The determinant of a 3 × 3 matrix

The determinant of the 3 × 3 matrix

$$A = \begin{pmatrix} a & b & c \\ d & e & f \\ g & h & i \end{pmatrix}$$

is defined as

$$\det A = a\begin{vmatrix} e & f \\ h & i \end{vmatrix} - b\begin{vmatrix} d & f \\ g & i \end{vmatrix} + c\begin{vmatrix} d & e \\ g & h \end{vmatrix}$$

$$= a(ei - fh) - b(di - fg) + c(dh - eg)$$

Notice that the first calculation in the determinant comes from a multiplied by the determinant of the 2 × 2 matrix derived from deleting the row and column in which a appears:

$$\begin{pmatrix} a & \blacksquare & \blacksquare \\ \blacksquare & e & f \\ \blacksquare & h & i \end{pmatrix}$$

The second calculation comes from $-b$ multiplied by the determinant of the 2 × 2 matrix derived from deleting the row and column in which b appears:

$$\begin{pmatrix} \blacksquare & b & \blacksquare \\ d & \blacksquare & f \\ g & \blacksquare & i \end{pmatrix}$$

The third calculation comes from c multiplied by the determinant of the 2 × 2 matrix derived from deleting the row and column in which c appears:

$$\begin{pmatrix} \blacksquare & \blacksquare & c \\ d & e & \blacksquare \\ g & h & \blacksquare \end{pmatrix}$$

Example 1

Given that $A = \begin{pmatrix} 1 & 2 & 4 \\ -2 & 5 & 1 \\ 3 & 1 & 2 \end{pmatrix}$, find det A.

$$\det A = 1\begin{vmatrix} 5 & 1 \\ 1 & 2 \end{vmatrix} - 2\begin{vmatrix} -2 & 1 \\ 3 & 2 \end{vmatrix} + 4\begin{vmatrix} -2 & 5 \\ 3 & 1 \end{vmatrix}$$

$$= (10 - 1) - 2(-4 - 3) + 4(-2 - 15)$$

$$= 9 + 14 - 68$$

$$= -45$$

Inverse of a 3 × 3 matrix

A 3×3 matrix A has an inverse providing det $A \neq 0$.

The same condition applies to 2×2 matrices.

Example 2

Given that $P = \begin{pmatrix} x & 4 & -1 \\ 2 & 0 & 1 \\ x & 1 & -2 \end{pmatrix}$, find det P.

Determine the value of x for which P^{-1} does not exist.

$$\det P = x \begin{vmatrix} 0 & 1 \\ 1 & -2 \end{vmatrix} - 4 \begin{vmatrix} 2 & 1 \\ x & -2 \end{vmatrix} + (-1) \begin{vmatrix} 2 & 0 \\ x & 1 \end{vmatrix}$$

$$= x(-1) - 4(-4 - x) - 1(2)$$

$$= -x + 16 + 4x - 2$$

$$\therefore \det P = 3x + 14$$

The inverse P^{-1} will not exist if det $P = 0$, so

$$3x + 14 = 0$$

$$\therefore x = -\frac{14}{3}$$

Finding the inverse of a 3×3 matrix is more complicated than finding the inverse of a 2×2 matrix. However, you can use your GDC to do it.

Example 3

Given that $A = \begin{pmatrix} 1 & 2 & 0 \\ -2 & 0 & 1 \\ 1 & 3 & 4 \end{pmatrix}$, and $B = \begin{pmatrix} 1 & 1 & 5 \\ -4 & -5 & -2 \\ -7 & 4 & 7 \end{pmatrix}$,

use the GDC to find A^{-1}.

Hence, find P such that $AP = B$.

Using the GDC to find A^{-1} gives

$$A^{-1} = \begin{pmatrix} -\frac{1}{5} & -\frac{8}{15} & \frac{2}{15} \\ \frac{3}{5} & \frac{4}{15} & -\frac{1}{15} \\ -\frac{2}{5} & -\frac{1}{15} & \frac{4}{15} \end{pmatrix}$$

$AP = B$. Multiplying both sides by A^{-1} gives

$$A^{-1}AP = A^{-1}B$$

$$(A^{-1}A)P = A^{-1}B$$

Since $A^{-1}A = I$,

$$IP = A^{-1}B$$

$$\therefore P = A^{-1}B$$

Therefore,

$$P = \begin{pmatrix} -\frac{1}{5} & -\frac{8}{15} & \frac{2}{15} \\ \frac{3}{5} & \frac{4}{15} & -\frac{1}{15} \\ -\frac{2}{5} & -\frac{1}{15} & \frac{4}{15} \end{pmatrix} \begin{pmatrix} 1 & 1 & 5 \\ -4 & -5 & -2 \\ -7 & 4 & 7 \end{pmatrix}$$

$$= \begin{pmatrix} 1 & 3 & 1 \\ 0 & -1 & 2 \\ -2 & 1 & 0 \end{pmatrix}$$

Solving three equations in three unknowns

When you have simultaneous equations in more than two unknowns, matrix methods can be a useful way to solve them.

Example 4

Given the simultaneous equations

$$x + 2y + 3z = 3$$
$$x - y + z = 2$$
$$2x + y + z = 8$$

write them in the matrix form $MX = C$, where M is a 3×3 matrix, $X = \begin{pmatrix} x \\ y \\ z \end{pmatrix}$ and $C = \begin{pmatrix} 3 \\ 2 \\ 8 \end{pmatrix}$.

Using the GDC, find M^{-1} and hence solve to find x, y and z.

· ·

Write the equations in matrix form:

$$\begin{pmatrix} 1 & 2 & 3 \\ 1 & -1 & 1 \\ 2 & 1 & 1 \end{pmatrix} \begin{pmatrix} x \\ y \\ z \end{pmatrix} = \begin{pmatrix} 3 \\ 2 \\ 8 \end{pmatrix}$$

So $M = \begin{pmatrix} 1 & 2 & 3 \\ 1 & -1 & 1 \\ 2 & 1 & 1 \end{pmatrix}$

From standard matrix algebra you know that

$$X = M^{-1}C$$

Using the GDC to find M^{-1} gives

$$M^{-1} = \begin{pmatrix} -\frac{2}{9} & \frac{1}{9} & \frac{5}{9} \\ \frac{1}{9} & -\frac{5}{9} & \frac{2}{9} \\ \frac{1}{3} & \frac{1}{3} & -\frac{1}{3} \end{pmatrix}$$

See page 329.

11 Use matrices to solve each of these sets of simultaneous equations.

a) $2x + 3y + z = 6$
$x + y + z = 2$
$3x - y + 2z = 3$

b) $3x - y + 2z = 15$
$2x + 5y + 6z = 29$
$x + y - 3z = 19$

c) $5x + y = 32$
$2x - 3y + z = 7$
$4x - 6y + 5z = 17$

d) $x + 3y - 2z = 15$
$2x - 5y - 3z = 6$
$2y - 5z = 14$

Summary

You should know how to ...

► Identify matrices.

▷ The order of a matrix indicates its number of rows and columns.

▷ The zero matrix **0** has all elements as zero.

▷ The identity matrix $I = \begin{pmatrix} 1 & 0 \\ 0 & 1 \end{pmatrix}$ for a 2×2 matrix or $\begin{pmatrix} 1 & 0 & 0 \\ 0 & 1 & 0 \\ 0 & 0 & 1 \end{pmatrix}$ for a 3×3 matrix

► Add and subtract matrices.

▷ To add or subtract two matrices, they must be of the same order.

► Multiply matrices.

▷ To multiply two matrices, the number of columns in the first matrix must equal the number of rows in the second matrix.

▷ For the identity matrix I and a matrix **M**, **MI = IM = M**

► Calculate determinants and inverses.

▷ For a 2×2 matrix $A = \begin{pmatrix} a & b \\ c & d \end{pmatrix}$

$\det A = |A| = \begin{vmatrix} a & b \\ c & d \end{vmatrix} = ad - bc$

$A^{-1} = \dfrac{1}{\det A} = \begin{pmatrix} d & -b \\ -c & a \end{pmatrix}$

▷ For a 3×3 matrix $A = \begin{pmatrix} a & b & c \\ d & e & f \\ g & h & i \end{pmatrix}$ $\det A = a\begin{vmatrix} e & f \\ h & i \end{vmatrix} - b\begin{vmatrix} d & f \\ g & i \end{vmatrix} + c\begin{vmatrix} d & e \\ g & h \end{vmatrix}$

Revision exercise 12

1 Let $A = \begin{pmatrix} \frac{1}{\sqrt{2}} & \frac{1}{\sqrt{2}} \\ -\frac{1}{\sqrt{2}} & \frac{1}{\sqrt{2}} \end{pmatrix}$

a) Find det A.

b) Find the matrix product $A\begin{pmatrix} \cos 135° \\ \sin 135° \end{pmatrix}$.

© IBO [1996]

2 Let the 2×2 matrix $M = \begin{pmatrix} a & \sqrt{1-a^2} \\ \sqrt{1-a^2} & -a \end{pmatrix}$, with $-1 \le a \le 1$.

Find the matrices a) M^2 b) M^3

© IBO [1997]

3 The matrices A, B and C are given by $A = \begin{pmatrix} 0 & -1 \\ 1 & 0 \end{pmatrix}$, $B = \begin{pmatrix} 1 & 0 & 0 \\ 0 & 0 & 1 \\ 0 & 1 & 0 \end{pmatrix}$,

$C = \begin{pmatrix} 1 & 0 & -1 \\ 0 & 2 & 3 \end{pmatrix}$.

a) Evaluate A^2 and B^2.

b) Solve for X the equation $AXB = C$.

© IBO [1998]

4 If $A = \begin{pmatrix} 2p & 3 \\ -4p & p \end{pmatrix}$ and det $A = 14$, find the possible values of p.

© IBO [1999]

5 A and B are 2×2 matrices, where $A = \begin{pmatrix} 5 & 2 \\ 2 & 0 \end{pmatrix}$ and $BA = \begin{pmatrix} 11 & 2 \\ 44 & 8 \end{pmatrix}$.

Find B.

© IBO [1999]

6 Let $A = \begin{pmatrix} 2 & a \\ -3 & a-3 \end{pmatrix}$.

a) Write down an expression for det A and simplify it.
   Hence find the value of a for which A does not have an inverse.

b) Find A^2 in terms of a. Hence find a value of a for which $A^2 = I$,
   where $I = \begin{pmatrix} 1 & 0 \\ 0 & 1 \end{pmatrix}$.

© IBO [1998]

7 A matrix X satisfies the equation $PX = Y$, where $P = \begin{pmatrix} 1 & 3 \\ 2 & 6 \end{pmatrix}$ and $Y = \begin{pmatrix} 5 \\ 10 \end{pmatrix}$.

a) Show that in the case when $X = \begin{pmatrix} 2 \\ 1 \end{pmatrix}$, the equation $PX = Y$ is
   satisfied.

b) In the case when $X = \begin{pmatrix} 8 \\ k \end{pmatrix}$, the equation is also satisfied.
   Find the value of k.

c) i) Calculate $P = \begin{pmatrix} 3t \\ -t \end{pmatrix}$, where t is any real number.

ii) Hence or otherwise show that if $X = \begin{pmatrix} 2 \\ 1 \end{pmatrix} + t\begin{pmatrix} 3 \\ -1 \end{pmatrix}$,

then the equation $PX = Y$ is always satisfied. © IBO[1998]

8 This question uses the matrices I, O and D, where

$$I = \begin{pmatrix} 1 & 0 \\ 0 & 1 \end{pmatrix}, O = \begin{pmatrix} 0 & 0 \\ 0 & 0 \end{pmatrix}, D = \begin{pmatrix} p & 0 \\ 0 & q \end{pmatrix}, \quad p, q \in \mathbb{R}.$$

Let $M = \begin{pmatrix} 3 & 5 \\ 1 & 2 \end{pmatrix}$.

a) Calculate det M and find M^{-1}.

Let $B = \begin{pmatrix} 7 & 10 \\ -3 & -4 \end{pmatrix}$.

b) The matrix D satisfies the matrix equation $M^{-1}DM = B$. Find p and q.

c) Show that B satisfies the equation $(B - pI)(B - qI) = 0$.

d) Hence, or otherwise, show that $B^{-1} = \frac{1}{2}(3I - B)$. © IBO[1997]

9 The matrix $A = \begin{pmatrix} 2 & 0 & 2 \\ 5 & 1 & 0 \\ -1 & 4 & a \end{pmatrix}$.

a) Find an expression in terms of a for det A.

b) Find the value of a for which A^{-1} does not exist.

c) Solve the equation $A\begin{pmatrix} x \\ y \\ z \end{pmatrix} = \begin{pmatrix} 1 \\ 2 \\ 3 \end{pmatrix}$ when $a = 0$,

giving your answers correct to 3 significant figures.

10 a) Show that the simultaneous equations

$$\begin{aligned} x - 2y + 3z &= 1 \\ x - y + 2z &= 2 \\ -2x + 4y - 5z &= 3 \end{aligned}$$

have a unique solution.

b) Find this solution.

13 Vectors

13.1 Introduction

> A quantity which is specified by a magnitude and a direction is called a **vector**.

To specify a displacement, you need to say 'how far' (magnitude) and 'in which direction'.
To specify a velocity, you need to say 'how fast' (magnitude) and 'in which direction'.

For example, displacement and velocity are both specified by a magnitude and a direction and are therefore examples of vector quantities.

A quantity which is specified by just its magnitude is called a **scalar**. For example, distance and speed are both fully specified by a magnitude and are therefore examples of scalar quantities.

A vector is represented by a straight line with an arrowhead.

In the diagram, the line OA represents a vector $\overrightarrow{OA}$. You can write:

$$\overrightarrow{OA} = \begin{pmatrix} 4 \\ 2 \end{pmatrix}$$

which means that to go from O to A you move 4 units in the positive x direction and 2 units in the positive y direction. This is called a **column vector**.

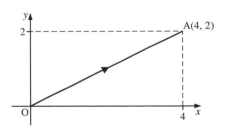

The magnitude or **modulus** of the vector $\overrightarrow{OA}$ is represented by the length OA and is denoted by $|\overrightarrow{OA}|$. In this example,

$$|\overrightarrow{OA}|^2 = 4^2 + 2^2 = 20$$
$$\therefore |\overrightarrow{OA}| = \sqrt{20} = 2\sqrt{5}$$

Vectors can also be written as lower case bold letters. For example, instead of $\overrightarrow{OA}$ you could write **v**. When writing vectors by hand, underline the letters to show they are vectors, for example $\underline{v}$.

Some properties of vectors

Two vectors are said to be equal if they have the same magnitude and direction.

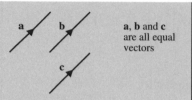

a, **b** and **c** are all equal vectors

If you multiply a vector by a scalar (number) the result is another vector.

2**a** and −3**a** are scalar multiples of **a**. Note that −3**a** is in the opposite direction to **a**.

a, 2**a** and −3**a** are parallel vectors.

Two vectors are parallel if one is a scalar multiple of the other.

Unit vectors

A **unit vector** is a vector of length **1**.

The standard unit vectors in two dimensions are

$$\mathbf{i} = \begin{pmatrix} 1 \\ 0 \end{pmatrix} \quad \text{and} \quad \mathbf{j} = \begin{pmatrix} 0 \\ 1 \end{pmatrix}$$

They can be represented diagrammatically as shown.

The vector $\overrightarrow{OA}$ can be written as

$$\overrightarrow{OA} = 4\mathbf{i} + 2\mathbf{j}$$

In three dimensions, the standard unit vectors are

$$\mathbf{i} = \begin{pmatrix} 1 \\ 0 \\ 0 \end{pmatrix} \qquad \mathbf{j} = \begin{pmatrix} 0 \\ 1 \\ 0 \end{pmatrix} \qquad \mathbf{k} = \begin{pmatrix} 0 \\ 0 \\ 1 \end{pmatrix}$$

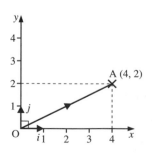

i, *j* and *k* are vectors along the *x*, *y* and *z* axes respectively.

For example, the column vector $\mathbf{v} = \begin{pmatrix} -4 \\ 2 \\ 3 \end{pmatrix}$ can be written as
$\mathbf{v} = -4\mathbf{i} + 2\mathbf{j} + 3\mathbf{k}$.

The magnitude of $\mathbf{v}$ is given by

$$|\mathbf{v}| = \sqrt{(-4)^2 + 2^2 + 3^2} = \sqrt{29}$$

However, a unit vector can be in any direction as Example 1 shows.

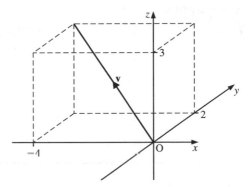

$\mathbf{v} = -4\mathbf{i} + 2\mathbf{j} + 3\mathbf{k}$

Example 1

Find the unit vector in the direction of $\mathbf{v} = 5\mathbf{i} - 2\mathbf{j} + 4\mathbf{k}$.

··

The magnitude of $\mathbf{v}$ is
$$|\mathbf{v}| = \sqrt{5^2 + (-2)^2 + 4^2} = \sqrt{45} = 3\sqrt{5}$$

If $\hat{\mathbf{v}}$ is a unit vector in the direction of $\mathbf{v}$:

$$\hat{\mathbf{v}} = \frac{5}{3\sqrt{5}}\mathbf{i} - \frac{2}{3\sqrt{5}}\mathbf{j} + \frac{4}{3\sqrt{5}}\mathbf{k}$$

The vector has length 1 and is in the direction of $\mathbf{v}$.

> Note that $\hat{\mathbf{v}}$ could be written as
> $$\frac{1}{3\sqrt{5}} \begin{pmatrix} 5 \\ -2 \\ 4 \end{pmatrix}.$$

13.2 Addition and subtraction of vectors

The diagram shows two possible paths that could be taken to travel from A to C.

One route is A to B then B to C. The other route is to go directly from A to C. This can be written as a vector equation:

$$\overrightarrow{AC} = \overrightarrow{AB} + \overrightarrow{BC}$$

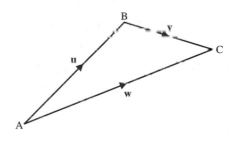

The vector $\overrightarrow{AC}$ is called the **resultant** of vectors $\overrightarrow{AB}$ and $\overrightarrow{BC}$.

Vectors may also be written as single letters in bold type, so let
$\mathbf{u} = \overrightarrow{AB}, \mathbf{v} = \overrightarrow{BC}$ and $\mathbf{w} = \overrightarrow{AC}$. Then:

$$\mathbf{w} = \mathbf{u} + \mathbf{v}$$

If $\overrightarrow{AB} = \mathbf{u}$, then $\overrightarrow{BA} = -\mathbf{u}$. In other words, the vector $\overrightarrow{AB}$ has the same magnitude as $\overrightarrow{BA}$ but is in the opposite direction.

If vectors are given in terms of the standard unit vectors, $\mathbf{i}, \mathbf{j}$ and $\mathbf{k}$, you add or subtract them by adding or subtracting the separate components.

Example 1

Given that $\overrightarrow{AB} = 3\mathbf{i} + 5\mathbf{j} - 4\mathbf{k}$ and $\overrightarrow{BC} = -\mathbf{i} + 4\mathbf{j} - \mathbf{k}$, find $\overrightarrow{AC}$.

$$\overrightarrow{AC} = \overrightarrow{AB} + \overrightarrow{BC}$$
$$= (3\mathbf{i} + 5\mathbf{j} - 4\mathbf{k}) + (-\mathbf{i} + 4\mathbf{j} - \mathbf{k})$$
$$\therefore \overrightarrow{AC} = 2\mathbf{i} + 9\mathbf{j} - 5\mathbf{k}$$

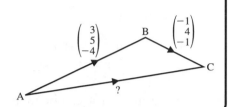

Example 2

Given that $\overrightarrow{BC} = 7\mathbf{i} - 2\mathbf{j} + \mathbf{k}$ and $\overrightarrow{AC} = \mathbf{i} - 6\mathbf{k}$, find $\overrightarrow{BA}$.

$$\overrightarrow{BA} = \overrightarrow{BC} + \overrightarrow{CA}$$
$$= \overrightarrow{BC} - \overrightarrow{AC}$$
$$= (7\mathbf{i} - 2\mathbf{j} + \mathbf{k}) - (\mathbf{i} - 6\mathbf{k})$$
$$\therefore \overrightarrow{BA} = 6\mathbf{i} - 2\mathbf{j} + 7\mathbf{k}$$

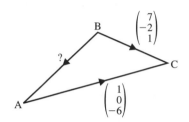

Example 3

Two vectors are given by $\mathbf{a} = 2\mathbf{i} - \mathbf{j} - \mathbf{k}$ and $\mathbf{b} = -\mathbf{i} + 3\mathbf{j} + 4\mathbf{k}$.
a) Find $\mathbf{a} + \mathbf{b}$ and $\mathbf{a} - \mathbf{b}$.
b) Draw a diagram showing $\mathbf{a} + \mathbf{b}$ and another showing $\mathbf{a} - \mathbf{b}$.

a) Adding the two vectors gives
$$\mathbf{a} + \mathbf{b} = (2\mathbf{i} - \mathbf{j} - \mathbf{k}) + (-\mathbf{i} + 3\mathbf{j} + 4\mathbf{k})$$
$$= \mathbf{i} + 2\mathbf{j} + 3\mathbf{k}$$
Subtracting the two vectors gives
$$\mathbf{a} - \mathbf{b} = (2\mathbf{i} - \mathbf{j} - \mathbf{k}) - (-\mathbf{i} + 3\mathbf{j} + 4\mathbf{k})$$
$$= 3\mathbf{i} - 4\mathbf{j} - 5\mathbf{k}$$
b) Adding $\mathbf{a}$ to $\mathbf{b}$ gives

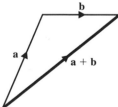

Subtracting $\mathbf{b}$ from $\mathbf{a}$ is the same as adding $-\mathbf{b}$:

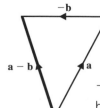

$-\mathbf{b}$ is the same magnitude as $\mathbf{b}$ but in the opposite direction.

Exercise 13A

1 Find the magnitude of each of these vectors.

a) $4\mathbf{i} + 3\mathbf{j}$ b) $5\mathbf{i} - 7\mathbf{j}$ c) $2\mathbf{i} - 2\mathbf{j} + \mathbf{k}$

d) $6\mathbf{i} - 3\mathbf{j} + 4\mathbf{k}$ e) $\begin{pmatrix} 12 \\ 5 \end{pmatrix}$ f) $\begin{pmatrix} 2 \\ -4 \end{pmatrix}$

g) $\begin{pmatrix} -9 \\ 7 \end{pmatrix}$ h) $\begin{pmatrix} 5 \\ -7 \\ 3 \end{pmatrix}$

2 Given $\mathbf{v} = x\mathbf{i} + 5\mathbf{j} - \sqrt{7}\mathbf{k}$ and $|\mathbf{v}| = 9$, find the possible values of the constant x.

3 Given that $|2\mathbf{i} + y\mathbf{j} - 4\mathbf{k}| = 6$, find the possible values of the constant y.

4 Find the possible values of the constant a such that $|a\mathbf{i} + 4a\mathbf{j} + 4\mathbf{k}| = 13$.

5 Find a unit vector in the direction of the vector $8\mathbf{i} - 6\mathbf{j}$.

6 Find a unit vector in the direction of $\mathbf{v} = 5\mathbf{i} - 8\mathbf{j}$.

7 Find a unit vector in the direction of the vector $\begin{pmatrix} -7 \\ 9 \end{pmatrix}$.

8 Find a unit vector in the direction of $\mathbf{v} = 3\mathbf{i} - 2\mathbf{j} + 5\mathbf{k}$.

9 Find a unit vector in the direction of the vector $\mathbf{i} - 3\mathbf{j} + 2\mathbf{k}$.

10 Find a unit vector in the direction of the vector $\begin{pmatrix} -3 \\ 12 \\ -4 \end{pmatrix}$.

11 Find a vector of magnitude 14 in the direction of the vector $6\mathbf{i} - 3\mathbf{j} + 2\mathbf{k}$.

12 Find a vector of magnitude $\sqrt{5}$ in the direction of $\mathbf{v} = 4\mathbf{i} - 8\mathbf{k}$.

13 Find a vector of magnitude $\sqrt{7}$ in the direction of the vector $\begin{pmatrix} 5 \\ -3 \\ 1 \end{pmatrix}$.

14 Given that $\overrightarrow{AB} = 2\mathbf{i} - 4\mathbf{j} + 5\mathbf{k}$ and $\overrightarrow{BC} = 3\mathbf{i} + 6\mathbf{j} - 2\mathbf{k}$, find $\overrightarrow{AC}$.

15 Given that $\overrightarrow{PQ} = -2\mathbf{i} + 3\mathbf{j} - 6\mathbf{k}$ and $\overrightarrow{QR} = 5\mathbf{i} - 7\mathbf{k}$, find $\overrightarrow{PR}$.

16 Given that $\overrightarrow{RS} = \begin{pmatrix} 3 \\ 6 \\ -8 \end{pmatrix}$ and $\overrightarrow{ST} = \begin{pmatrix} 5 \\ -5 \\ 0 \end{pmatrix}$, find $\overrightarrow{RT}$.

17 Given that $\overrightarrow{AB} = 5\mathbf{i} - 7\mathbf{j} - 2\mathbf{k}$ and $\overrightarrow{AC} = 2\mathbf{i} + 3\mathbf{j} - 2\mathbf{k}$, find $\overrightarrow{BC}$.

18 Given that $\overrightarrow{ML} = 10\mathbf{i} - 4\mathbf{j} - 6\mathbf{k}$ and $\overrightarrow{MN} = 7\mathbf{j} - 5\mathbf{k}$, find $\overrightarrow{NL}$.

19 Given that $\overrightarrow{PQ} = \begin{pmatrix} 5 \\ 2 \\ -8 \end{pmatrix}$ and $\overrightarrow{PR} = \begin{pmatrix} -2 \\ 5 \\ -6 \end{pmatrix}$, find $\overrightarrow{QR}$.

20 Given $\overrightarrow{AB} = x\mathbf{i} + 6\mathbf{j} + 4\mathbf{k}$, $\overrightarrow{BC} = 4\mathbf{i} + y\mathbf{j} - 3\mathbf{k}$ and
$\overrightarrow{AC} = -3\mathbf{i} + z\mathbf{k}$, find the values of the constants x, y and z.

13.3 Position vectors

The **position vector** of a point P with respect to a fixed origin O is
the vector $\overrightarrow{OP}$. This is not a free vector, since O is a fixed point.
You usually write

$$\overrightarrow{OP} = \mathbf{p}$$

Suppose the position vectors of two points P and Q are $\mathbf{p}$ and $\mathbf{q}$
respectively with respect to the origin O.
The diagram illustrates the two vectors.

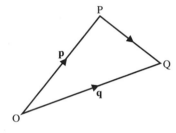

Then

$$\overrightarrow{PQ} = \overrightarrow{PO} + \overrightarrow{OQ}$$

$$= -\mathbf{p} + \mathbf{q}$$

$$\therefore \overrightarrow{PQ} = \mathbf{q} - \mathbf{p}$$

You can also show that $\overrightarrow{QP} = \mathbf{p} - \mathbf{q}$.

Example 1

The points A, B and C have position vectors $\mathbf{a}$, $\mathbf{b}$ and $\mathbf{c}$. Point P
is the mid-point of AB and point Q is the mid-point of BC.
Find

a) the position vectors of P and Q   b) $\overrightarrow{PQ}$

a) The diagram illustrates the situation.

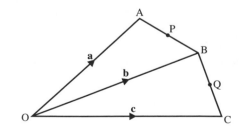

   The position vector of P is given by

$$\overrightarrow{OP} = \overrightarrow{OA} + \overrightarrow{AP}$$

$$= \overrightarrow{OA} + \tfrac{1}{2}\overrightarrow{AB}$$

$$= \mathbf{a} + \tfrac{1}{2}(\mathbf{b} - \mathbf{a})$$

$$= \tfrac{1}{2}\mathbf{a} + \tfrac{1}{2}\mathbf{b}$$

$$\therefore \overrightarrow{OP} = \tfrac{1}{2}(\mathbf{a} + \mathbf{b})$$

The position vector of Q is given by

$$\overrightarrow{OQ} = \overrightarrow{OB} + \overrightarrow{BQ}$$
$$= \overrightarrow{OB} + \tfrac{1}{2}\overrightarrow{BC}$$
$$= \mathbf{b} + \tfrac{1}{2}(\mathbf{c} - \mathbf{b})$$
$$= \tfrac{1}{2}\mathbf{b} + \tfrac{1}{2}\mathbf{c}$$
$$\therefore \overrightarrow{OQ} = \tfrac{1}{2}(\mathbf{b} + \mathbf{c})$$

> **Remember**
> You can multiply a vector by a scalar, such as $\tfrac{1}{2}$.
> The result is also a vector.

b) The vector $\overrightarrow{PQ}$ is given by

$$\overrightarrow{PQ} = \overrightarrow{OQ} - \overrightarrow{OP}$$
$$= \tfrac{1}{2}(\mathbf{b} + \mathbf{c}) - \tfrac{1}{2}(\mathbf{a} + \mathbf{b})$$
$$= \tfrac{1}{2}\mathbf{c} - \tfrac{1}{2}\mathbf{a}$$
$$\therefore \overrightarrow{PQ} = \tfrac{1}{2}(\mathbf{c} - \mathbf{a})$$

Example 2

The point O is the centre of the regular hexagon ABCDEF.
Given that $\overrightarrow{OA} = \mathbf{a}$ and $\overrightarrow{AB} = \mathbf{b}$, find

a) $\overrightarrow{OB}$ b) $\overrightarrow{BD}$ c) $\overrightarrow{CF}$

....

a) The diagram shows the hexagon ABCDEF.

$$\overrightarrow{OB} = \overrightarrow{OA} + \overrightarrow{AB}$$
$$- \mathbf{a} + \mathbf{b}$$

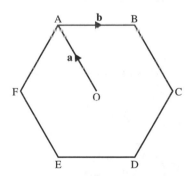

b) $\overrightarrow{BD} = \overrightarrow{BC} + \overrightarrow{CD}$. But $\overrightarrow{BC} = \overrightarrow{AO}$, since AO and BC are parallel and the same length, and $\overrightarrow{CD} = \overrightarrow{BO}$, since CD and BO are parallel and the same length. Therefore,

$$\overrightarrow{BD} = \overrightarrow{BC} + \overrightarrow{CD}$$
$$= \overrightarrow{AO} + \overrightarrow{BO}$$
$$= -\overrightarrow{OA} - \overrightarrow{OB}$$
$$= -\mathbf{a} - (\mathbf{a} + \mathbf{b})$$
$$= -2\mathbf{a} - \mathbf{b}$$

> **Remember:**
> Two vectors are equal if they have the same magnitude and direction.

c) $\overrightarrow{CF} = \overrightarrow{CD} + \overrightarrow{DE} + \overrightarrow{EF}$. You know that

$$\overrightarrow{CD} = \overrightarrow{BO} = -(\mathbf{a} + \mathbf{b})$$

and

$$\overrightarrow{DE} = \overrightarrow{BA} = -\mathbf{b}$$

and

$$\overrightarrow{EF} = \overrightarrow{OA} = \mathbf{a}$$

Therefore,

$$\overrightarrow{CF} = -(\mathbf{a} + \mathbf{b}) - \mathbf{b} + \mathbf{a}$$
$$= -2\mathbf{b}$$

This tells you that FC is parallel to AB, and twice the length of AB.

Example 3

OABC is a parallelogram with $\overrightarrow{OA} = \mathbf{a}$ and $\overrightarrow{OB} = \mathbf{b}$.
The point P lies on AB extended such that AB : BP = 2 : 1, and
the point Q lies on CB such that CQ : QB = 1 : 3.

a) Express each of these vectors in terms of
   a and **b**.

   i) $\overrightarrow{AB}$ ii) $\overrightarrow{AP}$

   iii) $\overrightarrow{OP}$ iv) $\overrightarrow{OQ}$

b) Hence show that $\overrightarrow{QP} = \frac{1}{4}\mathbf{a} + \frac{1}{2}\mathbf{b}$.

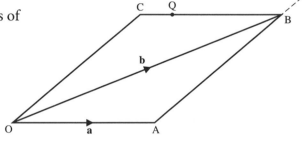

....

a) i) $\overrightarrow{AB} = \overrightarrow{OB} - \overrightarrow{OA}$

$$= \mathbf{b} - \mathbf{a}$$

ii) $\overrightarrow{AP} = \frac{3}{2}\overrightarrow{AB}$

$$= \frac{3}{2}(\mathbf{b} - \mathbf{a})$$

iii) $\overrightarrow{OP} = \overrightarrow{OA} + \overrightarrow{AP}$

$$= \mathbf{a} + \frac{3}{2}(\mathbf{b} - \mathbf{a})$$
$$= \frac{3}{2}\mathbf{b} - \frac{1}{2}\mathbf{a}$$

iv) $\overrightarrow{OQ} = \overrightarrow{OB} - \frac{3}{4}\overrightarrow{CB}$

$$= \mathbf{b} - \frac{3}{4}\mathbf{a}$$

b) $\overrightarrow{QP} = \overrightarrow{OP} - \overrightarrow{OQ}$

$$= \frac{3}{2}\mathbf{b} - \frac{1}{2}\mathbf{a} - (\mathbf{b} - \frac{3}{4}\mathbf{a})$$
$$= \frac{1}{4}\mathbf{a} + \frac{1}{2}\mathbf{b}$$

You can use vectors to show that three points lie in the same straight line, that is, they are **collinear**.

Example 4

The position vectors of the points, A, B and C are $2\mathbf{i} - \mathbf{j} + \mathbf{k}$, $3\mathbf{i} + 2\mathbf{j} - \mathbf{k}$ and $6\mathbf{i} + 11\mathbf{j} - 7\mathbf{k}$, respectively. Show that A, B and C are collinear.

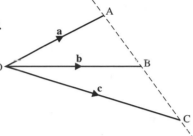

....

Let $\mathbf{a} = 2\mathbf{i} - \mathbf{j} + \mathbf{k}$, $\mathbf{b} = 3\mathbf{i} + 2\mathbf{j} - \mathbf{k}$ and $\mathbf{c} = 6\mathbf{i} + 11\mathbf{j} - 7\mathbf{k}$.

Then:

$$\overrightarrow{AB} = \mathbf{b} - \mathbf{a}$$

$$= \begin{pmatrix} 3 \\ 2 \\ -1 \end{pmatrix} - \begin{pmatrix} 2 \\ -1 \\ 1 \end{pmatrix}$$

$$\therefore \overrightarrow{AB} = \begin{pmatrix} 1 \\ 3 \\ -2 \end{pmatrix}$$

> The vectors are shown in column vector notation here.

and

$$\overrightarrow{BC} = \mathbf{c} - \mathbf{b}$$

$$= \begin{pmatrix} 6 \\ 11 \\ 7 \end{pmatrix} - \begin{pmatrix} 3 \\ 2 \\ 1 \end{pmatrix} = \begin{pmatrix} 3 \\ 9 \\ 6 \end{pmatrix} = 3\begin{pmatrix} 1 \\ 3 \\ 2 \end{pmatrix}$$

> Reduce the final vector to its simplest form.

$$\therefore \overrightarrow{BC} = 3\overrightarrow{AB}$$

$\overrightarrow{BC}$ is a multiple of $\overrightarrow{AB}$, which means that $\overrightarrow{BC}$ and $\overrightarrow{AB}$ are parallel. But $\overrightarrow{AB}$ and $\overrightarrow{BC}$ have a common point, namely B. Therefore, A, B and C are collinear.

Exercise 13B

....

1 OAB is a triangle with $\overrightarrow{OA} = \mathbf{a}$ and $\overrightarrow{OB} = \mathbf{b}$. P and Q are the mid-points of OA and AB, respectively.

a) Express each of these vectors in terms of **a** and **b**.

   i) $\overrightarrow{PA}$ ii) $\overrightarrow{AB}$ iii) $\overrightarrow{AQ}$ iv) $\overrightarrow{PQ}$

b) State two geometrical relationships connecting $\overrightarrow{OB}$ and $\overrightarrow{PQ}$.

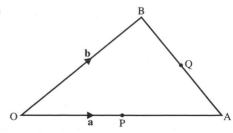

2 OABC is a rectangle with $\overrightarrow{OA} = \mathbf{a}$ and $\overrightarrow{OB} = \mathbf{b}$.
M is the mid-point of OC, and N is the point on CB such that
CN : NB = 2 : 1.
Express each of these vectors in terms of $\mathbf{a}$ and $\mathbf{b}$.

a) $\overrightarrow{OC}$ b) $\overrightarrow{ON}$ c) $\overrightarrow{MO}$ d) $\overrightarrow{MN}$

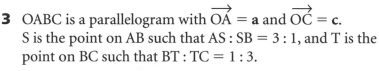

3 OABC is a parallelogram with $\overrightarrow{OA} = \mathbf{a}$ and $\overrightarrow{OC} = \mathbf{c}$.
S is the point on AB such that AS : SB = 3 : 1, and T is the
point on BC such that BT : TC = 1 : 3.

a) Express each of these vectors in terms of $\mathbf{a}$ and $\mathbf{c}$.

   i) $\overrightarrow{AC}$ ii) $\overrightarrow{SB}$ iii) $\overrightarrow{BT}$ iv) $\overrightarrow{ST}$

b) Explain why $\overrightarrow{ST}$ and $\overrightarrow{AC}$ are parallel, and state the value of
   the ratio ST : AC.

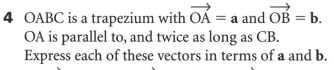

4 OABC is a trapezium with $\overrightarrow{OA} = \mathbf{a}$ and $\overrightarrow{OB} = \mathbf{b}$.
OA is parallel to, and twice as long as CB.
Express each of these vectors in terms of $\mathbf{a}$ and $\mathbf{b}$.

a) $\overrightarrow{CB}$ b) $\overrightarrow{BA}$ c) $\overrightarrow{CA}$ d) $\overrightarrow{CO}$

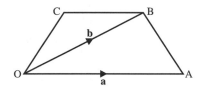

5

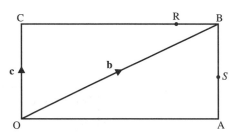

OABC is a rectangle with $\overrightarrow{OB} = \mathbf{b}$ and $\overrightarrow{OC} = \mathbf{c}$.
R is the point on CB such that CR : RB = 4 : 1, and S is the
point on AB such that AS : SB = 2 : 3.

a) Express each of these vectors in terms of $\mathbf{b}$ and $\mathbf{c}$.

   i) $\overrightarrow{OA}$ ii) $\overrightarrow{OR}$ iii) $\overrightarrow{OS}$ iv) $\overrightarrow{RS}$

b) Given also that M is the mid-point of RS, find an
   expression for $\overrightarrow{OM}$ in terms of $\mathbf{b}$ and $\mathbf{c}$.

6 OAB is a triangle with $\overrightarrow{OA} = \mathbf{a}$ and $\overrightarrow{OB} = \mathbf{b}$.
The point P lies on AB extended such that AB : BP = 1 : 3, and
the point Q lies on OA such that OQ : QA = 3 : 1.

a) Express each of these vectors in terms of $\mathbf{a}$ and $\mathbf{b}$.

   i) $\overrightarrow{AB}$ ii) $\overrightarrow{BP}$ iii) $\overrightarrow{OP}$ iv) $\overrightarrow{OQ}$

b) Hence show that $\overrightarrow{PQ} = \frac{15}{4}\mathbf{a} - 4\mathbf{b}$.

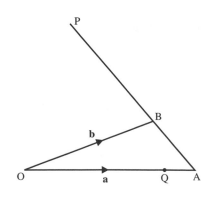

7 OABC is a trapezium with $\overrightarrow{OA} = \mathbf{a}$ and $\overrightarrow{OB} = \mathbf{b}$.

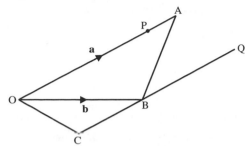

OA is parallel to, and three times as long as, CB.

The point P lies on OA such that OP : PA = 3 : 1, and the point Q lies on CB extended such that CB : BQ = 4 : 5.

a) Express each of these vectors in terms of **a** and **b**.

　i) $\overrightarrow{OP}$　　ii) $\overrightarrow{CB}$　　iii) $\overrightarrow{BQ}$

b) Show that $\overrightarrow{OC} = \overrightarrow{PQ}$.

8 Points P, Q and R have position vectors $\begin{pmatrix} 5 \\ 4 \\ 1 \end{pmatrix}$, $\begin{pmatrix} 7 \\ 5 \\ 4 \end{pmatrix}$ and $\begin{pmatrix} 11 \\ 7 \\ 10 \end{pmatrix}$, respectively.

a) Find $\overrightarrow{PQ}$ and $\overrightarrow{QR}$.

b) Deduce that P, Q and R are collinear and find the ratio PQ : QR.

9 The points A, B and C have coordinates $(1, -5, 6)$, $(3, -2, 10)$ and $(7, 4, 18)$ respectively. Show that A, B and C are collinear.

10 Show that the points $P(5, 4, -3)$, $Q(3, 8, -1)$ and $R(0, 14, 2)$ are collinear.

11 Given that $A(2, 13, -5)$, $B(3, y, -3)$ and $C(6, -7, z)$ are collinear, find the values of the constants y and z.

···

13.4 The scalar product

> The **scalar product** $\mathbf{v} \cdot \mathbf{w}$ of two vectors **v** and **w** is defined by
>
> 　$\mathbf{v} \cdot \mathbf{w} = |\mathbf{v}||\mathbf{w}|\cos\theta$
>
> where θ is the angle between the vectors.

The scalar product is also called the 'dot product' or 'inner product'.

❖ When the two vectors **v** and **w** are perpendicular, $\theta = 90°$ and $\cos 90° = 0$. Therefore, $\mathbf{v} \cdot \mathbf{w} = 0$.

❖ When the angle between the vectors **v** and **w** is acute, $\cos\theta > 0$ and therefore $\mathbf{v} \cdot \mathbf{w} > 0$.

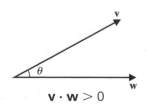

$\mathbf{v} \cdot \mathbf{w} > 0$

❖ When the angle between the vectors **v** and **w** is between 90°
and 180°, cos $\theta < 0$ and therefore **v** · **w** < 0.

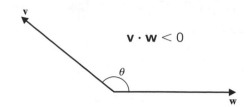

$$\mathbf{v} \cdot \mathbf{w} < 0$$

❖ **v** · **v** $= |v||v| \cos \theta$
$\qquad = v^2 \times 1$
$\qquad = v^2$
i · **i** $= 1$, because $|\mathbf{i}| = 1$

To calculate the scalar product **v** · **w**, let **v** $= v_1\mathbf{i} + v_2\mathbf{j}$ and
w $= w_1\mathbf{i} + w_2\mathbf{j}$. Then

$$\mathbf{v} \cdot \mathbf{w} = (v_1\mathbf{i} + v_2\mathbf{j}) \cdot (w_1\mathbf{i} + w_2\mathbf{j})$$
$$= v_1 w_1 \mathbf{i} \cdot \mathbf{i} + v_1 w_2 \mathbf{i} \cdot \mathbf{j} + v_2 w_1 \mathbf{i} \cdot \mathbf{j} + v_2 w_2 \mathbf{j} \cdot \mathbf{j}$$

> Expand the brackets as in algebra.

Now **i** · **i** = **j** · **j** $= 1$, and since **i** and **j** are perpendicular to each
other **i** · **j** = **j** · **i** $= 0$. Therefore

$$\mathbf{v} \cdot \mathbf{w} = v_1 w_1 + v_2 w_2$$

In three dimensions, the product is given by

$$\mathbf{v} \cdot \mathbf{w} = v_1 w_1 + v_2 w_2 + v_3 w_3$$

Example 1

Find the scalar product of each of these pairs of vectors.

a) $2\mathbf{i} + 3\mathbf{j}$ and $\mathbf{i} - 6\mathbf{j}$

b) $4\mathbf{i} - 2\mathbf{j} + \mathbf{k}$ and $2\mathbf{i} + \mathbf{j} - 3\mathbf{k}$

c) $\begin{pmatrix} 2 \\ -1 \\ 4 \end{pmatrix}$ and $\begin{pmatrix} -3 \\ 1 \\ 5 \end{pmatrix}$

....

a) $(2\mathbf{i} + 3\mathbf{j}) \cdot (\mathbf{i} - 6\mathbf{j}) = (2 \times 1) + (3 \times -6) = -16$

b) $(4\mathbf{i} - 2\mathbf{j} + \mathbf{k}) \cdot (2\mathbf{i} + \mathbf{j} - 3\mathbf{k})$
$\quad = (4 \times 2) + (-2 \times 1) + (1 \times -3)$
$\quad = 3$

c) $\begin{pmatrix} 2 \\ -1 \\ 4 \end{pmatrix} \cdot \begin{pmatrix} -3 \\ 1 \\ 5 \end{pmatrix} = (2 \times -3) + (-1 \times 1) + (4 \times 5) = 13$

You can use the scalar product to find the angle between two vectors.

Example 2

Find the angle between the vectors $\mathbf{v} = 2\mathbf{i} + \mathbf{j} + \mathbf{k}$ and
$\mathbf{w} = \mathbf{i} - \mathbf{j} + 3\mathbf{k}$.

....

$$\mathbf{v} \cdot \mathbf{w} = |\mathbf{v}||\mathbf{w}| \cos \theta \qquad\qquad [1]$$

Now

$$\mathbf{v} \cdot \mathbf{w} = (2 \times 1) + (1 \times -1) + (1 \times 3) = 4$$

We also have

$$|\mathbf{v}| = \sqrt{2^2 + 1^2 + 1^2} - \sqrt{6}$$

and

$$|\mathbf{w}| = \sqrt{1^2 + (-1)^2 + 3^2} = \sqrt{11}$$

Substituting into [1] gives

$$4 = \sqrt{6}\sqrt{11} \cos \theta$$

$$\therefore \cos \theta = \frac{4}{\sqrt{66}} \quad \text{and} \quad \theta = 60.5°$$

Example 3

Given that the two vectors $\mathbf{v} = (3t + 1)\mathbf{i} + \mathbf{j} - \mathbf{k}$ and
$\mathbf{w} = (t + 3)\mathbf{i} + 3\mathbf{j} - 2\mathbf{k}$ are perpendicular, find the possible
values of the constant t.

....

$\mathbf{v}$ and $\mathbf{w}$ are perpendicular, so $\mathbf{v} \cdot \mathbf{w} = 0$. That is:

$$(3t + 1)(t + 3) + (1 \times 3) + (-1 \times -2) = 0$$
$$3t^2 + 10t + 8 = 0$$
$$(3t + 4)(t + 2) = 0$$

Solving gives $t = -\frac{4}{3}$ or $t = -2$, so the possible values of t are
$-\frac{4}{3}$ and -2.

Exercise 13C

....

1 Given $\mathbf{a} = 3\mathbf{i} + 4\mathbf{j}$, $\mathbf{b} = \mathbf{i} - 3\mathbf{j}$ and $\mathbf{c} = 2\mathbf{i} + 5\mathbf{j}$, evaluate each of
these scalar products.

a) $\mathbf{a} \cdot \mathbf{b}$ b) $\mathbf{b} \cdot \mathbf{a}$ c) $\mathbf{a} \cdot \mathbf{c}$

d) $\mathbf{c} \cdot \mathbf{b}$ e) $\mathbf{a} \cdot \mathbf{a}$ f) $\mathbf{c} \cdot (\mathbf{a} + \mathbf{b})$

2 Given $\mathbf{x} = 2\mathbf{i} - 3\mathbf{j} + \mathbf{k}$, $\mathbf{y} = 5\mathbf{i} + 2\mathbf{j} - 7\mathbf{k}$ and $\mathbf{z} = \mathbf{i} - 4\mathbf{j} - 2\mathbf{k}$,
evaluate each of these scalar products.

a) $\mathbf{x} \cdot \mathbf{y}$ b) $\mathbf{y} \cdot \mathbf{x}$ c) $\mathbf{x} \cdot \mathbf{z}$

d) $\mathbf{z} \cdot \mathbf{z}$ e) $\mathbf{x} \cdot (\mathbf{y} + \mathbf{z})$ f) $\mathbf{y} \cdot (\mathbf{z} - \mathbf{x})$

3 Given

$$\mathbf{p} = \begin{pmatrix} -2 \\ 3 \end{pmatrix} \quad \mathbf{q} = \begin{pmatrix} 1 \\ 1 \end{pmatrix} \quad \text{and} \quad \mathbf{r} = \begin{pmatrix} 5 \\ -2 \end{pmatrix}$$

evaluate each of these scalar products.

a) $\mathbf{p} \cdot \mathbf{q}$ b) $\mathbf{q} \cdot \mathbf{r}$ c) $\mathbf{r} \cdot \mathbf{q}$

d) $\mathbf{q} \cdot \mathbf{q}$ e) $\mathbf{r} \cdot (\mathbf{q} + \mathbf{p})$ f) $\mathbf{p} \cdot (\mathbf{q} - \mathbf{r})$

4 Given

$$\mathbf{c} = \begin{pmatrix} 3 \\ 1 \\ -4 \end{pmatrix} \quad \mathbf{d} = \begin{pmatrix} -5 \\ -2 \\ 7 \end{pmatrix} \quad \text{and} \quad \mathbf{e} = \begin{pmatrix} 0 \\ 4 \\ -5 \end{pmatrix}$$

evaluate each of these scalar products.

a) $\mathbf{c} \cdot \mathbf{d}$ b) $\mathbf{d} \cdot \mathbf{e}$ c) $\mathbf{c} \cdot \mathbf{e}$

d) $\mathbf{d} \cdot (\mathbf{e} - \mathbf{c})$ e) $\mathbf{c} \cdot (\mathbf{c} + \mathbf{d})$ f) $\mathbf{e} \cdot (2\mathbf{c} - \mathbf{d})$

5 Decide which of these pairs of vectors are perpendicular, which are parallel, and which are neither perpendicular nor parallel.

a) $2\mathbf{i} + 8\mathbf{j}$ and $4\mathbf{i} - \mathbf{j}$

b) $3\mathbf{i} + 5\mathbf{j}$ and $6\mathbf{i} + 10\mathbf{j}$

c) $6\mathbf{i} - 8\mathbf{j} + 2\mathbf{k}$ and $9\mathbf{i} - 12\mathbf{j} + 3\mathbf{k}$

d) $5\mathbf{i} - 6\mathbf{j} + 2\mathbf{k}$ and $3\mathbf{i} + 2\mathbf{j} + \mathbf{k}$

e) $\begin{pmatrix} -3 \\ 1 \end{pmatrix}$ and $\begin{pmatrix} 6 \\ -2 \end{pmatrix}$ f) $\begin{pmatrix} 12 \\ 6 \end{pmatrix}$ and $\begin{pmatrix} 1 \\ -2 \end{pmatrix}$

g) $\begin{pmatrix} 3 \\ -1 \\ 4 \end{pmatrix}$ and $\begin{pmatrix} 9 \\ -3 \\ 12 \end{pmatrix}$ h) $\begin{pmatrix} 1 \\ 2 \\ 3 \end{pmatrix}$ and $\begin{pmatrix} 3 \\ 2 \\ 1 \end{pmatrix}$

6 Find the angle between each of these pairs of vectors, giving your answers correct to one decimal place.

a) $3\mathbf{i} - 4\mathbf{j}$ and $12\mathbf{i} + 5\mathbf{j}$

b) $\mathbf{i}$ and $\mathbf{i} + \mathbf{j}$

c) $2\mathbf{i} + \mathbf{j} - 2\mathbf{k}$ and $4\mathbf{i} - 3\mathbf{j} + 12\mathbf{k}$

d) $3\mathbf{i} - 5\mathbf{j} - 2\mathbf{k}$ and $\mathbf{i} - 6\mathbf{k}$

e) $\begin{pmatrix} 2 \\ -1 \end{pmatrix}$ and $\begin{pmatrix} 6 \\ 3 \end{pmatrix}$ f) $\begin{pmatrix} 3 \\ 5 \end{pmatrix}$ and $\begin{pmatrix} 2 \\ -3 \end{pmatrix}$

g) $\begin{pmatrix} -2 \\ 1 \\ 3 \end{pmatrix}$ and $\begin{pmatrix} 4 \\ -3 \\ 3 \end{pmatrix}$ h) $\begin{pmatrix} 3 \\ 0 \\ -1 \end{pmatrix}$ and $\begin{pmatrix} 2 \\ 5 \\ 0 \end{pmatrix}$

7 $\mathbf{a} = 4\mathbf{i} + 5\mathbf{j}, \mathbf{b} = x\mathbf{i} - 8\mathbf{j}$ and $\mathbf{c} = \mathbf{i} + y\mathbf{j}$.

a) Find the value of the constant x given that $\mathbf{a}$ and $\mathbf{b}$ are perpendicular.

b) Find the value of the constant y given that $\mathbf{a}$ and $\mathbf{c}$ are parallel.

8 $p = 6\mathbf{i} - \mathbf{j}$, $q = x\mathbf{i} + 2\mathbf{j}$ and $r = 2\mathbf{i} + y\mathbf{j}$.

a) Find the value of the constant x given that $\mathbf{p}$ and $\mathbf{q}$ are parallel.

b) Find the value of the constant y given that $\mathbf{p}$ and $\mathbf{r}$ are perpendicular.

9 Given that the vectors $2\mathbf{i} + t\mathbf{j} - 4\mathbf{k}$ and $\mathbf{i} - 3\mathbf{j} + (t - 4)\mathbf{k}$ are perpendicular, find the value of the constant t.

10 Given that $\begin{pmatrix} c \\ 2 + c \\ 3 \end{pmatrix}$ and $\begin{pmatrix} -1 \\ 3 \\ 4 - c \end{pmatrix}$ are perpendicular vectors, find the value of the constant c.

11 Find the possible values of the constant a, given that the vectors

$$a\mathbf{i} + 8\mathbf{j} + (3a + 1)\mathbf{k} \quad \text{and} \quad (a + 1)\mathbf{i} + (a - 1)\mathbf{j} - 2\mathbf{k}$$

are perpendicular.

12 Given that the vectors $\begin{pmatrix} t \\ 4 \\ 2t + 1 \end{pmatrix}$ and $\begin{pmatrix} t + 2 \\ 1 - t \\ -1 \end{pmatrix}$ are perpendicular, find the possible values of the constant t.

13 Find a unit vector which is perpendicular to both of the vectors $4\mathbf{i} + 2\mathbf{j} - 3\mathbf{k}$ and $2\mathbf{i} - 3\mathbf{j} + \mathbf{k}$.

14 In the triangle OAB, $\overrightarrow{OA} = \mathbf{a}$ and $\overrightarrow{OB} = \mathbf{b}$.

a) Show that $(\mathbf{a} - \mathbf{b}) \cdot (\mathbf{a} - \mathbf{b}) = \mathbf{a} \cdot \mathbf{a} + \mathbf{b} \cdot \mathbf{b} - 2\mathbf{a} \cdot \mathbf{b}$

b) Hence prove the cosine rule.

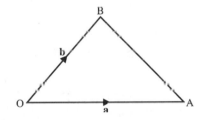

13.5 Vector equation of a line

Let $\mathbf{p}$ and $\mathbf{q}$ be the position vectors of two points P and Q with respect to an origin O. Let $\mathbf{r}$ be the position vector of a point R on the line PQ. Then

$$\overrightarrow{OR} = \overrightarrow{OP} + \overrightarrow{PR}$$
$$= \overrightarrow{OP} + t\overrightarrow{PQ}$$

where t is a scalar.

Therefore

$$\mathbf{r} = \mathbf{p} + t(\mathbf{q} - \mathbf{p}) \quad \text{or} \quad \mathbf{r} = \mathbf{p} + t\mathbf{d}, \quad \text{where } \mathbf{d} = \mathbf{q} - \mathbf{p}.$$

This is the vector equation of the line PQ.
The vector $(\mathbf{q} - \mathbf{p})$ is the direction vector of the line.

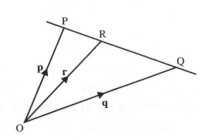

Each value of the parameter t corresponds to a point on the line PQ.

❖ When $t < 0$, point R is on the line QP produced.
❖ When $t = 0$, $\mathbf{r} = \mathbf{p}$, that is, R = P.
❖ When $0 < t < 1$, point R is between P and Q.
❖ When $t = 1$, $\mathbf{r} = \mathbf{q}$, that is, R = Q.
❖ When $t > 1$, point R is on the line PQ produced.

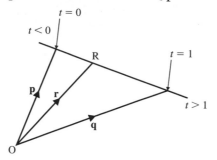

Example 1

Find the vector equation of the line passing through P(1, 3, 2) and Q(0, −1, 4). Does the point R(−2, 9, 1) lie on the line PQ?

. . . .

The position vectors of P and Q are given by

$$\mathbf{p} = \begin{pmatrix} 1 \\ 3 \\ 2 \end{pmatrix} \quad \text{and} \quad \mathbf{q} = \begin{pmatrix} 0 \\ -1 \\ 4 \end{pmatrix}$$

The vector equation of the line is given by

$$\mathbf{r} = \mathbf{p} + t(\mathbf{q} - \mathbf{p})$$

$$= \begin{pmatrix} 1 \\ 3 \\ 2 \end{pmatrix} + t\left(\begin{pmatrix} 0 \\ -1 \\ 4 \end{pmatrix} - \begin{pmatrix} 1 \\ 3 \\ 2 \end{pmatrix} \right)$$

$$\therefore \mathbf{r} = \begin{pmatrix} 1 \\ 3 \\ 2 \end{pmatrix} + t\begin{pmatrix} -1 \\ -4 \\ 2 \end{pmatrix}$$

This could be written in the form

$$\mathbf{r} = (1 - t)\mathbf{i} + (3 - 4t)\mathbf{j} + (2 + 2t)\mathbf{k}$$

If the point R(−2, 9, 1) lies on the line PQ, there will be a unique value of t for which

$$\begin{pmatrix} -2 \\ 9 \\ 1 \end{pmatrix} = \begin{pmatrix} 1 - t \\ 3 - 4t \\ 2 + 2t \end{pmatrix}$$

If $-2 = 1 - t$, then $t = 3$.
However, $t = 3$ does not satisfy $9 = 3 - 4t$.
Therefore, the point R(−2, 9, 1) does *not* lie on the line PQ.

Example 2

Two lines l and m have vector equations

$$\mathbf{r}_l = (2 - 3s)\mathbf{i} + (1 + s)\mathbf{j} + 4s\mathbf{k}$$

$$\mathbf{r}_m = (-1 + 3t)\mathbf{i} + 3\mathbf{j} + (7 - t)\mathbf{k}$$

respectively. Find

a) the position vector of their common point

b) the angle between the lines.

....

a) Rewriting the vector equations of the two lines in column form gives

$$\mathbf{r}_l = \begin{pmatrix} 2 - 3s \\ 1 + s \\ 4s \end{pmatrix} \quad \text{and} \quad \mathbf{r}_m = \begin{pmatrix} -1 + 3t \\ 3 \\ 7 - t \end{pmatrix}$$

At the point common to l and m,

$$\mathbf{r}_l = \mathbf{r}_m$$

$$\therefore \begin{pmatrix} 2 - 3s \\ 1 + s \\ 4s \end{pmatrix} = \begin{pmatrix} -1 + 3t \\ 3 \\ 7 - t \end{pmatrix}$$

Equating $\mathbf{i}$, $\mathbf{j}$ and $\mathbf{k}$ coefficients gives:

$\mathbf{i}$: $2 - 3s = -1 + 3t$ [1]

$\mathbf{j}$: $1 + s = 3$ [2]

$\mathbf{k}$: $4s = 7 - t$ [3]

From [2], $s = 2$. Substituting into [1] gives

$$2 - 3(2) = -1 + 3t$$

$$\therefore t = -1$$

Notice that $s = 2$, $t = -1$ also satisfies [3].

So, at the common point,

$$\mathbf{r}_l = \begin{pmatrix} 2 - 3(2) \\ 1 + 2 \\ 4(2) \end{pmatrix} = \begin{pmatrix} -4 \\ 3 \\ 8 \end{pmatrix}$$

The position vector of the common point is $-4\mathbf{i} + 3\mathbf{i} + 8\mathbf{k}$.

b) You know that

$$\mathbf{r}_l = \begin{pmatrix} 2 - 3s \\ 1 + s \\ 4s \end{pmatrix} = \begin{pmatrix} 2 \\ 1 \\ 0 \end{pmatrix} + s\begin{pmatrix} -3 \\ 1 \\ 4 \end{pmatrix}$$

and

$$\mathbf{r}_m = \begin{pmatrix} -1 + 3t \\ 3 \\ 7 - t \end{pmatrix} = \begin{pmatrix} -1 \\ 3 \\ 7 \end{pmatrix} + t\begin{pmatrix} 3 \\ 0 \\ -1 \end{pmatrix}$$

Therefore, the direction vector of line l is $-3\mathbf{i} + \mathbf{j} + 4\mathbf{k}$ and the direction vector of line m is $3\mathbf{i} - \mathbf{k}$. The angle between the lines l and m is the angle between these two direction vectors.

Let $\mathbf{u} = -3\mathbf{i} + \mathbf{j} + 4\mathbf{k}$ and $\mathbf{v} = 3\mathbf{i} - \mathbf{k}$. Then

$$|\mathbf{u}| = \sqrt{(-3)^2 + 1^2 + 4^2} = \sqrt{26}$$

and

$$|\mathbf{v}| = \sqrt{3^2 + (-1)^2} = \sqrt{10}$$

Using the scalar product,

$$\mathbf{u} \cdot \mathbf{v} = |\mathbf{u}||\mathbf{v}| \cos \theta$$

$$\therefore \quad \begin{pmatrix} -3 \\ 1 \\ 4 \end{pmatrix} \cdot \begin{pmatrix} 3 \\ 0 \\ -1 \end{pmatrix} = \sqrt{26}\,\sqrt{10}\,\cos\theta$$

$$-9 - 4 = \sqrt{260}\,\cos\theta$$

$$\therefore \quad \cos\theta = -\frac{13}{\sqrt{260}}$$

$$\therefore \quad \theta = 143.7°$$

The angle between the lines l and m is $143.7°$.

Exercise 13D

1 Find a vector equation for the line passing through the point $(4, 3)$ and parallel to the vector $\mathbf{i} - 2\mathbf{j}$.

2 Find a vector equation for the line passing through the point $(5, -2, 3)$ and parallel to the vector $4\mathbf{i} - 3\mathbf{j} + \mathbf{k}$.

3 Find a vector equation for the line passing through the point $(5, -1)$ and perpendicular to the vector $\mathbf{i} + \mathbf{j}$.

4 Find a vector equation for the line joining the points $(2, 6)$ and $(5, -2)$.

5 Find a vector equation for the line joining the points $(-1, 2, -3)$ and $(6, 3, 0)$.

6 Points A and B have coordinates $(4, 1)$ and $(2, -5)$, respectively. Find a vector equation for the line which passes through the point A, and which is perpendicular to the line AB.

7 Points P and Q have coordinates $(3, 5)$ and $(-3, -7)$, respectively. Find a vector equation for the line which passes through the point P, and which is perpendicular to the line PQ.

8 Find a vector equation for the perpendicular bisector of the points $(6, 3)$ and $(2, -5)$.

9 Find a vector equation for the perpendicular bisector of the points $(7, -1)$ and $(3, -3)$.

10 Three lines, l_1, l_2 and l_3, have equations

$$l_1: \quad \begin{pmatrix} x \\ y \end{pmatrix} = \begin{pmatrix} 1 \\ 1 \end{pmatrix} + s\begin{pmatrix} 3 \\ 1 \end{pmatrix}$$

$$l_2: \quad \begin{pmatrix} x \\ y \end{pmatrix} = \begin{pmatrix} 6 \\ -4 \end{pmatrix} + t\begin{pmatrix} 1 \\ 2 \end{pmatrix}$$

and

$$l_3: \quad \begin{pmatrix} x \\ y \end{pmatrix} = \begin{pmatrix} 12 \\ -8 \end{pmatrix} + u\begin{pmatrix} -1 \\ 6 \end{pmatrix}$$

Show that l_1, l_2 and l_3 are concurrent, and find the position vector of their point of intersection.

Lines are **concurrent** if they have a point in common, i.e. they pass through the same point.

11 Show that the lines $\mathbf{r}_1 = (6 - 2s)\mathbf{i} + (s - 5)\mathbf{j}$, $\mathbf{r}_2 = t\mathbf{i} + 3(1 - t)\mathbf{j}$ and $\mathbf{r}_3 = (5 - u)\mathbf{i} + (2u - 9)\mathbf{j}$ are concurrent, and find the position vector of their point of intersection.

12 Two lines, l_1 and l_2, have equations

$$l_1: \quad \begin{pmatrix} x \\ y \\ z \end{pmatrix} = \begin{pmatrix} 0 \\ -1 \\ -3 \end{pmatrix} + s\begin{pmatrix} 1 \\ 3 \\ 6 \end{pmatrix}$$

and

$$l_2: \quad \begin{pmatrix} x \\ y \\ z \end{pmatrix} = \begin{pmatrix} -2 \\ 1 \\ 1 \end{pmatrix} + t\begin{pmatrix} 1 \\ 1 \\ 2 \end{pmatrix}$$

a) Show that l_1 and l_2 are concurrent, and find the position vector of their point of intersection.

b) Find the angle between l_1 and l_2.

13 Two lines, l and m, have vector equations

$$\mathbf{r}_i = (3 - s)\mathbf{i} + (4s - 5)\mathbf{j} + (3s - 1)\mathbf{k}$$

and

$$\mathbf{r}_m = (2t - 5)\mathbf{i} + t\mathbf{j} + (2t - 1)\mathbf{k}$$

a) Show that l and m are concurrent, and find the position vector of their point of intersection.

b) Find the angle between l and m.

14 Points A, B and C have coordinates $(0, 5)$, $(9, 8)$ and $(4, 3)$, respectively.

a) Find a vector equation for the line, l, joining A and B.

b) Find a vector equation for the line, p, which passes through the point C and which is perpendicular to the line AB.

c) Find the coordinates of the point of intersection of the lines l and p.

d) Deduce that the perpendicular distance from the point C to the line AB is $\sqrt{10}$.

15 Points A, B and C have coordinates $(-1, -2)$, $(5, 10)$ and $(0, 5)$ respectively.

a) Find a vector equation for the line, l, joining A and B.

b) Find a vector equation for the line, p, which passes through the point C and which is perpendicular to the line AB.

c) Find the coordinates of the point of intersection of the lines l and p.

d) Deduce that the perpendicular distance from the point C to the line AB is $\sqrt{5}$.

16 The points A, B and C have position vectors $4\mathbf{i} + \mathbf{j} - 4\mathbf{k}$, $3\mathbf{i} + 2\mathbf{j} - 3\mathbf{k}$ and $2\mathbf{i} + 3\mathbf{j} - 5\mathbf{k}$, respectively.

Given that the angle between $\overrightarrow{AB}$ and $\overrightarrow{AC}$ is θ,

a) find the value of $\cos \theta$

b) deduce that $\sin \theta = \dfrac{\sqrt{6}}{3}$.

17 The points P, Q and R have position vectors $2\mathbf{i} + 5\mathbf{j} - 3\mathbf{k}$, $\mathbf{i} + 4\mathbf{j} - 2\mathbf{k}$ and $3\mathbf{i} + 3\mathbf{j} - 2\mathbf{k}$, respectively.

Given that the angle between $\overrightarrow{PQ}$ and $\overrightarrow{PR}$ is θ,

a) find the value of $\cos \theta$

b) deduce that $\sin \theta = \dfrac{\sqrt{7}}{3}$.

13.6 Applications of vectors

You have already seen that a quantity which has both magnitude and direction is called a vector. Consequently vectors are very applicable to everyday situations, such as the velocity of an aircraft. Directions are usually described using compass bearings.

Example 1

A tractor is moving across a field at $10\,\text{kmh}^{-1}$ on a bearing of $150°$. Express its velocity as a column vector.

The diagram illustrates the vector. The horizontal and vertical components of the vector can be obtained using trigonometry.

The horizontal component is represented by the length OA, where

$OA = 10 \sin 30°$

$= 5$

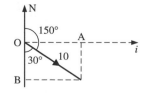

The vertical component is represented by the negative of the length OB, where

$$OB = 10 \cos 30°$$
$$= 5\sqrt{3}$$

Therefore the column vector is $\begin{pmatrix} 5 \\ -5\sqrt{3} \end{pmatrix}$.

You can write this as $5\mathbf{i} - 5\sqrt{3}\mathbf{j}$.

> Notice that the negative is important as it represents the direction of the vector (south).

Example 2

A pleasure boat leaves a harbour and its position (x, y) relative to the harbour at time t hours is given by the vector

$$\begin{pmatrix} x \\ y \end{pmatrix} = t\begin{pmatrix} 10 \\ 7 \end{pmatrix}$$

a) Find the speed of the boat.

The position (x, y) of a cargo ship at time t hours is given by the vector

$$\begin{pmatrix} x \\ y \end{pmatrix} = \begin{pmatrix} 12 \\ 12 \end{pmatrix} + t\begin{pmatrix} 4 \\ 1 \end{pmatrix}$$

b) Find the distance of the cargo ship from the harbour when the pleasure boat starts its journey.

c) If both the pleasure boat and the cargo ship continued on the same courses, after what period of time would they collide?

....

The speed of the boat is given by the magnitude of the vector $\begin{pmatrix} 10 \\ 7 \end{pmatrix}$. In other words

$$\text{speed} = \sqrt{10^2 + 7^2}$$
$$= \sqrt{149}$$

The pleasure boat starts its journey at $t = 0$. When $t = 0$ the cargo ship's position is given by

$$\begin{pmatrix} x \\ y \end{pmatrix} = \begin{pmatrix} 12 \\ 12 \end{pmatrix}$$

Therefore the distance of the cargo ship from the harbour at time $t = 0$ is given by

$$\sqrt{12^2 + 12^2} = 12\sqrt{2}$$

Given the pleasure boat and the cargo ship would collide at time t

$$t\begin{pmatrix} 10 \\ 7 \end{pmatrix} = \begin{pmatrix} 12 \\ 12 \end{pmatrix} + t\begin{pmatrix} 4 \\ 1 \end{pmatrix}$$

Therefore

$$10t = 12 + 4t \quad \text{and} \quad 7t = 12 + t$$

Therefore $t = 2$.

> The course of a ship is regularly updated to account for other sea traffic.

Exercise 13E

In all the questions a unit vector represents a displacement of 1 metre.

1 A man starts at the point $(5, 3)$ and walks such that his position (x, y) at a time t seconds is given by the formula

$$\begin{pmatrix} x \\ y \end{pmatrix} = \begin{pmatrix} 5 \\ 3 \end{pmatrix} + t\begin{pmatrix} 3 \\ 4 \end{pmatrix}$$

a) Where is the man after 10 seconds?

b) Find the speed of the man.

c) Obtain the equation of the man's path in the form
$ax + by + c = 0$

A woman starts at the point $(-19, 27)$ at the same time, and walks such that her position at a time t seconds is given by

$$\begin{pmatrix} x \\ y \end{pmatrix} = \begin{pmatrix} -19 \\ 27 \end{pmatrix} + t\begin{pmatrix} 4 \\ 3 \end{pmatrix}$$

d) Show that the man and the woman will meet.

e) Find their common position at that time.

2 A bird starts at the point $(2, 1, 5)$ and flies such that its position (x, y, z) at a time t seconds is given by the formula

$$\begin{pmatrix} x \\ y \\ z \end{pmatrix} = \begin{pmatrix} 2 \\ 1 \\ 5 \end{pmatrix} + t\begin{pmatrix} 4 \\ 3 \\ 12 \end{pmatrix}$$

a) Find the speed of the bird.

A second bird sets off from the point $(-10, 49, 17)$ at the same time, and flies such that its position at a time t seconds is given by the formula

$$\begin{pmatrix} x \\ y \\ z \end{pmatrix} = \begin{pmatrix} -10 \\ 49 \\ 17 \end{pmatrix} + t\begin{pmatrix} 5 \\ -1 \\ 11 \end{pmatrix}$$

b) Show that the two birds will meet, and find the time at which this will happen.
Find also their common position at this time.

3 A toy boat is involved in a navigation exercise on a lake. It starts at the origin, O, and travels for 12 seconds with velocity $\begin{pmatrix} 5 \\ 6 \end{pmatrix}$ to a point A.

a) Find the coordinates of A.

At A it changes velocity to $\begin{pmatrix} 1 \\ -4 \end{pmatrix}$ and travels for 20 seconds to a point B.

b) Find the coordinates of B.

From B it returns to O.

c) Given that the journey from B to O takes 8 seconds, find, in vector form, an expression for its velocity on that final stage.

Summary

You should know how to ...

► Use properties of vectors.
  ▷ The magnitude of vector $\overrightarrow{PQ} = a\mathbf{i} + b\mathbf{j} + c\mathbf{k}$ is given by $|\overrightarrow{PQ}| = \sqrt{(a^2 + b^2 + c^2)}$
  ▷ The vectors $\overrightarrow{AB}$ and $\overrightarrow{CD}$ are parallel if $\overrightarrow{AB} = k\overrightarrow{CD}$.

► Calculate a scalar (dot) product.
  ▷ $\mathbf{v} \cdot \mathbf{w} = |\mathbf{v}||\mathbf{w}| \cos \theta$

    where θ is the angle between the vectors

  ▷ $\mathbf{v}$ and $\mathbf{w}$ are perpendicular if $\mathbf{v} \cdot \mathbf{w} = 0$

► Understand and use the vector equation of a line.
  ▷ $\mathbf{r} = \mathbf{p} + t\mathbf{d}$

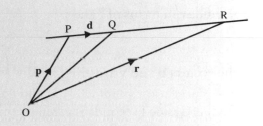

Revision exercise 13

1 A cuboid has dimensions 10 cm, 15 cm and 22 cm and one vertex is at the origin.

a) Write down the position vectors of the other seven vertices.

b) Find the length of a diagonal of the cuboid.

2 The triangle ABC is defined by the following information.

$$\overrightarrow{OA} = \begin{pmatrix} 2 \\ -3 \end{pmatrix}, \quad \overrightarrow{AB} = \begin{pmatrix} 3 \\ 4 \end{pmatrix}, \quad \overrightarrow{AB} \cdot \overrightarrow{BC} = 0,$$

$\overrightarrow{AC}$ is parallel to $\begin{pmatrix} 0 \\ 1 \end{pmatrix}$

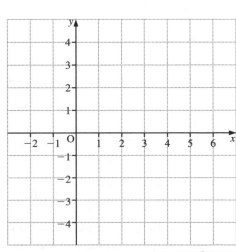

a) On a copy of the grid shown, draw an accurate diagram of triangle ABC.

b) Write down the vector $\overrightarrow{OC}$.

© IBO [2000]

3 The vectors $\begin{pmatrix} 2x \\ x-3 \end{pmatrix}$ and $\begin{pmatrix} x+1 \\ 5 \end{pmatrix}$ are perpendicular for two values of x.

  a) Write down the quadratic equation which the two values of x must satisfy.

  b) Find the two values of x.   © IBO[2001]

4 The vectors **i** and **j** are unit vectors along the x- and y-axes respectively. The vectors $\mathbf{u} = -\mathbf{i} + 2\mathbf{j}$ and $\mathbf{v} = 3\mathbf{i} + 5\mathbf{j}$ are given.

  a) Find $\mathbf{u} + 2\mathbf{v}$ in terms of **i** and **j**.

  A vector **w** has the same direction as $\mathbf{u} = -\mathbf{i} + 2\mathbf{j}$, and has a magnitude of 26.

  b) Find **w** in terms of **i** and **j**.   © IBO[1999]

5 The vectors **u** and **v** are given by $\mathbf{u} = \begin{pmatrix} 6 \\ 4 \\ 0 \end{pmatrix}$ and $\mathbf{v} = \begin{pmatrix} 1 \\ 1 \\ 2 \end{pmatrix}$.

  a) Find $|\mathbf{u} + \mathbf{v}|$.

  b) Find a and b such that $a\mathbf{u} + b\mathbf{v} = \begin{pmatrix} 8 \\ 6 \\ 4 \end{pmatrix}$.

6 The line L passes through the points $(1, 2, 3)$ and $(-1, 0, 4)$.

  Find an equation of L in the form $\begin{pmatrix} x \\ y \\ z \end{pmatrix} = \begin{pmatrix} a \\ b \\ c \end{pmatrix} + t\begin{pmatrix} c \\ d \\ e \end{pmatrix}$ where t

  is any real number.

7 a) Find the scalar product of the vectors $\begin{pmatrix} 60 \\ 25 \end{pmatrix}$ and $\begin{pmatrix} -30 \\ 40 \end{pmatrix}$.

  b) Two markers are at the points $P(60, 25)$ and $Q(-30, 40)$.
    A surveyor stands at $O(0, 0)$ and looks at marker P.
    Find the angle she turns through to look at marker Q.   © IBO[2003]

8 Two lines L_1 and L_2 have the following equations.

$$L_1: \mathbf{r} = \begin{pmatrix} 2 \\ 2 \\ 2 \end{pmatrix} + s\begin{pmatrix} 1 \\ 2 \\ 3 \end{pmatrix} \quad \text{and} \quad L_2: \mathbf{r} = \begin{pmatrix} 1 \\ 0 \\ -1 \end{pmatrix} + t\begin{pmatrix} 0 \\ 5 \\ 1 \end{pmatrix}$$

  a) Calculate the angle between L_1 and L_2.

  b) Find the coordinates of the point where the two lines intersect.

c) Write down an equation for line L_3 which is parallel to L_1 and passes through the point P$(1, 5, 0)$.

d) Show that P also lies on L_2.

e) Hence state the point of intersection of L_2 and L_3 and the angle between them.

9 In this question, a unit vector represents a displacement of 1 metre.

A miniature car moves in a straight line, starting at the point $(2, 0)$. After t seconds, its position, (x, y), is given by the vector equation

$$\begin{pmatrix} x \\ y \end{pmatrix} = \begin{pmatrix} 2 \\ 0 \end{pmatrix} + t \begin{pmatrix} 0.7 \\ 1 \end{pmatrix}$$

a) How far from the point $(0, 0)$ is the car after 2 seconds?

b) Find the speed of the car.

c) Obtain the equation of the car's path in the form $ax + by = c$.

Another miniature vehicle, a motorcycle, starts at the point $(0, 2)$, and travels in a straight line with constant speed. The equation of its path is

$$y = 0.6x + 2, \quad x \geqslant 0$$

Eventually, the two miniature vehicles collide.

d) Find the coordinates of the collision point.

e) If the motorcycle left point $(0, 2)$ at the same moment that the car left point $(2, 0)$, find the speed of the motorcycle.   © *IBO* [2001]

10 Three of the coordinates of the parallelogram STUV are S$(-2, -2)$, T$(7, 7)$, U$(5, 15)$.

a) Find the vector $\overrightarrow{ST}$ and hence the coordinates of V.

b) Find a vector equation of the line (UV) in the form $\mathbf{r} = \mathbf{p} + s\mathbf{d}$ where $s \in \mathbb{R}$.

c) Show that the point E with position vector $\begin{pmatrix} 1 \\ 11 \end{pmatrix}$ is on the line (UV), and find the value of s for this point.

The point W has position vector $\begin{pmatrix} a \\ 17 \end{pmatrix}$, $a \in \mathbb{R}$.

d) i) If $|\overrightarrow{EW}| = 2\sqrt{13}$, show that one value of a is -3 and find the other possible value of a.

ii) For $a = -3$, calculate the angle between $\overrightarrow{EW}$ and $\overrightarrow{ET}$.   © *IBO* [2002]

11 The diagram shows the positions of towns O, A, B and X.

Town A is 240 km East and 70 km North of O.
Town B is 480 km East and 250 km North of O.
Town X is 339 km East and 238 km North of O.

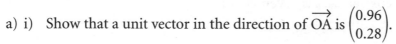

diagram
not to scale

An airplane flies at a constant speed of 300 km h^{-1} from O
towards A.

a) i) Show that a unit vector in the direction of $\overrightarrow{OA}$ is $\begin{pmatrix} 0.96 \\ 0.28 \end{pmatrix}$.

   ii) Write down the velocity vector for the airplane in the

   form $\begin{pmatrix} v_1 \\ v_2 \end{pmatrix}$.

   iii) How long does it take for the airplane to reach A?

At A the airplane changes direction so it now flies towards B.
The angle between the original direction and the new direction
is θ as shown in the second diagram. This diagram also shows the
point Y, between A and B, where the airplane comes closest to X.

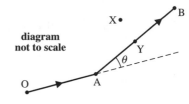

diagram
not to scale

b) Use the scalar product of two vectors to find the value of θ in
   degrees.

c) i) Write down the vector $\overrightarrow{AX}$.

   ii) Show that the vector $\mathbf{n} = \begin{pmatrix} -3 \\ 4 \end{pmatrix}$ is perpendicular to $\overrightarrow{AB}$.

   iii) Calculate the distance XY.

d) How far is the airplane from A when it reaches Y?   © *IBO*[2001]

14 Probability

Suppose a fair die is rolled. Then there are six possible outcomes,

    1   2   3   4   5   6

Suppose you want to know the likelihood of getting a 3.
There is only one way of getting a 3.

Therefore the probability of getting a 3 is one out of six, or $\frac{1}{6}$.

> The list of all possible outcomes is called the **sample space**, and those outcomes which meet the particular requirement are called the **event**.

In the diagram, the rectangle U represents the sample space, and the oval represents the event A. The probability of A is defined by

$$P(A) = \frac{n(A)}{n(U)}$$

where $n(A)$ is the number of elements in the set A, and $n(U)$ is the number of elements in the whole sample space U. That is

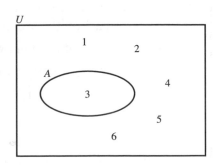

> The probability of $A = \dfrac{\text{number of ways in which } A \text{ can happen}}{\text{number of possible outcomes}}$

In the original example, A is the event that the score is 3, so $n(A) = 1$ and $n(U) = 6$. Therefore

$$P(A) = \frac{n(A)}{n(U)} = \frac{1}{6}$$

as shown earlier.

Example 1

Assuming that births are equally likely on any day of the week, find the probability that the next person you meet was born on a weekday.

....

Let A be the event that the day is a weekday – Monday, Tuesday, Wednesday, Thursday or Friday. So $n(A) = 5$.
There are seven days in a week, so $n(U) = 7$.

Hence

$$P(A) = \frac{n(A)}{n(U)} = \frac{5}{7}$$

Example 2

Two fair dice are thrown. Find the probability that the total of the scores on the two dice is five.

There are 36 possible outcomes:

$$[1, 1] \quad [1, 2] \quad [1, 3] \quad [1, 4] \quad [1, 5] \quad [1, 6]$$
$$[2, 1] \quad [2, 2] \quad [2, 3] \quad [2, 4] \quad [2, 5] \quad [2, 6]$$
$$[3, 1] \quad [3, 2] \quad [3, 3] \quad [3, 4] \quad [3, 5] \quad [3, 6]$$
$$[4, 1] \quad [4, 2] \quad [4, 3] \quad [4, 4] \quad [4, 5] \quad [4, 6]$$
$$[5, 1] \quad [5, 2] \quad [5, 3] \quad [5, 4] \quad [5, 5] \quad [5, 6]$$
$$[6, 1] \quad [6, 2] \quad [6, 3] \quad [6, 4] \quad [6, 5] \quad [6, 6]$$

where, for example, $[1, 2]$ means the outcome of getting a 1 on the first die and a 2 on the second die.

Let A be the event that the *total score is five*.

There are four ways of getting a *total score of five*:

$$[1, 4] \, [2, 3] \, [3, 2] \, [4, 1]$$

So $n(A) = 4$, and $n(U) = 36$.

Hence, $P(A) = \dfrac{n(A)}{n(U)} = \dfrac{4}{36} = \dfrac{1}{9}$

14.1 Combined events

Now consider two events, A and B. Two possible outcomes are

$A \cap B$, which means that A and B **both** occur

and $A \cup B$, which means that A occurs **or** B occurs.

Note that $A \cup B$ includes the case when A and B both occur.

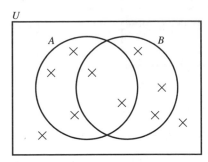

From elementary set theory, you know that

$$n(A \cup B) = n(A) + n(B) - n(A \cap B)$$

So:

$$P(A \cup B) = P(A) + P(B) - P(A \cap B)$$

Playing cards

In many countries games are played with playing cards.

An ordinary pack consists of 52 cards.

Cards are in four different suits; hearts, clubs, diamonds, and spades. In each suit there are 13 cards, labelled Ace, 2, 3, 4, …, 10, Jack, Queen, and King.

A typical suit of hearts is illustrated. The diamonds and hearts are usually coloured red, and the clubs and spades are coloured black.

Example 1

A card is selected at random from an ordinary pack of 52 cards. Find the probability that the card is

a) a king   b) a heart   c) the king of hearts

d) either a king or a heart.

··

Let K denote the event that the card is a king, and let H denote the event that the card is a heart.

a) $P(K) = \frac{4}{52} = \frac{1}{13}$

b) $P(H) = \frac{13}{52} = \frac{1}{4}$

c) The event 'choosing the king of hearts' is written as $K \cap H$. So:

$$P(K \cap H) = \tfrac{1}{52}$$

> There is only one king of hearts in the pack.

d) Choosing the *king* or a *heart* is denoted by the event $K \cup H$,

$$P(K \cup H) = P(K) + P(H) - P(K \cap H)$$

$$\therefore P(K \cup H) = \tfrac{4}{52} + \tfrac{13}{52} - \tfrac{1}{52}$$

$$= \tfrac{16}{52} = \tfrac{4}{13}$$

Mutually exclusive events

Two events, A and B, are said to be **mutually exclusive** when they have *no outcome in common*. In other words, when $n(A \cap B) = 0$. The Venn diagram illustrates such a case.

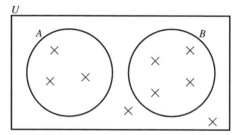

A and *B* do not intersect, so $P(A \cap B) = 0$.

When A and B are mutually exclusive,

$$P(A \cup B) = P(A) + P(B)$$

Example 2

Given that the events A and B are mutually exclusive with $P(A) = \tfrac{3}{10}$ and $P(B) = \tfrac{2}{5}$, find the value of $P(A \cup B)$.

....

$$P(A \cup B) = P(A) + P(B) - P(A \cap B)$$

However, $P(A \cap B) = 0$, so:

$$P(A \cup B) = \tfrac{3}{10} + \tfrac{2}{5} = \tfrac{7}{10}$$

Exhaustive events

Two events, A and B, are said to be **exhaustive** if together they include *possible outcomes* in the sample space. In other words, when $A \cup B = U$. The Venn diagram illustrates such a case.

When A and B are exhaustive,

$$P(A \cup B) = 1$$

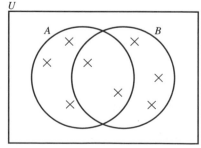

There are no outcomes that are not included in *A* or *B*.

Example 3

Given $P(X) = \frac{4}{5}$, $P(Y) = \frac{1}{2}$ and $P(X \cap Y) = \frac{3}{10}$, show that the events X and Y are exhaustive.

....

$$P(X \cup Y) = P(X) + P(Y) - P(X \cap Y)$$

Therefore,

$$P(X \cup Y) = \frac{4}{5} + \frac{1}{2} - \frac{3}{10} = 1$$

Hence X and Y are exhaustive.

Complementary events

The **complement** of an event A (written A') consists of all outcomes in the sample space which are *not* contained in A. Notice that A and A' are both mutually exclusive:

$$P(A' \cup A) = P(A') + P(A)$$

and exhaustive:

$$P(A' \cup A) = 1$$

Therefore,

$$P(A') + P(A) = 1$$
$$\therefore P(A') = 1 - P(A)$$

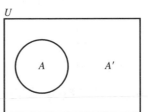

A' is the event 'not A'.
$A \cup A' = U$

Example 4

Given $P(A) = 0.55$, $P(A \cup B) = 0.7$ and $P(A \cap B) = 0.2$, find $P(B')$.

....

$$P(A \cup B) = P(A) + P(B) - P(A \cap B)$$
$$\therefore P(B) = 0.7 + 0.2 - 0.55 = 0.35$$
$$P(B') = 1 - P(B)$$
$$\therefore P(B') = 1 - 0.35 = 0.65$$

Example 5

Given $P(G') = 5x$, $P(H) = \frac{3}{5}$, $P(G \cup H) = 8x$ and $P(G \cap H) = 3x$, find the value of x.

....

$$P(G) = 1 - P(G') = 1 - 5x$$
$$P(G \cup H) = P(G) + P(H) - P(G \cap H)$$
$$\therefore 8x = (1 - 5x) + \frac{3}{5} - 3x$$
$$16x = \frac{8}{5}$$
$$\therefore x = \frac{1}{10}$$

Exercise 14A

1 A card is selected at random from a pack of 52 cards.
Find the probability that the card is

a) black

b) an honour [aces, kings, queens and jacks are honours]

c) a black honour

d) either black or an honour.

2 In a bag are 100 discs numbered 1 to 100. A disc is selected at random from the bag.
Find the probability that the number on the selected disc is

a) even

b) a multiple of five

c) a multiple of ten

d) either even or a multiple of five.

3 Two fair dice are thrown. Find the probability that

a) at least one of the dice shows a *four*

b) the sum of the scores on the two dice is *nine*

c) one of the dice shows a *four* and the other shows a *five*

d) either at least one of the dice shows a *four* or the total of the scores on the two dice is *nine*.

4 In a class half the pupils study Mathematics, a third study English and a quarter study both Mathematics and English.
Find the probability that a pupil selected at random from the class studies either Mathematics or English.

5 Given $P(A) = \frac{3}{5}$, $P(B) = \frac{2}{3}$ and $P(A \cap B) = \frac{1}{2}$, find the value of $P(A \cup B)$.

6 Given $P(X) = 0.37$, $P(Y) = 0.48$ and $P(X \cup Y) = 0.69$, find the value of $P(X \cap Y)$.

7 Given $P(A) = \frac{7}{10}$, $P(A \cup B) = \frac{9}{10}$ and $P(A \cap B) = \frac{3}{20}$, find the value of $P(B)$.

8 Given $P(F) = 4x$, $P(G) = \frac{1}{3}$, $P(F \cap G) = x$ and $P(F \cup G) = 8x$, find the value of x.

9 Given that the events X and Y are mutually exclusive with $P(X) = \frac{4}{7}$ and $P(Y) = \frac{1}{3}$, find the value of $P(X \cup Y)$.

10 Given $P(S) = 0.34$, $P(T) = 0.49$ and $P(S \cup T) = 0.83$, show that the events S and T are mutually exclusive.

11 Given that the events M and N are mutually exclusive with $P(M) = 3x$, $P(N) = 4x$ and $P(M \cup N) = 1 - x$, find the value of x.

12 Given that the events A and B are exhaustive with $P(A) = \frac{2}{3}$ and $P(B) = \frac{3}{4}$, find the value of $P(A \cap B)$.

13 Given $P(X) = \frac{5}{8}$, $P(Y) = \frac{11}{12}$ and $P(X \cap Y) = \frac{13}{24}$, show that the events X and Y are exhaustive.

14 Given that the events S and T are exhaustive with $P(S) = x$, $P(T) = 3x$ and $P(S \cap T) = 1 - 5x$, find the value of x.

15 When a roulette wheel is spun, the score will be a number from 0 to 36 inclusive. Each score is equally likely.
Find the probability that the score is
a) an even number
b) a multiple of 3
c) a multiple of 6
d) an odd number which is not a multiple of 3.

16 As a result of a survey of the households in a town, it is found that 80% have a video recorder and 24% have satellite television. Given that 15% have both a video recorder and satellite television, find the proportion of households with neither a video recorder nor satellite television.

17 The children at a party were asked about their pets. Two-thirds had a dog and three-quarters had a cat. Given that half the children had both a cat and a dog, calculate the probability that a child selected at random is found to have neither a cat nor a dog.

18 Given $P(C) = 0.44$, $P(C \cap D) = 0.21$ and $P(C \cup D) = 0.03$, find $P(D')$.

19 Given $P(G') = 3x$, $P(H) = 4x$, $P(G \cap H) = \frac{1}{4}$ and $P(G \cup H) = 9x$, find the value of x.

20 Given $P(A) = 0.6$, $P(A \cap B') = 0.4$ and $P(A \cup B) = 0.85$, find the value of $P(B)$.

21 Show that $P(A \cup B \cup C) = P(A) + P(B) + P(C) - P(A \cap B) - P(B \cap C) - P(C \cap A) + P(A \cap B \cap C)$.

14.2 Conditional probability

Suppose you are considering two related events, A and B, and you are told that B has occurred. This information may influence the likelihood of A occurring. For example, if you select a card at random from a pack of 52, then the probability that it will be a heart is $\frac{13}{52} = \frac{1}{4}$.

However, if you are given the additional information that the card selected is red, then this probability is increased to $\frac{13}{26} = \frac{1}{2}$.

The probability that event A will occur given that event B has already occurred is given by

$$\frac{n(A \cap B)}{n(B)}$$

This probability, which is denoted by $P(A \mid B)$, is illustrated in the Venn diagram.

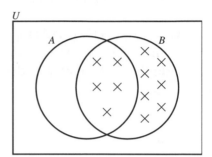

$$P(A \mid B) = \frac{P(A \cap B)}{P(B)}$$

Going back to the pack of cards, A is the event *obtain a heart* and B is the event *obtain a red card*, so $A \cap B$ is the event *obtain a heart* (all hearts being red).

Now $P(A \cap B) = \frac{13}{52}$ and $P(B) = \frac{26}{52}$. Therefore,

$$P(A \mid B) = \frac{\frac{13}{52}}{\frac{26}{52}} = \frac{1}{2}$$

Example 1

Two fair dice are thrown. Find the probability that one of the dice shows a *four* given that the total on the two dice is *ten*.

....

Let F denote the event that one of the dice shows a four and T denote the event that the total on the two dice is ten.

$$
\begin{array}{cccccc}
[1, 1] & [1, 2] & [1, 3] & [1, 4] & [1, 5] & [1, 6] \\
[2, 1] & [2, 2] & [2, 3] & [2, 4] & [2, 5] & [2, 6] \\
[3, 1] & [3, 2] & [3, 3] & [3, 4] & [3, 5] & [3, 6] \\
[4, 1] & [4, 2] & [4, 3] & [4, 4] & [4, 5] & [4, 6] \\
[5, 1] & [5, 2] & [5, 3] & [5, 4] & [5, 5] & [5, 6] \\
[6, 1] & [6, 2] & [6, 3] & [6, 4] & [6, 5] & [6, 6]
\end{array}
$$

Then $F \cap T$ means $[4, 6]$ or $[6, 4]$, and T means $[4, 6]$ or $[5, 5]$ or $[6, 4]$.

Therefore,

$$P(F \cap T) = \frac{2}{36} \quad \text{and} \quad P(T) = \frac{3}{36}$$

Then: $P(F \mid T) = \dfrac{P(F \cap T)}{P(T)}$

$$\therefore P(F \mid T) = \frac{\frac{2}{36}}{\frac{3}{36}} = \frac{2}{3}$$

Example 2

Given $P(A) = \frac{1}{2}$, $P(A \mid B) = \frac{1}{4}$ and $P(A \cup B) = \frac{2}{3}$, find $P(B)$.

....

$$P(A \mid B) = \frac{P(A \cap B)}{P(B)}$$

$$\therefore \frac{1}{4} = \frac{P(A \cap B)}{P(B)}$$

$$\therefore P(A \cap B) = \tfrac{1}{4}P(B) \qquad\qquad [1]$$

$$P(A \cup B) = P(A) + P(B) - P(A \cap B)$$

So: $\quad \frac{2}{3} = \frac{1}{2} + P(B) - P(A \cap B)$

$$\therefore \tfrac{1}{6} = P(B) - P(A \cap B) \qquad\qquad [2]$$

Eliminating $P(A \cap B)$ between [1] and [2] gives

$$\tfrac{1}{6} = P(B) - \tfrac{1}{4}P(B)$$

$$\tfrac{3}{4}P(B) = \tfrac{1}{6}$$

$$\therefore P(B) = \tfrac{2}{9}$$

Independence

Two events A and B are said to be **independent** if

$$P(A \mid B) = P(A)$$

In this case:

$$\frac{P(A \cap B)}{P(B)} = P(A)$$

$$\therefore P(A \cap B) = P(A) \times P(B)$$

Example 3

A card is picked at random from a pack of 52 and a fair die is thrown. Find the probability that the card is the ace of spades and the die shows a three.

....

You could choose to list all the 52×6 possible outcomes, but it is easier to use independence and write

$$P(\text{ace of spades and a two}) = P(\text{ace of spades}) \times P(\text{two})$$

$$= \tfrac{1}{52} \times \tfrac{1}{6}$$

$$= \tfrac{1}{312}$$

Tree diagrams

Sometimes, rather than list all the possible outcomes, it is easier to view probabilities on a **tree diagram**. The next example shows how this is done.

Example 4

A fair coin is thrown twice. Find the probability that the result is a head and a tail, in either order.

$\cdots\cdots\cdots\cdots\cdots\cdots\cdots\cdots\cdots\cdots\cdots\cdots\cdots\cdots\cdots\cdots$

You could just calculate this simple example like this:

Without using a tree diagram, and using independence to multiply probabilities, and mutual exclusivity to add them,

$$\text{P(head and tail)} = \text{P}(HT) + \text{P}(TH)$$

$$= \tfrac{1}{2} \times \tfrac{1}{2} + \tfrac{1}{2} \times \tfrac{1}{2}$$

$$= \tfrac{1}{4} + \tfrac{1}{4}$$

$$= \tfrac{2}{4} = \tfrac{1}{2}$$

> The outcomes *HT* and *TH* are mutually exclusive.

> The events *H* and *T* are independent, so multiply the probabilities.

Alternatively, you can draw a tree diagram like this:

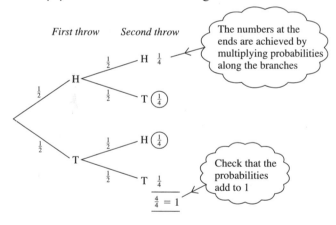

> The numbers at the ends are achieved by multiplying probabilities along the branches

> Check that the probabilities add to 1

You want all those combinations of branches which include one head and one tail.

To calculate P(*HT*), multiply the probabilities along the branches.

For example,

$$\text{P}(HT) = \tfrac{1}{2} \times \tfrac{1}{2} = \tfrac{1}{4}$$

You can now easily see that the probability you want is the sum of those circled on the diagram: HT and TH. That is:

$$\tfrac{1}{4} + \tfrac{1}{4} = \tfrac{2}{4} = \tfrac{1}{2}$$

Example 4 is a very simple one. However, tree diagrams can be very useful in more complicated problems.

Example 5

A bag contains four red discs and five blue discs.
Three discs are selected at random, without replacing them.
Find the probably that two are red and the other is blue.

....

The probability of the first disc being red is $\frac{4}{9}$. That leaves 8
discs, of which 3 are red. The probability of the second disc
being red is thus $\frac{3}{8}$. The probability of the third disc being blue
is $\frac{5}{7}$, as there are still 5 blue discs left.

So $P(RRB) = \frac{4}{9} \times \frac{3}{8} \times \frac{5}{7}$ and so on.

So:

$$P(\text{two red and one blue}) = P(RRB) + P(RBR) + P(BRR)$$
$$= \left(\frac{4}{9} \times \frac{3}{8} \times \frac{5}{7}\right) + \left(\frac{4}{9} \times \frac{5}{8} \times \frac{3}{7}\right)$$
$$+ \left(\frac{5}{9} \times \frac{4}{8} \times \frac{3}{7}\right)$$
$$= \frac{60}{504} + \frac{60}{504} + \frac{60}{504}$$
$$= \frac{5}{14}$$

Alternatively, you can use a tree diagram:

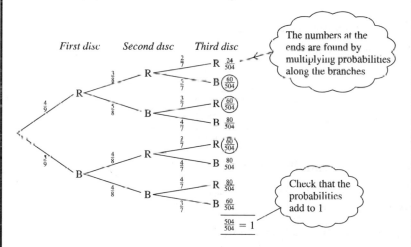

The numbers at the ends are found by multiplying probabilities along the branches

Check that the probabilities add to 1

You want all those combinations of branches which include
two red discs and one blue disc: for example, *RRB*.

To calculate P(*RRB*), multiply the probabilities along the
branches:

$$P(RRB) = \frac{4}{9} \times \frac{3}{8} \times \frac{5}{7} = \frac{60}{504}$$

The probability you want is the sum of those circled on the
diagram: *RRB*, *RBR* and *BRR*. That is:

$$\frac{60}{504} + \frac{60}{504} + \frac{60}{504} = \frac{5}{14}$$

Exercise 14B

In questions **1** to **6** two fair dice are rolled, and the outcome is the total of the scores on each of the two dice. It will help if you first draw a grid to illustrate the 36 possible outcomes for the total score.

1 Find the probability that one of the dice shows a *two* given that the total on the two dice is *six*.

2 Find the probability that one of the dice shows a *three* given that the total on the two dice is *seven*.

3 Find the probability that the total score is *eight* given that at least one of the dice shows a *two*.

4 Find the probability that the scores on each of the two dice are the same given that the total on the two dice is *four*.

5 Find the probability that the total on the two dice is *eight* given that neither die shows a *five*.

6 Find the probability that neither dice shows a *four* given that the total on the two dice is *nine*.

7 Three fair coins are tossed.
a) List the eight possible outcomes.
b) Find the probability that all three coins show *heads* given that there is an odd number of *heads* showing.

8 Four fair coins are tossed.
a) List the 16 possible outcomes.
b) Find the probability that the coins show two *heads* and two *tails* given that there is at least one *tail*.

9 A computer randomly chooses two different numbers from the list 1, 2, 3, 4, 5.
a) Draw a grid to show the 20 possible outcomes.
b) Find the probability that both numbers are odd given that neither is a *four*.

10 Five identical cards are labelled A, B, C, D, E. The cards are placed in a box, and a card is selected at random. This card is replaced in the box, and a second random selection is made.
a) Draw a grid to show the 25 possible outcomes.
b) Given that neither card has a letter which appears in the word VICTORY, calculate the probability that both are vowels.

11 Given $P(A \cap B) = \frac{1}{3}$ and $P(B) = \frac{3}{5}$, find $P(A \mid B)$.

12 Given $P(A \mid B) = \frac{1}{5}$ and $P(B) = \frac{1}{2}$, find $P(A \cap B)$.

13 Given $P(A \mid B) = \frac{5}{6}$, $P(A) = \frac{3}{4}$ and $P(B) = \frac{2}{3}$, find $P(A \cup B)$.

14 Given $P(A) = \frac{2}{5}$, $P(B) = \frac{7}{10}$ and $P(A \cup B) = \frac{4}{5}$, find $P(A \mid B)$.

15 Given $P(A) = \frac{2}{3}$, $P(B) = \frac{4}{9}$ and $P(A \mid B) = \frac{3}{4}$, find $P(A \cup B)$.

16 Given $P(A) = 0.2$, $P(A \mid B) = 0.3$ and $P(A \cup B) = 0.4$, find $P(B)$.

17 Given $P(A) = \frac{1}{3}$, $P(B) = \frac{3}{8}$ and $P(A \cup B) = \frac{7}{12}$, show that A and B are independent.

18 Given $P(A) = x$, $P(B) = 2x$, $P(A \cup B) = 1 - x$ and $P(A \mid B) = \frac{1}{5}$, find the value of x.

19 A bag contains seven red cards and three blue cards. Two cards are selected at random.

a) Copy and complete the tree diagram of possible outcomes.

b) Find the probability that both cards are the same colour.

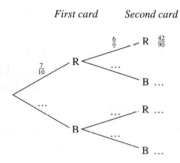

20 Another bag also contains seven red cards and three blue cards. This time a card is selected and replaced. A second card is then selected.

a) Copy and complete the tree diagram of possible outcomes.

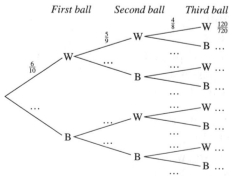

b) Find the probability that the cards are a different colour.

21 A box contains six white balls and four black balls. Three balls are selected at random.

a) Copy and complete the tree diagram of possible outcomes.

b) Find the probability that two of the selected balls are white and the other is black.

22 Two sacks, *A* and *B*, each contain a mixture of plastic and leather rugby balls. Sack *A* contains four plastic balls and two leather balls, and sack *B* contains three plastic balls and five leather balls. A sack is selected at random and a ball is taken from it.

a) Represent this information on a tree diagram.

b) Calculate the probability that the ball is leather.

c) Given that the ball is leather, calculate the probability that it came from sack *A*.

23 A girl cannot decide what to wear for a party. She will wear either trousers or a skirt. She has two wardrobes. In wardrobe *A* are two pairs of trousers and three skirts, and in wardrobe *B* are five pairs of trousers and one skirt. She decides to select a wardrobe at random and then randomly select one item from that wardrobe.

Draw a tree diagram and find the probability that

a) she wears trousers

b) she selected wardrobe *A*, given that she wears trousers.

24 An eccentric mathematics teacher decides to award a prize to a pupil selected from one of his three classes. Class 1 has five boys and seven girls; class 2 has eight boys and two girls; and class 3 has three boys and three girls. He decides to award the prize by selecting a class at random and then randomly selecting a pupil from that class.

Use a tree diagram to find the probability that

a) the selected pupil is a boy

b) the selected pupil is from class 3, given that the selected pupil is a boy.

25 A bag contains eleven discs numbered 1 to 11. Two discs are selected at random from the bag. Given that the sum of the selected numbers is even, use a tree diagram to find the probability that the number on each of the selected discs is odd.

26 All of the pupils in two classes sit the same examination. In class A, 15 out of 20 pupils pass; in class B, 8 out of 12 pupils pass. A pupil is selected at random from the 32 pupils who have taken the examination. Given that she passes, use a tree diagram to find the probability that she was in class A.

27 In a given week, 200 people take a driving test in a given centre. Of the 120 who pass, 50 were taking the test for the first time; and of those who failed, 60 were taking the test for the first time. Estimate the probability of passing the driving test at a first attempt at this centre.

Summary

You should know how to ...

▶ Calculate the probability of an event.

    ▷ For an event A, $P(A) = \dfrac{n(A)}{n(U)}$ where $n(A)$ is the set of outcomes comprising event A, and $n(U)$ is the total set of all possible outcomes.

▶ Calculate probabilities associated with more than one event.

    ▷ $P(A \cup B) = P(A) + P(B) - P(A \cap B)$ ▷ If A and B are mutually exclusive,

$$P(A \cup B) = P(A) + P(B)$$

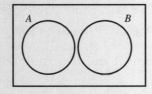

$A \cap B$

    ▷ $P(A) + P(A') = 1$, where A' is the complementary event to A (not A).

    ▷ $P(A \mid B) = \dfrac{P(A \cap B)}{P(B)}$ where $P(A \mid B)$ is a conditional probability

    ▷ If A and B are independent events

$$P(A \mid B) = P(A), \quad \text{and} \quad P(A \cap B) = P(A) \times P(B)$$

Revision exercise 14

1 In a survey, 100 students were asked, 'Do you prefer to watch television or play sport?' Of the 46 boys in the survey, 33 said they would choose sport, while 29 girls made this choice.

By copying and completing this table, find the probability that

	Boys	Girls	Total
Television			
Sport	33	29	
Total	46		100

    a) a student selected at random prefers to watch television

    b) a student prefers to watch television, given that the student is a boy.   © *IBO* [2000]

2 Two ordinary, six-sided dice are rolled and the total scored is noted.

    a) Copy and complete the tree diagram by entering probabilities and listing outcomes.

    b) Find the probability of getting one or more sixes.

Outcomes

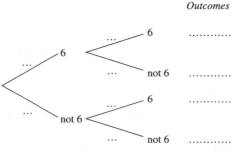

© *IBO* [2000]

3 This Venn diagram shows a sample space U and events A and B.

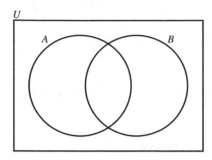

$n(U) = 36, n(A) = 11, n(B) = 6$ and $n(A \cup B)' = 21$.

a) On a copy of the diagram, shade the region $(A \cup B)'$.

b) Find i) $n(A \cap B)$ ii) $P(A \cap B)$.

c) Explain why events A and B are not mutually exclusive.   © IBO [2000]

4 A bag contains 10 red balls, 10 green balls and 6 white balls. Two balls are drawn at random from the bag without replacement. What is the probability that they are different colours?   © IBO [2001]

5 A box contains 22 red apples and 3 green apples. Three apples are selected at random, one after the other without replacement.

a) The first two apples are green. What is the probability that the third apple is red?

b) What is the probability that exactly two of the three apples are red?   © IBO [2002]

6 The events A and B are independent and $P(A \cap B) = 0.6$, $P(B) = 0.8$. Find

a) $P(A \mid B)$ b) $P(A \mid B')$ © IBO [1998]

7 For the events A and B, $P(A) = 0.3$ and $P(B) = 0.4$.

a) Find $P(A \cup B)$ if A and B are independent events.

b) Find $P(A' \cap B')$ if A and B are mutually exclusive events.   © IBO [1996]

8 In a certain country, 40% of the population own a car and of these 90% are male. The two sexes are equally represented in the population.

a) Find the probability that a randomly selected person is a male car owner.

b) What percentage of women do not own a car?   © IBO [1998]

9 In a school of 88 boys, 32 study Economics (E), 28 study History (H) and 39 do not study either subject. This information is represented in the Venn diagram.

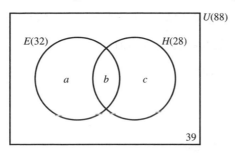

a) Calculate the values of a, b and c.

b) A student is selected at random.

    i) Calculate the probability that he studies both Economics and History.

    ii) Given that he studies Economics, calculate the probability that he does not study History.

c) A group of three students is selected at random from the school.

    i) Calculate the probability that none of these students studies Economics.

    ii) Calculate the probability that at least one of these students studies Economics.

© *IBO*[2003]

10 A student sits three examinations in Physics, Chemistry and French. From past experience she estimates that the probability of passing Physics is 0.4. If she passes Physics the probability of passing Chemistry is 0.6, otherwise it is 0.3. Passing French is independent of her performance in the other two subjects and has a probability of 0.45.

a) Draw a tree diagram to represent this information.

b) Find the probability that she

    i)   passes all three subjects

    ii) passes exactly one subject

    iii) passes at least one subject.

Stem and leaf diagrams

Tally charts are not always suitable if the range of possible values is very large. In these cases a **stem and leaf diagram** summarises the data better. The stem represents the most significant figure and the leaves are the less significant figures. For example, the ages of people staying in a hotel on one particular night are:

10, 13, 12, 20, 9, 21, 14, 32, 21, 6,
10, 56, 41, 4, 51, 12, 33, 8, 31, 23

This can be summarised as: If you order the data you get:

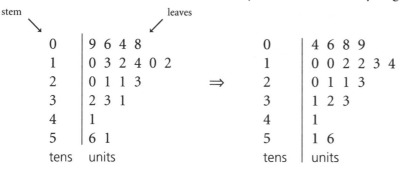

stem	leaves
0	9 6 4 8
1	0 3 2 4 0 2
2	0 1 1 3
3	2 3 1
4	1
5	6 1
tens	units

$\Rightarrow$

0	4 6 8 9
1	0 0 2 2 3 4
2	0 1 1 3
3	1 2 3
4	1
5	1 6
tens	units

> The stem is the tens and the leaves are the units.
> 0|9 means 9
> 3|2 means 32

Bar charts

You can represent the number of cars sold per week by a garage over a 20-week period, as recorded in the tally chart on page 385, in a bar chart. The lengths of the bars are proportional to the number of observations of each outcome.

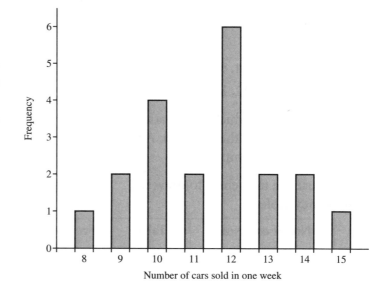

Line graphs

In a line graph, the data points are plotted on axes and joined by straight lines. The car sales data are shown here as a line graph.

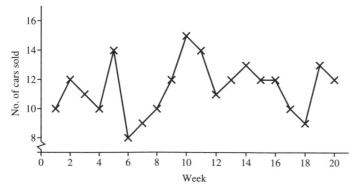

> The line graph shows how sales vary over time.

Pictograms

Pictograms are bar charts in which the bars are replaced by symbols (pictures) representing the subject of the data.
They are less precise than bar charts.

Example 1

The table shows January sales figures for four manufacturers of cars.

Manufacturer	Jaguar	Porsche	Saab	Volvo
Number of cars	1550	466	925	2050

Represent the data using a suitable pictogram.

Using a picture of a car to represent 500 car sales, the pictogram is as shown.

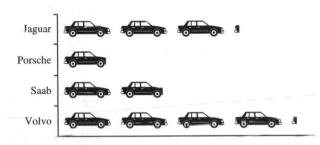

Pie charts

In a pie chart, the areas of the portions of the pie are in proportion to the quantities being represented.

Example 2

The table shows the number of 240 students who achieved grades A to E in their mathematics examinations in a particular year.

Grade	A	B	C	D	E
Number of students	35	62	71	54	18

Represent these data in a pie chart.

First convert the frequencies to angles. These are the angles that represent each grade as a proportion of the pie chart.

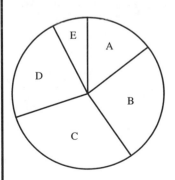

Grade	Number of students	Angle
A	35	$\frac{35}{240} \times 360° = 52.5°$
B	62	$\frac{62}{240} \times 360° = 93°$
C	71	$\frac{71}{240} \times 360° = 106.5°$
D	54	$\frac{54}{240} \times 360° = 81°$
E	18	$\frac{18}{240} \times 360° = 27°$
Total	240	360°

Exercise 15A

1 Here are the points scored by the final 30 competitors in a talent contest:

> 74, 75, 77, 82, 80, 77, 79, 79, 81, 82, 82, 82, 79, 80, 78,
> 81, 86, 79, 80, 81, 81, 83, 83, 84, 81, 84, 80, 80, 77, 81

a) By first drawing a tally chart, obtain a frequency table.

b) Represent these scores on a bar chart.

2 Here are the numbers of puppies in 40 different litters.

> 8, 6, 7, 2, 10, 5, 7, 3, 3, 4, 8, 8, 4, 5, 7, 5, 6, 5, 9, 5,
> 6, 7, 7, 5, 7, 4, 7, 8, 8, 8, 5, 12, 4, 9, 7, 10, 5, 12, 2, 6

a) By first drawing a tally chart, obtain a frequency table.

b) Represent these scores on a bar chart.

3 The bar chart shows the colours of the cars in a car park.

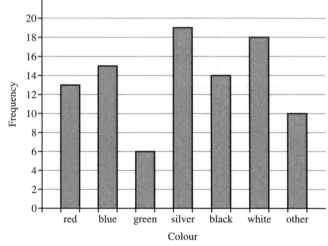

a) How many cars were in the car park?

b) What percentage of the cars were silver?

4 The pictogram shows the number of e-mails received in a given week by each of five employees of a company.

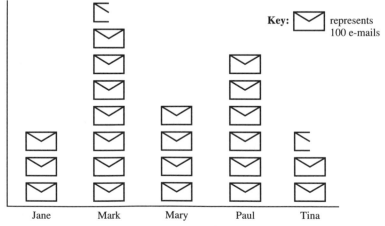

a) How many e-mails did Mark receive?

b) How many e-mails were received in total by the five employees?

5 Here are the marks of a group of students in a Chemistry exam.

45, 56, 67, 73, 82, 91, 67, 85, 94, 88, 65, 76, 48, 62, 84,
57, 74, 88, 73, 93, 82, 45, 58, 72, 81, 60, 91, 73, 77, 81

Construct a stem and leaf diagram to summarise these marks.

6 The points scored by a rugby team in all its matches in one season are given here.

23, 32, 15, 27, 0, 29, 34, 46, 17, 8,
33, 63, 24, 12, 48, 37, 26, 0, 36, 41

Construct a stem and leaf diagram to summarise these results.

7 The table shows the number of tubers under 50 potato plants.

Number of tubers	0	1	2	3	4	5	6	7
Frequency	6	2	5	8	14	9	4	2

Construct a line graph to summarise these results.

8 The frequency table shows the age at which a group of mothers gave birth to their first child.

Age	16–20	21–25	26–30	31–35	36–40	41–45
Frequency	7	12	9	7	3	2

Using mid-interval values, construct a line graph to summarise these results.

> The mid-interval value for the class 16–20 is 18.

9 The shoe sizes of the children in a class are:

4, 5, 7, 6, 4, 6, 7, 7, 8, 5, 4, 7, 6, 4, 7, 8, 6, 5, 7, 4

Construct a pie chart to represent these data.

10 In a drawing competition all the exhibits are awarded a grade from A to E. The results are summarised in the pie chart.

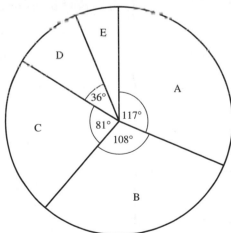

a) Given that 13 drawings were awarded an A, how many were awarded a C?

b) How many drawings were in the competition altogether?

15.3 Mode, median, mean

There are three common ways of measuring the typical, or **average**, value of a set of data. These averages are called the mean, median and mode.

The mode

The **mode** of a set of data is the single value that occurs most often. If two outcomes occur with the greatest frequency then there is no unique mode. The data are described as being **bimodal**. If there are three or more such outcomes then the data are described as being multimodal.

Example 1

A fair die is thrown ten times and the following results obtained:

5, 4, 5, 1, 3, 5, 6, 2, 1, 4

What is the modal score?

···

The score 5 occurs the most (three times) and therefore the modal score is 5.

In the case of continuous data, the modal class is the class which has the greatest frequency.

Example 2

The results of a survey of the age of 110 cars passing a particular point are given in the table.

Age (years)	1–2	2–3	3–4	4–5	5–6
Frequency	15	27	36	21	11

What is the modal class?

···

The greatest frequency is 36, corresponding to 3–4 years. The modal class is 3–4 years.

> When the data are grouped into classes (see page 392), you can't tell exactly what the mode is, only which class it is in.

The median

The **median** is the middle value when all the observations or outcomes are arranged in order of magnitude.

Example 3

Find the median of these sets of data.

a) In an office block the amount spent on lunch by a cross-section of office workers on a particular Friday was recorded as €4, €14, €2, €6, €6, €4, €24, €10, €12, to the nearest euro.

b) The ages of university students in a tutorial group were recorded as 20, 23, 18, 19, 28, 26, 22, 18.

···

a) Arrange in order of magnitude:

   2, 4, 4, 6, 6, 10, 12, 14, 24

   The middle value is 6 and therefore the median is €6.

b) Arrange in order of magnitude:

   18, 18, 19, 20, 22, 23, 26, 28

   In this case no one number is in the middle as there is an even number of observations. The observations 20 and 22 are in the middle. The median is taken to be

   $\frac{1}{2}(20 + 22) = 21$

The mean

The **mean** is the sum of all the observed values divided by the total number of observations. It is written as $\bar{x}$ and calculated by the formula:

$$\bar{x} = \frac{1}{n}\sum x$$

where x_i represents the observed values/outcomes.

> This is also called the arithmetic mean.

Example 4

Calculate the mean of the set of data 3.7, 4.2, 4.0, 2.5 (cm).

....

The mean is given by

$$\bar{x} = \frac{3.7 + 4.2 + 4 + 2.5}{4} = \frac{14.4}{4} = 3.6$$

The mean of the data is 3.6 cm.

The mean of a frequency distribution

Very often, data are presented in the form of a **frequency distribution**. For example, the lengths of stems of a group of 35 plants were recorded, to the nearest 5 cm, and the data recorded in a table.

Length (x cm)	10	15	20	25
Frequency (f)	6	12	13	4

The table shows that there were 6 plants whose stems were 10 cm long, to the nearest 5 cm. There were 12 plants whose stems were 15 cm long, to the nearest 5 cm, and so on.

Notice that the total number of observations/outcomes is the total of the frequencies:

   $n = 6 + 12 + 13 + 4 = 35$

To find the sum of the observed values, draw the table vertically and add another column:

Length (x cm)	Frequency (f)	Total for each group ($x \times f$)
10	6	$10 \times 6 = 60$
15	12	$15 \times 12 = 180$
20	13	$20 \times 13 = 260$
25	4	$25 \times 4 = 100$
	$\Sigma f_i = 35$	$\Sigma f_i x_i = 600$

Therefore

$$\bar{x} = \frac{60 + 180 + 260 + 100}{35} = \frac{600}{35} = 17.1 \text{ (3 s.f.)}$$

For a frequency distribution, the mean is given by

$$\bar{x} = \frac{\Sigma fx}{\Sigma f}$$

There is an important distinction between the values x representing observations of an entire population and the observations x representing a sample. For example, the observations may be the set of measurements relating to the entire workforce in a factory – the whole population. Alternatively, the observations may relate to only 10 employees and will be used as a sample from which conclusions can be drawn about the entire workforce.

Grouped frequency tables

When you have a lot of numerical data, it is helpful to group it into classes or **class intervals**. Each class interval lies between an **upper class boundary** and a **lower class boundary**.

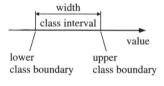

For example, consider the following class intervals representing lengths in cm:

0–10   10–20   20–30   30–40

For the class interval 30–40, 30 is the lower class boundary, 40 is the upper class boundary, and the class width is 10.

Example 5

The heights of flowers in a flower bed were recorded and the grouped frequency table summarises the results.

Length (cm)	0–10	10–20	20–30	30–40
Number of flowers	28	34	17	6

Find an estimate of the mean.

Redraw the table vertically and add additional columns for further calculations. You do not know the actual height of every flower, only that 28 of them have heights in the range 0–10 cm. So to calculate an approximate value for the mean you must use the mid-point value of each interval.

Length (cm)	Mid-point (x)	No. of flowers (f)	$x \times f$
0–10	5	28	140
10–20	15	34	510
20–30	25	17	425
30–40	35	6	210
		$n = 85$	$\Sigma f_i x_i = 1285$

The mean is given by

$$\bar{x} = \frac{\Sigma f_i x_i}{n} = \frac{140 + 510 + 425 + 210}{85} = \frac{1285}{85} = 15.1$$

The mean length is 15.1 cm.

For grouped data, the formula for the mean is

$$\bar{x} = \frac{\Sigma fx}{n}$$

Exercise 15B

1 A statistician throws a fair die 20 times and records the following results.

   4, 2, 1, 6, 4, 5, 3, 4, 6, 2, 3, 2, 2, 4, 6, 3, 5, 1, 2, 3

What is the modal score?

2 Each of the children in a class was asked to name the day of the week on which they were born. The results are given in the table.

Day	Mon	Tue	Wed	Thu	Fri	Sat	Sun
Frequency	2	3	2	5	4	1	6

What is the modal day?

3 Each of the passengers on a bus was asked to name a favourite colour of the rainbow. The results are given in the bar chart.

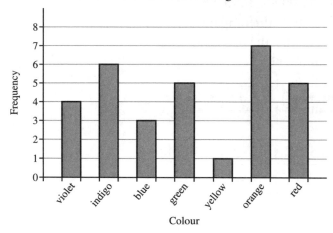

What is the modal colour?

4 Fifteen boxes of matches were purchased, and the number of matches in each box was recorded. The results are:

52, 48, 49, 51, 51, 47, 50, 55, 48, 52, 52, 46, 53, 47, 46

What is the median number of matches?

5 The masses, in kg, of the 11 players in a football team are:

67, 74, 59, 82, 76, 58, 72, 66, 85, 67, 63

What is the median mass?

6 The masses, in kg, of the eight oarsmen in an VIII are:

76, 82, 83, 78, 92, 85, 98, 75

What is the median mass?

7 An athlete records his time over 200 m over five successive races. The times, in seconds, are:

24.2, 25.1, 22.7, 23.5, 24.0

What is his mean time?

8 The four children in a family have ages 5 years, 6 years, 11 years and 14 years. Calculate their mean age.

9 The heights, in cm, of the seven policemen on duty in a police station are:

169, 183, 171, 178, 184, 172, 189

What is the mean height?

10 Four apples have a mean mass of 124 g. Another apple is added to the original four, and the mean mass of the five apples is now 132 g. What is the mass of the additional apple?

11 The table shows the salaries of 80 secretaries in a company.

Salary (€)	8000–10 000	10 000–12 000	12 000–15 000	15 000–20 000
Frequency	45	18	12	5

Calculate an estimate of the mean salary.

12 The table shows the masses of 50 chickens on a supermarket shelf.

Mass (kg)	2.2–2.4	2.4–2.6	2.6–3.2	3.2–3.8	3.8–4.6
Frequency	12	14	11	9	4

Calculate an estimate of the mean mass.

13 The table shows the IQs of 70 students.

IQ	80–100	100–120	120–130	130–140	140–160
Frequency	6	22	25	14	3

Calculate an estimate of the mean IQ.

14 The bar chart shows the speeds of cars on a stretch of road.

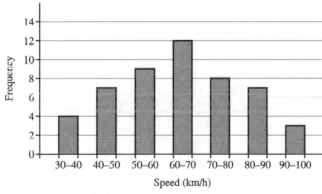

a) How many cars were recorded?

b) Calculate an estimate of the mean speed.

15.4 Measures of dispersion

Range

Look at these two sets of data:

2, 3, 4, 5, 6   [1]

−3, 2, 3, 5, 13   [2]

They both have a mean of 4. However, you can see that data set [2] is more spread out than data set [1]. The mean doesn't tell you this.

To represent the data more accurately, you need the mean plus a measure of the spread or dispersion of the data. One simple measure of dispersion is the **range**.

The range of a set of data is the highest value minus the lowest value.

In this case the range of set [1] is $6 - 2 = 4$.
The range of set [2] is $13 - (-3) = 16$.

The range is easy to calculate, but there are other measures of spread that are more useful.

Mean deviation squared from the mean

A common measure of spread is the mean of the sum of the squares of the deviations from the mean:

$$\frac{\Sigma(x - \bar{x})^2}{n} = \frac{(x_1 - \bar{x})^2 + (x_2 - \bar{x})^2 + \ldots + (x_n - \bar{x})^2}{n}$$

$x - \bar{x}$ is the deviation, or difference, of the value x from the mean value, $\bar{x}$.

The more variation in the data, the larger the value of $\frac{\Sigma(x - \bar{x})^2}{n}$.

For data set [1] in the previous example:

$$\Sigma(x - \bar{x})^2 = (2 - 4)^2 + (3 - 4)^2 + (4 - 4)^2 + (5 - 4)^2 + (6 - 4)^2$$

$$= 10$$

$$\therefore \frac{\Sigma(x - \bar{x})^2}{n} = \frac{10}{5} = 2$$

For data set [2]:

$$\Sigma(x - \bar{x})^2 = (-3 - 4)^2 + (2 - 4)^2 + (3 - 4)^2 + (5 - 4)^2 + (13 - 4)^2$$

$$= 136$$

$$\therefore \frac{\Sigma(x - \bar{x})^2}{n} = \frac{136}{5} = 27.2$$

The value 27.2 compared with the value 2 clearly shows a greater variation in data set [2].

Variance and standard deviation

The mean of the deviations from the mean squared is called the **variance**, usually written as σ^2.

σ is the Greek letter sigma.

The variance is calculated as:

$$\sigma^2 = \frac{1}{n}\Sigma(x - \bar{x})^2$$

Sometimes it is easier to calculate a variance using the alternative formula:

$$\sigma^2 = \frac{1}{n}\Sigma x^2 - \left(\frac{\Sigma x}{n}\right)^2$$

This is often more useful when performing calculations without the use of the GDC.

> The **standard deviation**, σ, is defined as the square root of the variance:
>
> $$\sigma = \sqrt{\frac{1}{n}\Sigma x^2 - \left(\frac{\Sigma x}{n}\right)^2} = \sqrt{\frac{1}{n}\Sigma(x-\bar{x})^2}$$

Example 1

The heights of eight plants are measured 3 months after they are fed with a new plant food.

Calculate the mean and standard deviation of the heights:

　　30 cm, 17 cm, 32 cm, 25 cm, 31 cm, 28 cm, 35 cm, 26 cm

Putting the data into the GDC and calculating the mean gives 28 cm.

The GDC will give you the standard deviation 5.15, from which you can calculate the variance as 26.5.

Using the formula would give the mean as:

$$\frac{\Sigma x}{n} = \frac{30 + 17 + 32 + \ldots + 26}{8} = 28$$

The variance is:

$$\sigma^2 = \frac{1}{n}\Sigma x^2 - \left(\frac{\Sigma x}{n}\right)^2$$

$$= \frac{1}{8}(30^2 + 17^2 + 32^2 + \ldots + 26^2) - 28^2$$

$$= 26.5$$

Therefore the standard deviation is $\sigma = \sqrt{26.5} = 5.15$

The mean height is 28 cm and the standard deviation is 5.15 cm.

> Ensure that your GDC is in statistical mode.

Example 2

The heights, in metres, of 10 children of a particular age in a school were recorded as

　　1.0, 1.2, 1.3, 1.1, 1.2, 1.4, 1.1, 0.9, 1.3, 1.2

Calculate a) the mean　b) the standard deviation.

Inputting the data into the GDC gives the following results:

a) the mean is 1.17 m

b) $\sigma = 0.142$ m

Standard deviation for frequency distributions

In this case the x values are the mid-points of the class intervals. The formula for variance becomes

$$\sigma^2 = \frac{1}{n}\Sigma fx^2 - \left(\frac{\Sigma fx}{n}\right)^2$$

where f represents the frequency of the x observation, and $n = \Sigma f$.

Example 3

A restaurant serves a variety of wines of different prices. In a week chosen at random the numbers of bottles of wine sold were recorded by price and the results are shown in the table.

€x	$0 \leqslant x < 5$	$5 \leqslant x < 10$	$10 \leqslant x < 15$	$15 \leqslant x < 20$	$20 \leqslant x < 25$
Number of bottles of wine sold	27	15	8	3	1

Determine the mean and the standard deviation.

Extending the table to include the mid-values gives:

€x	Mid-value	No. of bottles of wine sold
0–5	2.5	27
5–10	7.5	15
10–15	12.5	8
15–20	17.5	3
20–25	22.5	1

The mid-point of the class 10–15 is

$$\frac{10 + 15}{2} = 12.5$$

The mean is given by

$$\bar{x} = \frac{\Sigma fx}{n} = \frac{(27 \times 2.5) + (15 \times 7.5) + \dots + (1 \times 22.5)}{54} = 6.57$$

The variance is given by

$$\sigma^2 = \frac{1}{n}\Sigma fx^2 - \left(\frac{\Sigma fx}{n}\right)^2$$

$$= \frac{1}{54}[(27 \times 2.5^2) + (15 \times 7.5^2) + \dots + (1 \times 22.5^2)] - (6.57)^2$$

$$= 25.1$$

Therefore the standard deviation is $\sigma = \sqrt{25.1} = 5.01$

The mean price is €6.57 and the standard deviation is €5.01.

Exercise 15C

1 Use your GDC to calculate the mean and standard deviation of each of these sets of numbers.

a) 3, 5, 7, 8, 9, 10

b) 5, 5, 7, 8, 9, 10, 12

c) 0, 0, 1, 3, 5, 6, 8, 9

d) −2, −1, 3, 4, 6

2 Use your GDC to estimate the mean and standard deviation for each of these frequency distributions.

a)

x	0–10	10–20	20–30	30–40
Frequency	4	6	7	3

b)

x	6–10	11–15	16–20	21–25	26–30	31–35
Frequency	2	3	8	6	5	1

c)

x	20–22	22–24	24–26	26–28	28–30
Frequency	8	14	14	10	4

3 The table shows the prices of the 25 cars which are for sale in a garage.

Price ($s)	1000–2000	2000–4000	4000–6000	6000–10 000	10 000–15 000
Frequency	6	7	5	4	3

a) Use your GDC to calculate an estimate of the mean and the standard deviation of the prices of the cars in the garage.

A second garage has 30 cars for sale at the following prices.

Price ($s)	1000–2000	2000–4000	4000–6000	6000–10 000	10 000–15 000
Frequency	8	13	6	2	1

b) Use your GDC to calculate an estimate of the mean and the standard deviation of the prices of the cars in this garage.

c) Comment on the difference in the cars for sale in the two garages.

4 The bar chart shows the heights of players in a basketball squad.

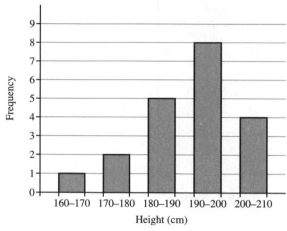

a) How many players are in the squad?

b) Calculate an estimate of the mean height.

c) Calculate an estimate of the standard deviation.

5 a) Calculate the mean and standard deviation of these numbers:

1, 2, 3, 4, 5

b) Deduce the mean and standard deviation of each data set:

i) 11, 12, 13, 14, 15

ii) 45, 46, 47, 48, 49

iii) 10, 20, 30, 40, 50

iv) 43, 46, 49, 52, 55

15.5 Cumulative frequency

The **cumulative frequency** is the total frequency up to a particular value or class boundary. The following example illustrates how to construct a cumulative table and then draw a **cumulative frequency curve**.

Example 1

The heights to the nearest centimetre of a type of plant were recorded 6 months after planting. The frequency distribution is shown in the table.

Height (cm)	Frequency
3–5	1
6–8	3
9–11	6
12–14	10
15–17	12
18–20	4

Show these results on a cumulative frequency curve.

The interval 6–8 ranges from 5.5 to 8.5.

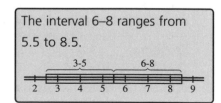

Construct the cumulative frequency table, including the upper class boundary.

Height (cm)	Upper class boundary	Frequency	Cumulative frequency
3–5	5.5	1	1
6–8	8.5	3	4
9–11	11.5	6	10
12–14	14.5	10	20
15–17	17.5	12	32
18–20	20.5	4	36

> $1 + 3 = 4$
> $4 + 6 = 10$
> $10 + 10 = 20$
> and so on.

Plot the cumulative frequencies against the corresponding upper class boundaries to give the cumulative frequency curve shown.

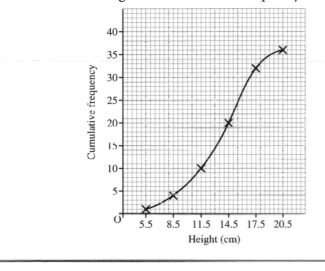

> Join the points with a curve.

Median, quartiles and interquartile range

Cumulative frequency provides a convenient way of estimating the median. In Example 1, the middle plant is roughly the 18th, since $36 \div 2 = 18$. From the curve, read down from 18 on the vertical axis to give:

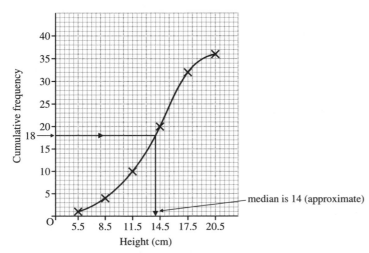

median is 14 (approximate)

This tells you that there are 18 plants with height less than 14 cm.

The range of values includes any extreme measurements included in the data. It is useful to discard any such extreme measurements. The most common way of doing this is to exclude the top and bottom quarters of the distribution.

The lower point, chosen so that one quarter of the values are less than or equal to it, is called the **first quartile** and denoted as Q_1.

> The first quartile Q_1 is also called the **lower quartile**.

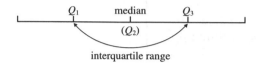

The upper point, with one quarter of the values greater than it, is the **third quartile** and is denoted as Q_3.
(Q_2 is the median.)

> The third quartile Q_3 is also called the **upper quartile**.

The difference between the two $(Q_3 - Q_1)$ is called the **interquartile range**.

Look again at the cumulative frequency polygon for the heights of plants.

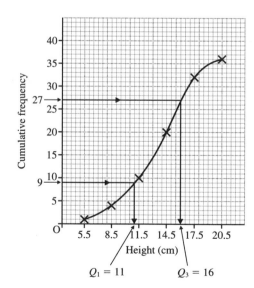

The quartiles are approximately the heights of the 9th $(36 \div 4)$ plant and the 27th $(3 \times 36 \div 4)$ plant. From the cumulative curve you see that these values are 11 and 16.

Therefore the interquartile range is $16 - 11 = 5$.

Percentiles

The distribution may be split into a greater number of parts.
A very large sample may be split into 100 parts called **percentiles**.
If it is divided into ten parts they are called **deciles**.

In Example 1, the 3rd decile is found by calculating
$\frac{3}{10} \times 36 = 10.8$, and projecting down on the cumulative
frequency curve as shown.

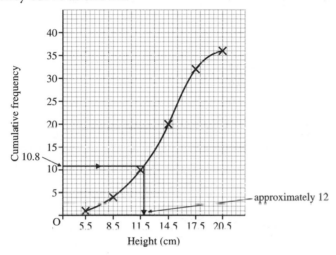

The 3rd decile is approximately 12.

approximately 12

> The 3rd decile is also called the 30th percentile.

Instead of forming a curve, you can join the points with straight
lines to make a **cumulative frequency polygon**.

Example 2

In a cricket match the 40 completed innings gave this
distribution of scores.

Score	0–9	10–19	20–29	30–39	40–49
Frequency	8	10	6	5	6

Score	50–59	60–69	70–79	80–89	90–99
Frequency	2	0	2	0	1

a) Show these results on a cumulative frequency polygon.

b) Use your polygon to estimate the median score.

c) From your polygon find the upper and lower quartiles, and
hence estimate the interquartile range.

d) The cricket club decide to award prizes to the top 20% of
players in the match. What score should be used as a
minimum to award prizes?

a) The cumulative frequencies are shown together with the upper class boundaries.

Score	Upper boundary	Frequency	Cumulative frequency
0–9	9.5	8	8
10–19	19.5	10	18
20–29	29.5	6	24
30–39	39.5	5	29
40–49	49.5	6	35
50–59	59.5	2	37
60–69	69.5	0	37
70–79	79.5	2	39
80–89	89.5	0	39
90–99	99.5	1	40

The cumulative frequency polygon is shown.

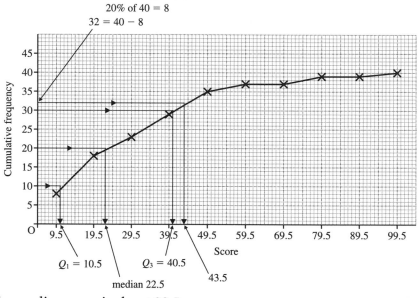

b) The median score is about 22.5.

c) $Q_3 - Q_1 = 40.5 - 10.5 = 30$

d) Score of 43.5 should be used as a minimum.

Box and whisker plots

A box and whisker plot is a useful way of representing data.
It shows at a glance the median and quartiles and gives an idea of the spread of the data.

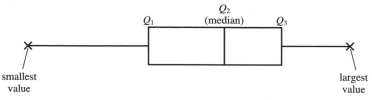

The box represents the central 50% of the data ($Q_3 - Q_1$), and the 'whiskers' show the smallest and largest values.

Example 3

Illustrate the data from Example 2 using a box and whisker plot.

$Q_1 = 10.5$, $Q_2 = 22.5$, $Q_3 = 43.5$, lowest value $= 0$,
highest value $= 99.5$

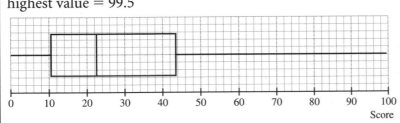

> Note that the box and whisker plot clearly illustrates where the central 50% of the data lies relative to the smallest and largest values.

Exercise 15D

1 The masses of 800 eggs were recorded and the results are summarised in the table.

Mass (grams)	50–55	55–60	60–65	65–70	70–75	75–80
Frequency	88	152	212	234	78	36

a) Show these results on a cumulative frequency curve

b) Use your curve to estimate the median mass of an egg.

c) From your curve read off the upper quartile and lower quartile, and hence estimate the interquartile range.

d) Represent this information on a box and whisker plot.

2 The heights of 600 trees in a garden centre were recorded and the results are summarised in the table.

Height (cm)	140–150	150–160	160–170	170–180	180–190	190–200	200–210
Frequency	26	38	56	85	125	212	58

a) Show these results on a cumulative frequency curve.

b) Use your curve to estimate the median height of a tree.

c) Use your curve to read off the upper quartile and lower quartile, and hence estimate the interquartile range.

d) Represent this information on a box and whisker plot.

3 The marks of 120 students in a Physics exam were recorded, and the results are summarised in the table.

Mark	0–20	21–40	41–60	61–80	81–100
Frequency	6	11	24	62	17

a) Show these results on a cumulative frequency curve.

b) Given that the pass mark is 46, estimate the percentage of students who pass.

c) If the chief examiner decides that only 76% of the pupils should pass, to what should he raise the pass mark?

4 The lifetimes of 400 electric lightbulbs were recorded, and the results are summarised in the table.

Lifetime (h)	700–800	800–900	900–1000	1000–1100	1100–1200
Frequency	62	116	98	76	48

a) Show these results on a cumulative frequency polygon.

b) Use your polygon to estimate the median lifetime of a lightbulb.

c) Use your polygon to read off the upper quartile and lower quartile, and hence estimate the interquartile range.

5 As part of a survey, the times of arrival of 160 trains were recorded, and the number of minutes that each train was late is recorded in the table.

Minutes late	0	0–5	5–10	10–20	20–30	30–60
Frequency	24	34	32	28	20	22

a) Show these results on a cumulative frequency polygon.

b) Use your polygon to estimate the 2nd decile, or 20th percentile.

c) Use your polygon to estimate the 7th decile, or 70th percentile.

6 A survey of the ages of 240 cars in a car park gives these results.

Age (years)	0–2	2–4	4–8	8–12	12–16
Frequency	114	36	48	28	14

a) Show these results on a cumulative frequency polygon.

b) Use your polygon to estimate the 42nd percentile.

c) Use your polygon to estimate the 84th percentile.

7 Two hundred students each sat a Maths exam and an English exam. Their marks are summarised in the table.

Mark	0–20	21–40	41–60	61–80	81–100
Frequency (Maths)	14	30	76	56	24
Frequency (English)	6	8	18	122	46

a) Show these results on two cumulative frequency curves, drawn on the same set of axes.

b) Use your curves to read off the median mark and the interquartile range for each of the two exams.

c) Comment on the differences between the marks achieved in each of the two exams.

15.6 Histograms

Histograms are used to illustrate grouped or continuous data. Although histograms may look the same as bar charts, there are some important differences.

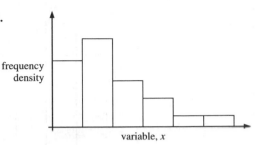

frequency density

variable, x

In the case of a bar chart the variable axis represents discrete data and is therefore simply divided into spaces.

A histogram represents continuous data and therefore the variable axis is a continuous number line.

In a histogram, rectangles representing frequency may have different widths. The area of each rectangle is proportional to the class frequency.

It is useful to work out the **frequency density** for each class. This is defined as

> The Mathematics Standard Level syllabus only deals with histograms with equal class intervals.

frequency density $= \dfrac{f}{w}$

where f = frequency and w = class width.

When all the class widths are equal, a histogram is easy to construct.

Example 1

100 students take a mathematics test which has a maximum possible score of 40. The results are shown in the table.

Mark	1–5	6–10	11–15	16–20	21–25	26–30	31–35	36–40
Number of students	3	10	22	28	20	10	2	5

Show the results on a histogram.

....

Work out the frequency densities:

Mark	Class width (w)	No. of students (f)	Frequency density ($f \div w$)
1–5	5	3	0.6
6–10	5	10	2
11–15	5	22	4.4
16–20	5	28	5.6
21–25	5	20	4
26–30	5	10	2
31–35	5	2	0.4
36–40	5	5	1

Note: The class widths are calculated as $5.5 - 0.5 = 5$, $10.5 - 5.5 = 5$ etc.

The heights of the sections of the histogram are in proportion to the frequency densities. In this case it makes sense to use 0.5–5.5, 5.5–10.5, etc. on the horizontal axis due to fractional marks and rounding.

Constructing the histogram gives:

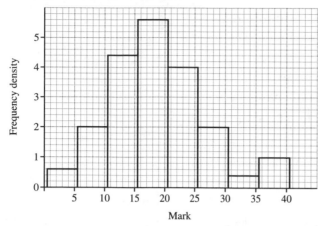

The next example illustrates how to construct a histogram with unequal class widths.

Example 2

A survey of family salaries in a holiday resort revealed the following results.

Salary (€)	Number of families
20 000–30 000	2
30 000–40 000	6
40 000–60 000	18
60 000–80 000	15
80 000–100 000	3
100 000–140 000	4

Note: Histograms with unequal class width will not be examined. This example is for interest only.

Show these results on a histogram.

···

Notice in this case that the class intervals are unequal. The first two are €10 000, the third, fourth and fifth are €20 000 and the sixth is €40 000.

Calculating the frequency densities gives:

Salary (€)	Class width (w)	No. of families (f)	Frequency density (f ÷ w)
20 000–30 000	10 000	2	0.0002
30 000–40 000	10 000	6	0.0006
40 000–60 000	20 000	18	0.0009
60 000–80 000	20 000	15	0.00075
80 000–100 000	20 000	3	0.00015
100 000–140 000	40 000	4	0.0001

Constructing the histogram gives:

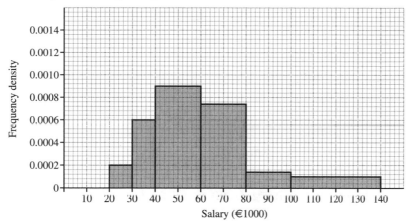

Notice that it is the area, not the height, of each bar that represents the frequency.

Exercise 15E

1 A teacher records the Intelligence Quotients of the 24 children in her class. The results are:

>125, 132, 117, 98, 151, 146, 133, 106, 114, 128, 126, 141,
>137, 122, 152, 108, 137, 121, 142, 102, 148, 133, 136, 95

a) Construct a frequency table of these results with classes 90–99, 100–109, …

b) Show these results on a histogram.

2 A biologist records the length of 20 worms. The results in mm are:

>152, 183, 125, 88, 194, 164, 129, 182, 173, 137, 91,
>102, 148, 162, 156, 183, 146, 117, 166, 172

a) Construct a frequency table of these results with classes 80–99, 100–119, …

b) Show these results on a histogram.

3 Sixteen athletes record their times to run 400 m. The results, in seconds, are:

>62, 58, 72, 66, 63, 78, 67, 56, 78, 83, 69, 75, 71, 68, 61, 73

a) Construct a frequency table of these results with classes 55–59, 60–64, …

b) Show these results on a histogram.

4 A farmer records the masses of 30 piglets at birth. The results in kg are:

> 3.12, 4.23, 3.67, 4.16, 4.45, 3.56, 3.92, 4.16, 5.02, 3.86, 4.39,
> 5.13, 4.58, 3.74, 4.12, 3.77, 4.32, 4.73, 3.36, 4.55, 5.23, 4.34,
> 4.61, 3.98, 4.32, 3.80, 3.54, 4.52, 4.68, 3.74

a) Construct a frequency table of these results with classes 3.00–3.49, 3.50–3.99, …

b) Show these results on a histogram.

5 The marks obtained by 100 students in a Science exam are given in the table.

Marks	0–19	20–39	40–59	60–79	80–99
Frequency	8	18	26	42	6

a) Show these results on a histogram.

b) Calculate an estimate of the mean mark.

6 The heights of the 34 players in a rugby match are given in the table.

Height (cm)	160–169	170–179	180–189	190–199
Frequency	6	13	11	4

a) Show these results on a histogram.

b) Calculate an estimate of the mean height.

7 On a section of road with a 50 km h^{-1} speed limit, 40 cars are caught on a speed camera. Their speeds are given in the table.

Speed (km h^{-1})	51–60	61–70	71–80	81–90	91–100	101–110
Frequency	13	10	8	5	3	1

a) Show these results on a histogram.

b) Calculate an estimate of the mean amount by which the cars were exceeding the speed limit.

15.7 Random variables

A **random variable**, denoted by X, is a measurable quantity which can take any value or range of values. Its value is the result of a random observation or experiment. Actual measured values are represented by x.

A **discrete random variable** means that a list of its possible numerical values could be made. That is, the data only has certain discrete values.

Probability distributions

Suppose you throw a fair coin twice.

You can draw a probability tree diagram to show this.

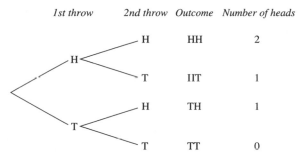

If you define the random variable X as 'the number of heads obtained', you can summarise the information in a table:

x	0	1	2
$P(X = x)$	$\frac{1}{4}$	$\frac{1}{2}$	$\frac{1}{4}$

> $P(X = x)$ is the probability that the number of heads is x.

Example 1

For each of these examples, draw a table of possible values of x, together with the associated probability $P(X = x)$.

a) A box contains 2 red marbles and 6 green marbles.
   Two marbles are chosen at random with replacement, and
   X is the number of green marbles obtained.

b) A fair die has faces labelled 1, 1, 1, 2, 3, 3 and X is the score
   when the die is thrown.

c) Two fair dice are thrown and X is the difference between the
   higher score and the lower score.

....

a) The table shows the possible ways of choosing green
   marbles together with the associated probabilities.

Marble 1	Marble 2	Number of green marbles	Probability
R	R	0	$\frac{2}{8} \times \frac{2}{8} = \frac{1}{16}$
R	G	1	$\frac{2}{8} \times \frac{6}{8} = \frac{3}{16}$
G	R	1	$\frac{6}{8} \times \frac{2}{8} = \frac{3}{16}$
G	G	2	$\frac{6}{8} \times \frac{6}{8} = \frac{9}{16}$

> As the marbles are replaced, each event is independent. So you multiply the probabilities (see page 375).

x	0	1	2
$P(X = x)$	$\frac{1}{16}$	$\frac{3}{16} + \frac{3}{16} = \frac{6}{16}$	$\frac{9}{16}$

Notice that $\frac{1}{16} + \frac{6}{16} + \frac{9}{16} = 1$, which shows that all possible combinations have been considered for the random variable X.

b) The probability distribution for X is:

x	1	2	3
$P(X = x)$	$\frac{3}{6}$	$\frac{1}{6}$	$\frac{2}{6}$

c) In this case the possible outcomes can be best summarised in a table as shown.

2nd die

	1	2	3	4	5	6
1	0	1	2	3	4	5
2	1	0	1	2	3	4
3	2	1	0	1	2	3
4	3	2	1	0	1	2
5	4	3	2	1	0	1
6	5	4	3	2	1	0

1st die

There are 36 possible outcomes. The probability distribution is summarised in this table.

x	0	1	2	3	4	5
$P(X = x)$	$\frac{6}{36}$	$\frac{10}{36}$	$\frac{8}{36}$	$\frac{6}{36}$	$\frac{4}{36}$	$\frac{2}{36}$

Expectation

The mean value of the random variable X is called the **expected value** of X and is written $E(X)$.

The expected value of X is

$$E(X) = \Sigma x\, P(X = x)$$

Example 2

A fair coin is thrown twice and X is the number of heads obtained. Find $E(X)$.

. . . .

The table below gives the possible outcomes of throwing a fair coin twice, together with the associated probabilities.

1st throw	2nd throw	Number of heads (x)	Probability
T	T	0	$\frac{1}{2} \times \frac{1}{2} = \frac{1}{4}$
T	H	1	$\frac{1}{2} \times \frac{1}{2} = \frac{1}{4}$
H	T	1	$\frac{1}{2} \times \frac{1}{2} = \frac{1}{4}$
H	H	2	$\frac{1}{2} \times \frac{1}{2} = \frac{1}{4}$

This is summarised as:

x	0	1	2
$P(X = x)$	$\frac{1}{4}$	$\frac{1}{4} + \frac{1}{4} = \frac{1}{2}$	$\frac{1}{4}$

Therefore,

$E(X) = \Sigma x\, P(X = x)$

$= \left(0 \times \frac{1}{4}\right) + \left(1 \times \frac{1}{2}\right) + \left(2 \times \frac{1}{4}\right)$

$= 1$

Therefore the expected number of heads is 1.

Example 3

The random variable X can only take the values 1, 2 and 3.
Given that the value 3 is twice as likely as each of the values 1
and 2, and values 1 and 2 are equally likely,

a) draw a table of possible values of x together with $P(X = x)$

b) determine the expectation of X.

....

a) Let $P(X = 1) = P(X = 2) = p$, then
$P(X = 3) = 2p$. The table of possible
values and probabilities is shown on the right.

x	1	2	3
$P(X = x)$	p	p	$2p$

Since X can only take the values 1, 2 and 3,

$p + p + 2p = 1$

$\therefore p = \frac{1}{4}$

The completed table is shown on the right.

b) $E(X) = \left(1 \times \frac{1}{4}\right) + \left(2 \times \frac{1}{4}\right) + \left(3 \times \frac{1}{2}\right)$

$= \frac{9}{4}$

x	1	2	3
$P(X = x)$	$\frac{1}{4}$	$\frac{1}{4}$	$\frac{1}{2}$

Example 4

The random variable X has a probability distribution given by

$P(X = x) = \dfrac{x}{k}, \quad x = 1, 2, 3, 4$

a) Find the value of the constant k.

b) Calculate $E(X)$.

....

a)

x	1	2	3	4
$P(X = x)$	$\frac{1}{k}$	$\frac{2}{k}$	$\frac{3}{k}$	$\frac{4}{k}$

You know that

$$\frac{1}{k} + \frac{2}{k} + \frac{3}{k} + \frac{4}{k} = 1$$

$$\frac{10}{k} = 1$$

$$\therefore k = 10$$

b) $E(X) = \left(1 \times \frac{1}{10}\right) + \left(2 \times \frac{2}{10}\right) + \left(3 \times \frac{3}{10}\right) + \left(4 \times \frac{4}{10}\right)$

$\quad = \frac{1}{10} + \frac{4}{10} + \frac{9}{10} + \frac{16}{10}$

$\quad = 3$

Exercise 15F

In questions **1** to **7**, for each random variable X, draw a table of possible values, x, together with the matching probabilities, $P(X = x)$.

1 A fair coin is thrown three times, and X is the number of heads obtained.

2 A bag contains four red discs and five blue discs. Two discs are taken out at random without replacement, and X is the number of red discs obtained.

3 Three cards are selected at random without replacement, from a pack of 52, and X is the number of spades.

4 Two fair dice are rolled and X is the sum of the scores on the two faces.

5 In a box are four cards numbered 5, 10, 15 and 20. Two cards are taken out at random without replacement, and X is the total of the numbers on the two cards.

6 A fair die is thrown and X is the square of the number scored.

7 A fair coin is thrown three times and X is the cube of the number of heads showing.

8 Find the value of p in each of these probability distributions. Calculate also the value of $E(X)$.

a)
x	1	2	3
$P(X = x)$	0.2	0.3	p

b)
x	2	3	4	5
$P(X = x)$	p	$\frac{1}{5}$	$\frac{1}{5}$	$\frac{2}{5}$

c)

x	0	1	2	3	4
P(X = x)	0.15	p	p	0.22	0.31

d)

x	−2	0	1	3
P(X = x)	$\frac{2}{9}$	$\frac{1}{9}$	p	$\frac{2}{9}$

e)

x	−4	−1	2	5	8
P(X = x)	0.12	0.24	p	2p	0.07

f)

x	5	7	9	11	13
P(X = x)	p	p	p	0.12	0.04

9 X has probability distribution $P(X = x) = kx$, for
$x = 1, 2, 3, 4, 5$.

a) Find the value of the constant k.

b) Calculate $E(X)$.

10 X has probability distribution $P(X = x) = kx^2$, for
$x = 1, 2, 3, 4, 5, 6$.

a) Find the value of the constant k.

b) Calculate $E(X)$.

11 X has probability distribution $P(X = x) = \dfrac{k}{x}$, for $x = 1, 2, 3, 4$.

a) Find the value of the constant k.

b) Calculate $E(X)$.

12 A probability distribution is given by
$P(X = x) = A(1 + x)(2 + x)$, for $x = 0, 1, 2, 3$.

a) Find the value of the constant A.

b) Calculate $E(X)$.

13 A probability distribution is given by
$P(X = x) = C(4 - x)(5 - x)$, for $x = 0, 1, 2, 3, 4, 5, 6$.

a) Find the value of the constant C.

b) Calculate $E(X)$.

14

x	1	2	3	4	5
P(X = x)	0.1	0.2	a	b	0.2

Given that $E(X) = 3.1$, calculate the values of a and b.

15.8 The binomial distribution

A **binomial distribution** is a discrete distribution defined by two parameters, 'the number of trials', n, and the 'probability of a success', p.

You write $X \sim B(n, p)$, to indicate that the discrete random variable X is binomially distributed.

Here is an illustration of the binomial distribution.

Annie is trying her skills at a shooting range. In a round she has three shots and the probability of a success on any shot is $\frac{1}{4}$. Annie is interested in the number of 'hits' or successes in any round and lists the possibilities in a table.

Possibilities	Number of successes, i.e. hits	Probability
HHH	3	$\frac{1}{4} \times \frac{1}{4} \times \frac{1}{4} = \frac{1}{64}$
HHM	2	$\frac{1}{4} \times \frac{1}{4} \times \frac{3}{4} = \frac{3}{64}$
HMH	2	$\frac{1}{4} \times \frac{3}{4} \times \frac{1}{4} = \frac{3}{64}$
MHH	2	$\frac{3}{4} \times \frac{1}{4} \times \frac{1}{4} = \frac{3}{64}$
HMM	1	$\frac{1}{4} \times \frac{3}{4} \times \frac{3}{4} = \frac{9}{64}$
MHM	1	$\frac{3}{4} \times \frac{1}{4} \times \frac{3}{4} = \frac{9}{64}$
MMH	1	$\frac{3}{4} \times \frac{3}{4} \times \frac{1}{4} = \frac{9}{64}$
MMM	0	$\frac{3}{4} \times \frac{3}{4} \times \frac{3}{4} = \frac{27}{64}$

H = hit
M = miss

Let X be the number of hits, i.e. the number of successes. X has the distribution shown:

x	0	1	2	3
$P(X = x)$	$\frac{27}{64}$	$3 \times \frac{9}{64}$	$3 \times \frac{3}{64}$	$\frac{1}{64}$

In this example, note that the outcomes of the 'experiment' only comprise success (hit) and failure (miss). Also note that we have assumed that the probability of a success was the same for each trial and that each trial was independent of the others.

The distribution of X is a binomial distribution.

The binomial distribution function is given by
$$P(X = r) = {}^nC_r \, p^r (1 - p)^{n-r}$$
where p is the probability of success and ${}^nC_r = \binom{n}{r} = \frac{n!}{r!(n-r)!}$

You met ${}^nC_r$ in Chapter 6 when you studied binomial expansions. You can use your GDC to work it out.

It is common to write $q = 1 - p$, which gives

$$P(X = r) = {}^nC_r \, p^r q^{n-r}$$

Notice that ${}^nC_r$ is the coefficient of $p^r q^{n-r}$ in the binomial

expansion of $(p + q)^n$ (see page 201).

Example 1

A fair coin is thrown 10 times. Find the probability that 6 heads will occur.

....

Let success be obtaining a head, then $p = \frac{1}{2}$ and $n = 10$.

Let the random variable X be the number of heads obtained, so $X \sim B(10, \frac{1}{2})$. Therefore

$$P(X = r) = {}^{10}C_r \left(\tfrac{1}{2}\right)^r \left(\tfrac{1}{2}\right)^{10-r}$$

Therefore

$$P(X = 6) = {}^{10}C_6 \left(\tfrac{1}{2}\right)^6 \left(\tfrac{1}{2}\right)^4 = 0.205$$

Use your GDC to work out ${}^{10}C_6$.

Example 2

The probability that a particular page in a mathematics book contains a misprint is 0.2. Find the probability that of 12 pages in the book

a) four of them contain a misprint

b) fewer than two of them contain a misprint.

....

Let success be a 'misprint on a page', then $p = 0.2$ and $n = 12$.

Let the random variable X be the number of pages containing a misprint, so $X \sim B(12, 0.2)$.

$$P(X = r) = {}^{12}C_r (0.2)^r (0.8)^{12-r}$$

a) $P(X = 4) = {}^{12}C_4 (0.2)^4 (0.8)^8 = 0.133$

b) $P(X < 2) = P(X = 0) + P(X = 1)$.

$$P(X = 0) = {}^{12}C_0 (0.2)^0 (0.8)^{12} = 0.0687$$

$$P(X = 1) = {}^{12}C_1 (0.2)(0.8)^{11} = 0.2062$$

Therefore $P(X < 2) = 0.0687 + 0.2062 = 0.275$

Returning to Annie practising her shooting, you know that her probability of successfully hitting a target is $\frac{1}{4}$.

If the probability of Annie hitting a target is p then in n attempts she would expect to hit the target $n \times p$ times. This leads to the generalisation for the mean of the binomial distribution $B(n, p)$.

> This means that out of every 4 attempts Annie would expect to hit the target on 1 occasion. In 8 attempts she would expect to hit the target twice and so on.

> For a binomial distribution $X \sim B(n, p)$, the mean value of X is given by np. This is also the expected value of X.

Example 3

An unbiased die is thrown 24 times. Find the expected number of threes thrown.

....

Let success be obtaining a three, then $p = \frac{1}{6}$ and $n = 24$.

Let the random variable X be the number of threes obtained, so $X \sim B(24, \frac{1}{6})$.

The expected number of threes is given by the mean,

$$np = 24 \times \frac{1}{6} = 4$$

The expected number of threes is 4.

Example 4

a) A bag contains 2 red counters and 3 green counters. Two counters are drawn at random from the bag without replacement. Find the probability that exactly 2 green counters are drawn.

b) Find the probability that out of 15 attempts of drawing two counters (replaced after each two drawn) more than 12 of them result in exactly 2 green counters being drawn.

c) What is the expected number of attempts in which exactly 2 green counters will be drawn?

....

a) $P(\text{exactly 2 green}) = \frac{3}{5} \times \frac{2}{4} = \frac{3}{10}$

b) Let a success be choosing 2 green counters, then $p = \frac{3}{10}$.

Let the random variable X be the number of attempts in which 2 green counters are drawn. Then $X \sim B(15, \frac{3}{10})$ and

$$P(X = r) = {}^{15}C_r \left(\frac{3}{10}\right)^r \left(\frac{7}{10}\right)^{15-r}$$

You require $P(X > 12) = P(X = 13) + P(X = 14) + P(X = 15) = 8.72 \times 10^{-6}$.

c) The expected number of attempts in which exactly 2 green counters will be drawn is given by $15 \times \frac{3}{10} = 4.5$.

Exercise 15G

1 A fair six-sided die has the faces numbered from 1 to 6. The die is thrown four times. Calculate the probability of exactly two fives.

2 A coin is biased in such a way that the probability it lands heads is $\frac{1}{3}$. The coin is thrown six times. Calculate the probability of exactly four heads.

3 One in every five cars on the road is red. Calculate the probability that exactly one out of the next four cars to pass my house is red.

4 There are 20 children in a class. Calculate the probability that exactly four of them were born on a Sunday.

5 In the game of roulette a ball rolls into one of 37 slots. All the slots are equally likely. Eighteen of the slots are red, eighteen are white, and one is green. The ball is rolled ten times. Calculate the probability of exactly six reds.

6 Three out of every four adult males have dark eyes. Calculate the probability that exactly eight of the adult males in a soccer team of eleven have dark eyes.

7 40% of the books in a library are paperbacks. I select twelve books at random. Calculate the probability that exactly five are paperbacks.

8 15% of cars on the road have faulty brakes. As part of a survey, 25 cars are randomly stopped and tested. Calculate the probability that exactly four have faulty brakes.

9 30% of the population have had a particular vaccination. Sixteen people are selected at random. Calculate the probability that exactly three have had the vaccination.

10 When a darts player attempts to hit the centre he has a 0.7 probability of success. He makes 50 attempts at the centre. Calculate the probability that he is successful with exactly 40 of these attempts.

11 A machine produces ornamental plates. It is known that 10% of the plates are cracked. A random sample of 12 plates is selected.

  a) Calculate the probability that
    i)   none is cracked
    ii) exactly one is cracked
    iii) exactly two are cracked
    iv) there are fewer than three cracked plates.

  b) Calculate the expected number of cracked plates in the sample.

15.9 The normal distribution

In Exercise 15E you were asked to plot some histograms. These histograms represented a small sample from a large population, for example the length of 20 worms or the masses of 30 piglets. Rough sketches of two such histograms are shown below.

If, instead of restricting yourself to small samples of size 20 or 30, you take samples of size 1000 or 10 000, the histograms would look much more like this:

Notice that the distribution is symmetrical, and bell-shaped, with fewer observations as you move away from the central value. These are properties of the **Normal Distribution**, which is the most important distribution in statistics.

A **normal distribution** is a continuous distribution defined by two parameters, the mean μ and the variance σ^2.

Because of the symmetrical shape of the normal curve, the mean is equal to the mode and the median.

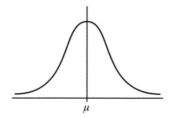

A normal distribution has a bell-shaped curve.

The normal distribution is written as $X \sim N(\mu, \sigma^2)$, to indicate that the continuous random variable X is normally distributed with mean μ and variance σ^2.

The standard normal distribution

The **standard normal distribution** has mean 0 and variance 1. In this case the random variable is called Z.

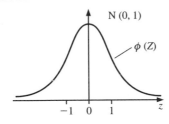

The standard normal distribution

> Note that the area under the curve is 1.

The probability density function for Z is usually denoted by ϕ. The distribution function is denoted by Φ.

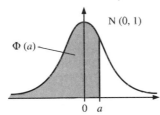

You can use tables of the normal distribution function together with a GDC to calculate areas under the N(0, 1) curve.

> Tables of the normal distribution are provided in the formula book, and are duplicated here on page 468.

Example 1

Use your GDC to find each of the shaded areas under the N(0, 1) curve shown.

a)
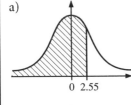

b)
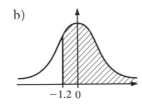

c)
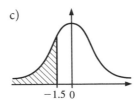

d)
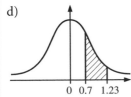

e)
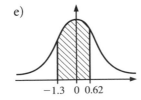

a) $\Phi(2.55) = 0.99461$

b) Because the normal curve is symmetrical, the two areas shown are equal:

Therefore the required area is $\Phi(1.2) = 0.8849$.

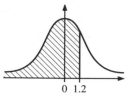 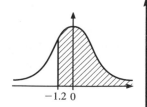

c) Due to symmetry the two areas shown are equal:

Therefore

$$\Phi(-1.5) = 1 - \Phi(1.5)$$
$$= 1 - 0.9332$$
$$= 0.0668$$

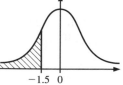 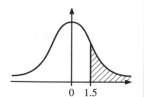

d) The required area is the difference between the areas shown.

$$\Phi(1.23) - \Phi(0.7) = 0.8907 - 0.7580$$
$$= 0.1327$$

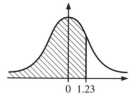 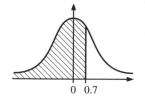

e) The required area is the difference between the areas shown.

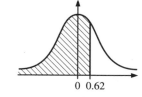 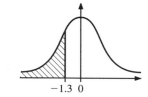

$$\Phi(0.62) - \Phi(-1.3) = \Phi(0.62) - (1 - \Phi(1.3))$$
$$= 0.7324 - 0.0968$$
$$= 0.6356$$

Example 2

Use your GDC to find each of the values a.

a)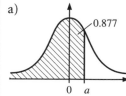

b)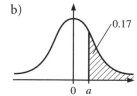

c)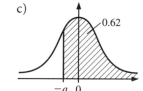

d)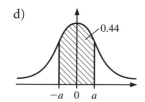

....

a) $\Phi(a) = 0.877$

$\therefore a = 1.16$

b) $1 - \Phi(a) = 0.17$

$\Phi(a) = 1 - 0.17 = 0.83$

$\therefore a = 0.95$

c)

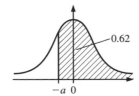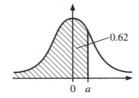

By symmetry, the two shaded areas are equal.
$\Phi(a) = 0.62$
$\therefore a = 0.31$

d) By symmetry

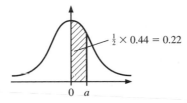

Therefore
$\Phi(a) = 0.5 + 0.22 = 0.72$
$\therefore a = 0.58$

The area under the normal curve represents a **probability**.

If $Z \sim N(0, 1)$ then $\Phi(a)$ represents the probability $P(z < a)$.

Example 3

Given that $Z \sim N(0, 1)$, find each of these probabilities:
a) $P(z < 2.68)$
b) $P(z > -1.41)$
c) $P(0.2 < z < 1.71)$
d) $P(|z| < 1.1)$

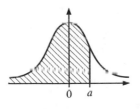

a) $P(z < 2.68) = \Phi(2.68)$
$= 0.9963$

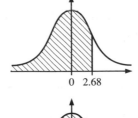

b) $P(z > -1.41) = \Phi(1.41)$
$= 0.9207$

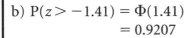

c) $P(0.2 < z < 1.71) = \Phi(1.71) - \Phi(0.2)$
$= 0.9564 - 0.5793$
$= 0.3771$

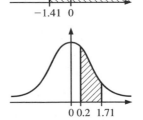

d) $P(|z| < 1.1) = P(-1.1 < z < 1.1)$
$$= \Phi(1.1) - \Phi(-1.1)$$
$$= \Phi(1.1) - (1 - \Phi(1.1))$$
$$= 0.8643 - 0.1357$$
$$= 0.7286$$

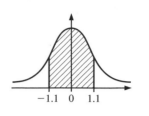

Probabilities for other normal distributions

You now know what to do with the standard normal distribution – but few normal distributions will actually have a mean of zero and a standard deviation of 1. However it is not difficult to transform any set of normal data into the standard normal.

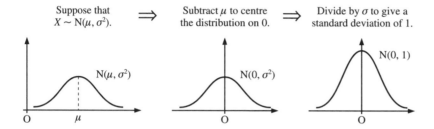

Suppose that $X \sim N(\mu, \sigma^2)$. $\Rightarrow$ Subtract μ to centre the distribution on 0. $\Rightarrow$ Divide by σ to give a standard deviation of 1.

Suppose the random variable $X \sim N(\mu, \sigma^2)$. In such cases the transformation

$$Z = \frac{X - \mu}{\sigma}$$

is used to give a normal distribution with a mean of 0 and a standard deviation of 1. You write $Z \sim N(0, 1)$.

The transformation combines a translation $(-\mu)$ and a change of scale (dividing by σ).

Example 4

Given that the random variable $X \sim N(2, 3^2)$, determine $P(X < 3)$.

3 in standard units is $Z = \dfrac{3 - 2}{3} = \dfrac{1}{3}$.

$$P(X < 3) = P\left(Z < \frac{1}{3}\right)$$
$$= \Phi(0.\dot{3})$$
$$= 0.6306$$

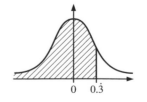

c)

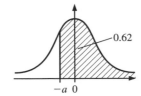 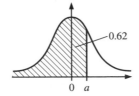

By symmetry, the two shaded areas are equal.

$\Phi(a) = 0.62$

$\therefore a = 0.31$

d) By symmetry

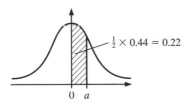

Therefore

$\Phi(a) = 0.5 + 0.22 = 0.72$

$\therefore a = 0.58$

The area under the normal curve represents a **probability**.

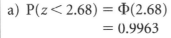 If $Z \sim N(0, 1)$ then $\Phi(a)$ represents the probability $P(z < a)$.

Example 3

Given that $Z \sim N(0, 1)$, find each of these probabilities:

a) $P(z < 2.68)$

b) $P(z > -1.41)$

c) $P(0.2 < z < 1.71)$

d) $P(|z| < 1.1)$

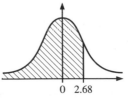

....

a) $P(z < 2.68) = \Phi(2.68)$

$\qquad\qquad = 0.9963$

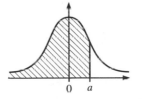

b) $P(z > -1.41) = \Phi(1.41)$

$\qquad\qquad = 0.9207$

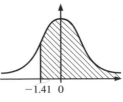

c) $P(0.2 < z < 1.71) = \Phi(1.71) - \Phi(0.2)$

$\qquad\qquad = 0.9564 - 0.5793$

$\qquad\qquad = 0.3771$

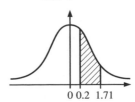

d) $P(|z| < 1.1) = P(-1.1 < z < 1.1)$

$\qquad\qquad\quad = \Phi(1.1) - \Phi(-1.1)$

$\qquad\qquad\quad = \Phi(1.1) - (1 - \Phi(1.1))$

$\qquad\qquad\quad = 0.8643 - 0.1357$

$\qquad\qquad\quad = 0.7286$

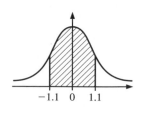

Probabilities for other normal distributions

You now know what to do with the standard normal distribution – but few normal distributions will actually have a mean of zero and a standard deviation of 1. However it is not difficult to transform any set of normal data into the standard normal.

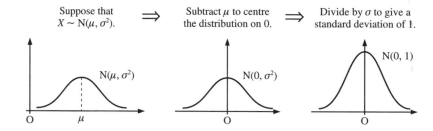

| Suppose that $X \sim N(\mu, \sigma^2)$. | $\Rightarrow$ | Subtract μ to centre the distribution on 0. | $\Rightarrow$ | Divide by σ to give a standard deviation of 1. |

Suppose the random variable $X \sim N(\mu, \sigma^2)$. In such cases the transformation

$$Z = \frac{X - \mu}{\sigma}$$

is used to give a normal distribution with a mean of 0 and a standard deviation of 1. You write $Z \sim N(0, 1)$.

The transformation combines a translation $(-\mu)$ and a change of scale (dividing by σ).

Example 4

Given that the random variable $X \sim N(2, 3^2)$, determine $P(X < 3)$.

3 in standard units is $Z = \dfrac{3 - 2}{3} = \dfrac{1}{3}$.

$P(X < 3) = P\left(Z < \dfrac{1}{3}\right)$

$\qquad\qquad = \Phi(0.\dot{3})$

$\qquad\qquad = 0.6306$

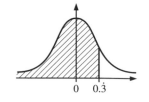

a) 945 in standard units is $Z = \dfrac{945 - 950}{10} = -0.5$

$$P(V < 945) = P(Z < -0.5)$$
$$= 1 - \Phi(0.5)$$
$$= 1 - 0.6915$$
$$= 0.3085$$

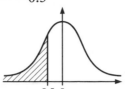

b) Let m be the maximum volume that a carton must contain to be accepted. M in standard units is $\dfrac{M - 950}{10}$.

$$P(V > m) = 0.06$$
$$P\left(Z > \dfrac{m - 950}{10}\right) = 0.06$$

Now $1 - \Phi(a) = 0.06$
$$\Phi(a) = 1 - 0.06$$
$$= 0.94$$
$$\therefore a = 1.555$$

Therefore
$$\dfrac{m - 950}{10} = 1.555$$
$$m = 965.55$$

The maximum volume is 966 ml, to the nearest ml.

Example 11

A machine produces components whose lengths are normally distributed with a mean of 20 cm. Given that 8% of the components produced by the machine have a length greater than 20.5 cm, find the standard deviation.

. . . .

Let L be the length of components, then $L \sim N(20, \sigma^2)$.

You know that
$$P(L > 20.5) = 0.08$$
$$\therefore P\left(Z > \dfrac{20.5 - 20}{\sigma}\right) = 0.08$$

Now
$$1 - \Phi(a) = 0.08$$
$$\Phi(a) = 0.92$$
$$\therefore a = 1.4053$$

20.5 in standard units is $\dfrac{20.5 - 20}{\sigma}$.

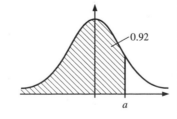

$$\dfrac{20.5 - 20}{\sigma} = 1.4053$$
$$\therefore \sigma = 0.356$$

Exercise 15H

In questions **1** to **3** the random variable Z has a standard normal distribution with mean zero and variance 1.

1 Use your GDC to find each of these shaded areas.

a)
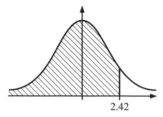
2.42

b)
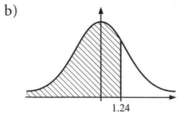
1.24

c)
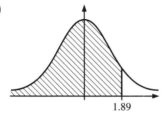
1.89

d)
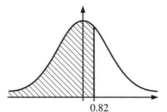
0.82

e)
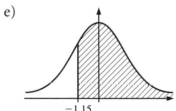
−1.15

f)
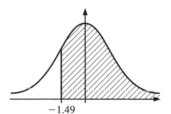
−1.49

g)
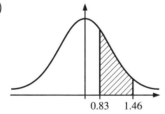
0.83   1.46

h)
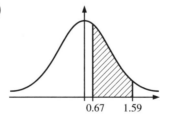
0.67   1.59

i)
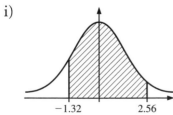
−1.32   2.56

j)
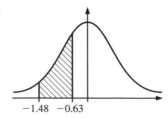
−1.48 −0.63

k)
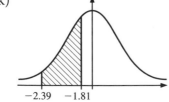
−2.39   −1.81

l)
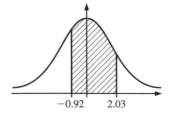
−0.92   2.03

2 Use your GDC to find each of these values of *a*.

a)

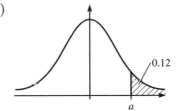

b)

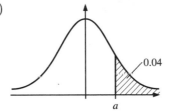

c)

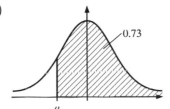

d)

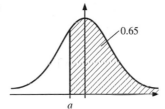

e)

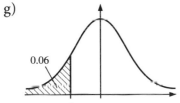

f)

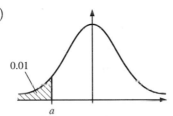

g)

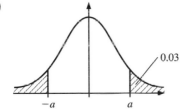

h)

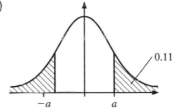

i)

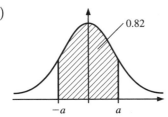

j)

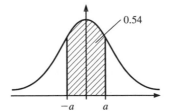

k)

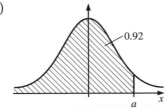

l)

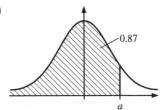

3 Find each of these probabilities, for $Z \sim N(0, 1)$.

a) $P(Z < 1.3)$ b) $P(Z < 2.18)$

c) $P(Z > 1.53)$ d) $P(Z > 1.9)$

e) $P(Z < -0.8)$ f) $P(Z > -1.64)$

g) $P(Z > -2.27)$ h) $P(0.2 < Z < 1.4)$

i) $P(1.13 < Z < 1.72)$ j) $P(-0.3 < Z < 0.4)$

k) $P(-2.02 < Z < 1.29)$ l) $P(|Z| < 1.23)$

4 Given that $X \sim N(14, 4^2)$, calculate these probabilities.

a) $P(X > 19)$ b) $P(X < 15)$

c) $P(X < 12)$ d) $P(X > 7)$

5 Given that $X \sim N(22, 5^2)$, calculate these probabilities.

a) $P(X < 27)$ b) $P(X < 10)$

c) $P(X > 15)$ d) $P(X > 31)$

6 Given that $X \sim N(42, 10^2)$, calculate these probabilities.

a) $P(43 < X < 56)$ b) $P(46 < X < 57)$

c) $P(40 < X < 50)$ d) $P(35 < X < 45)$

7 Given that $X \sim N(37, 8^2)$, calculate these probabilities.

a) $P(23 < X < 43)$ b) $P(23 < X < 35)$

c) $P(15 < X < 55)$ d) $P(|X - 37| < 8)$

8 Given that $X \sim N(-3, 5^2)$, calculate these probabilities.

a) $P(-5.7 < x < -3.1)$ b) $P(-1.1 < X < 2.6)$

c) $P(|X| < 2.2)$ d) $P(|X| > 5)$

9 The mass of an adult male is normally distributed with a mean of 79.4 kg and a standard deviation of 20 kg.
Find the probability that a randomly selected adult male has a mass of at least 65 kg.

10 The recorded speeds of cars on a section of motorway are normally distributed with a mean of 134.7 km h^{-1} and a standard deviation of 18 km h^{-1}. Calculate the percentage of cars which are travelling at less than 120 km h^{-1}.

11 The mid-day temperature in Gaiole in July is normally distributed with a mean of 35.8 °C and a standard deviation of 4 °C. Calculate the probability that it will be between 30 °C and 40 °C at mid-day in Gaiole on 12th July next year.

12 The length of earthworms is known to be normally distributed with a mean of 138 mm and a standard deviation of 40 mm. Calculate the proportion of earthworms which are between 100 mm and 200 mm in length.

13 Chicken eggs have a mean mass of 63.2 g and a standard deviation of 10 g. Eggs under 50 g are classified as *small*. Those between 50 g and 70 g are classified as *medium* and those above 70 g are classified as *large*.

a) What percentage of chicken eggs are classified as *small*?

b) What percentage of chicken eggs are classified as *medium*?

c) What percentage of chicken eggs are classified as *large*?

14 Conifers in a nursery have a mean height of 153 cm and a standard deviation of 50 cm. Those under 120 cm are classified as *short*. Those between 120 cm and 180 cm are classified as *medium* and those above 180 cm are classified as *tall*.

a) What percentage of conifers are classified as *short*?

b) What percentage of conifers are classified as *medium*?

c) What percentage of conifers are classified as *tall*?

15 The heights of candidates for the post of air hostess are normally distributed with a mean of 172 m and a standard deviation of 9 cm. 8% of candidates are rejected for being too tall. Calculate the critical height for an air hostess.

16 The speeds of lorries through a village are normally distributed with a mean of 65 km h^{-1} and a standard deviation of 12 km h^{-1}. Given that 10.56% of lorries are breaking the speed limit, what is that speed limit?

17 The masses of jars of home-made sweets are normally distributed with a mean of 245 g and a standard deviation of 23 g. Those jars with a mass below L grams are rejected. Given that 5% of jars are rejected, calculate the value of L.

18 The average daily temperature during the winter months is normally distributed with a mean of 12 °C and a standard deviation of 8 °C. The government decide to make *cold weather payments* when this average falls below a certain value, T°C. Given the government make cold weather payments on 1.5% of days in the winter months, calculate the value of T.

19 The time for an athlete to run 100 m is normally distributed with a mean of 12.4 seconds, and unknown standard deviation, σ seconds. Given the athlete records a time of more than 13.1 s in 4% of runs, calculate the value of σ.

20 The volume of beer in a glass is normally distributed with mean 523 cm^3, and unknown standard deviation, σ cm^3. Given that 20% of glasses have less than 500 cm^3 of beer, calculate the value of σ.

21 6% of the babies to be born in a hospital have a mass of over 3200 g, and 14% have a mass of less than 2600 g. Given the masses are normally distributed, calculate the mean and the standard deviation.

Summary

You should know how to ...

▶ Draw appropriate statistical diagrams, including:
  ▷ stem and leaf diagrams   ▷ cumulative frequency graphs
  ▷ bar charts   ▷ box and whisker plots
  ▷ line graphs   ▷ histograms
  ▷ pictograms   ▷ pie charts

▶ Calculate measures of location.
  ▷ The mode is the value that occurs most often.
  ▷ The median is the middle value in order of size.
  ▷ The mean $\bar{x}$ is given by the formula:

$$\bar{x} = \frac{1}{n} \Sigma x, \quad \text{or} \quad \bar{x} = \frac{\Sigma fx}{\Sigma f} \text{ for a frequency distribution.}$$

▶ Calculate measures of dispersion.
  ▷ The variance σ^2 is given by the formula:

$$\sigma^2 = \frac{1}{n} \Sigma(x - \bar{x})^2, \quad \text{or} \quad \sigma^2 = \frac{1}{n} \Sigma x^2 - \left(\frac{\Sigma x}{n}\right)^2$$

  ▷ The standard deviation σ is the square root of the variance.

  ▷ For a frequency distribution:

$$\sigma^2 = \frac{1}{n} \Sigma fx^2 - \left(\frac{\Sigma fx}{n}\right)^2$$

▶ Estimate quartiles and percentiles for large sets of data.

▶ Calculate with discrete random variables.
  ▷ The expected value of X is
$$E(X) = \Sigma x\, P(X = x)$$

▶ Use the binomial distribution
  ▷ $P(X = r) = {}^nC_r p^r (1 - p)^{n - r}$, where $X \sim B(n, p)$.
  ▷ The mean value of X is np.

▶ Use the normal distribution
  ▷ The standard normal distribution is $Z \sim N(0, 1)$, where the mean is 0 and the variance is 1.
  ▷ For a normal distribution $X \sim N(\mu, \sigma^2)$, $Z = \dfrac{X - \mu}{\sigma}$.

Revision exercise 15

1 Given the following frequency distribution, find
   a) the median
   b) the mean.

Number (x)	1	2	3	4	5	6
Frequency (f)	5	9	16	18	20	7

© IBO[2001]

2 Three positive integers a, b and c, where $a < b < c$, are such that their median is 11, their mean is 9 and their range is 10. Find the value of a.

© IBO[2002]

3 From January to September the mean number of car accidents per month was 630. From October to December the mean was 810 accidents per month.

What was the mean number of car accidents per month for the whole year?

© IBO[2002]

4 A supermarket records the amount of money d spent by customers in their store during a busy period. The results are as follows:

Money in $ (d)	0–20	20–40	40–60	60–80	80–100	100–120	120–140
Number of customers (n)	24	16	22	40	18	10	4

   a) Find an estimate for the mean amount of money spent by the customers, giving your answer to the nearest dollar ($).

   b) Copy and complete the following cumulative frequency table and use it to draw a cumulative frequency graph.
   Use a scale of 2 cm to represent $20 on the horizontal axis, and 2 cm to represent 20 customers on the vertical axis.

Money in $ (d)	<20	<40	<60	<80	<100	<120	<140
Number of customers (n)	24	40					

   c) The time t (minutes) spent by customers in the store may be represented by the equation

$$t = 2d^{\frac{2}{3}} + 3$$

   i) Use this equation and your answer to part a) to estimate the mean time in minutes spent by customers in the store.
   ii) Use the equation and the cumulative frequency graph to estimate the number of customers who spent more than 37 minutes in the store.

© IBO[2000]

5 A taxi company has 200 taxi cabs. The cumulative frequency curve shows the fares in dollars ($) taken by the cabs on a particular morning.

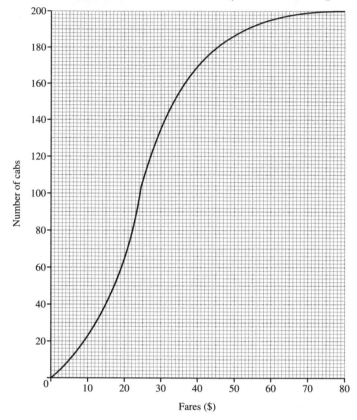

Fares ($)

a) Use the curve to estimate

  i) the median fare

  ii) the number of cabs in which the fare taken is $35 or less.

The company charges 55 cents per kilometre for distance travelled. There are no other charges. Use the curve to answer the following.

b) On that morning, 40% of the cabs travel less than a km. Find the value of a.

c) What percentage of the cabs travel more than 90 km on that morning?

© IBO [2002]

6 The diagram represents the lengths, in cm, of 80 plants grown in a laboratory.

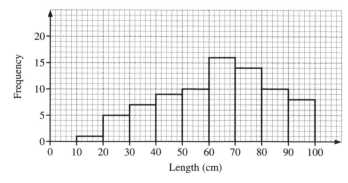

Length (cm)

a) How many plants have lengths in cm between

   i) 50 and 60

   ii) 70 and 90?

b) Calculate estimates for the mean and the standard deviation of the lengths of the plants.

c) Explain what feature of the diagram suggests that the median is different from the mean.

d) The following is an extract from the cumulative frequency table.

Length in cm less than	Cumulative frequency
...	...
50	22
60	32
70	48
80	62
...	...

Use the information in the table to estimate the median.
Give your answer to **two** significant figures. © *IBO* [*Spec.*]

7 The ages of people living in a certain street were recorded in years. The minimum age was 1 year and the maximum age was 91 years. The median age was found to be 42 years, and the quartiles were 25 and 60 years.

a) Draw a box and whisker plot to illustrate this information.

For a second street the data were presented in the following box and whisker plot.

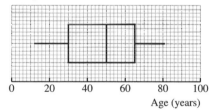

b) State the range and interquartile range for this street.

c) Write two statements comparing the distribution of ages in the two streets.

8 A factory makes computer components.
Over a long period it is found that 7% of components are faulty.
A sample of 30 components is tested.
Find the probability that

a) exactly two components are faulty   b) none are faulty

c) two or fewer are faulty   d) more than two are faulty.

9 An airline knows from past experience that the probability of a person booking a seat and then not turning up is 0.04. A small plane has 50 seats and 55 bookings are made.

a) A binomial distribution is used to model this situation. What assumption must be made? Comment on how reasonable this assumption is.

b) What is the expected number of no-shows?

c) What is the probability that more people turn up than there are seats available?

10 The masses of a group of students are normally distributed with mean 65 kg and standard deviation 10 kg.

a) Find the probability that a student, chosen at random, has a mass which is
i) less than 60 kg
ii) more than 70 kg
iii) between 60 and 70 kg.

b) Three students are chosen at random. Find the probability that only one of them has a mass of more than 70 kg.

11 A daily rail service operates to Euphoria City. The trains arrive T minutes after 08:00, where T is a random variable with a normal distribution. The mean value of T is 27, and there is a 5% probability that a train arrives after 08:38.

a) Show that the standard deviation of T is approximately 6.7.

b) Find the probability that a train arrives before 08:20.

The trains are scheduled to arrive at 08:23.

c) Show that approximately 20% of the arrivals are within 2 minutes of the scheduled time.

© IBO [1998]

12 The graph shows a normal curve for the random variable X, with mean μ and standard deviation σ.

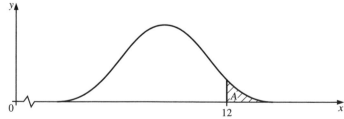

It is known that $P(X \geqslant 12) = 0.1$.

a) The shaded region A is the region under the curve where $x \geqslant 12$. Write down the area of the shaded region A.

It is also known that $P(X \leqslant 8) = 0.1$.

b) Find the value of μ, explaining your method in full.

c) Show that $\sigma = 1.56$ to an accuracy of three significant figures.

d) Find $P(X \leqslant 11)$.

© IBO [1999]

Practice Papers

Practice Paper One

1 A geometric series has first term a and common ratio $\frac{1}{3}$.
  - a) Find an expression in terms of a for
    - i) the nth term
    - ii) the sum of the first ten terms.
  - b) The sum to infinity of this series is 30. Find the value of the first term.

2 Find: a) $\dfrac{\mathrm{d}}{\mathrm{d}x}\left(\dfrac{x^4 + 2}{x^3}\right)$ b) $\displaystyle\int \dfrac{x^4 + 2}{x^3}\,\mathrm{d}x$

3 Consider the function $f(x) = x^2 - 8x + 15$.
  - a) Write the function in the form $f(x) = (x - a)^2 + b$.
  - b) State the co-ordinates of the maximum point on the graph of $y = f(x)$.
  - c) Find the equation of the line of symmetry of the graph.
  - d) Find the co-ordinates of the points where the graph crosses the x-axis.

4 The figure shows part of the graph of $y = f(x)$ where $f(x) = x^2$.

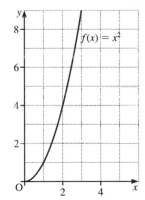

Describe fully the transformations which map this graph onto the graphs of

  a) $y = 3 + f(x)$ b) $y = f(x + 3)$ c) $y = f^{-1}(x)$

5 The population of a certain country t years after 2000 can be modelled by the function $P = Ae^{kt}$. In 2000 the population was 4.5 million and the growth rate was expected to be constant at 1.5% per annum.
  - a) Write down the values of A and k.
  - b) In what year would the population first exceed 10 million?

6 A particle passes a fixed point O with velocity 40 ms^{-1}. It travels in a straight line and its acceleration t seconds after passing O is given by $a = -4t\,\text{ms}^{-2}$.
a) Find an expression for its velocity in terms of t.
b) Find its velocity 5 seconds after passing O
c) Find an expression for its distance from O after t seconds.

7 The diagram shows a square based pyramid ABCDV of side 10 cm and height 4 cm. O is the centre of the base.

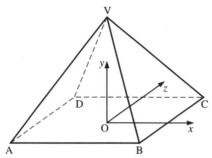

a) Find the position vectors of V and B
b) Find $|\overrightarrow{BV}|$.

8 Show that the points on the curve $y = \dfrac{x}{3} - \dfrac{3}{x}$ at which the gradient is $\frac{2}{3}$ satisfy the equation $x^2 - 9 = 0$.
Hence find the co-ordinates of these points.

9 A slice of angle 0.5 radians is cut from a circular cake. The cake has a radius of 10 cm and is 8 cm thick. Find the volume of the slice.

10 Consider the equation $\tan x = 4$.
a) How many solutions does this equation have for $-360° \leqslant x \leqslant 360°$?
b) Find the largest of these solutions.

11 a) Expand $\left(x - \dfrac{3}{x}\right)^4$ and simplify each term.
b) Find the constant term when $(1 + x)\left(x - \dfrac{3}{x}\right)^4$ is expanded.

12 A set of data has median 15 and is distributed symmetrically about this value. The interquartile range is 10 and the range is 26. Draw a box and whisker plot to represent this data.

13 A six-sided die is biased so that the probability of a prime number is twice that of a non-prime number.
a) Find the probability distribution of the scores.
b) How many times would you expect to get a prime number in 900 rolls of the dice?

14 The random variable X is normally distributed with mean 30 and variance σ^2. Also $P(X > 33) = 0.115$.
Find σ.

15 The function f is given by $f(x) = x + \dfrac{3}{x}, x \neq 0$.

a) Sketch the graph of this function for $-5 \leqslant x \leqslant 5$

b) Shade the area represented by $\left| \displaystyle\int_{-2}^{-1} f(x)\, dx \right|$.

c) Which of the following integrals represents an area under this curve?

    i) $\displaystyle\int_{0}^{2} f(x)\, dx$ ii) $\displaystyle\int_{-2}^{2} f(x)\, dx$ iii) $\displaystyle\int_{1}^{3} f(x)\, dx$

d) Find the area between the curve and the line $y = 4$.

Practice Paper Two

1 The function f is defined by $f \colon x \to 3x + 2$.

a) Using equal scales on both axes sketch the graph of $y = f(x)$.

b) On the same diagram sketch the graph of $y = f^{-1}(x)$.

c) Find an expression for $f^{-1}(x)$.

The function g is defined by $g \colon x \to x^2 - 1$.

d) Find an expression for $f \circ g(x)$.

e) Show that the equation $f \circ g(x) = f(x)$ has two real solutions.

2 Part A

The speeds in $km\ h^{-1}$ of cars passing a point on a highway are recorded in the following table.

Speed v	$v \leqslant 60$	$60 < v \leqslant 70$	$70 < v \leqslant 80$	$80 < v \leqslant 90$	$90 < v \leqslant 100$
Number of cars	0	7	25	63	70
Speed v	$100 < v \leqslant 110$	$110 < v \leqslant 120$	$120 < v \leqslant 130$	$130 < v \leqslant 140$	$v > 140$
Number of cars	71	39	20	5	0

a) Calculate an estimate of the mean speed of the cars.

b) The following table gives some of the cumulative frequencies for the information above.

Speed v	$v \leqslant 60$	$v \leqslant 70$	$v \leqslant 80$	$v \leqslant 90$	$v \leqslant 100$
Number of cars	0	7	32	95	a
Speed v	$v \leqslant 110$	$v \leqslant 120$	$v \leqslant 130$	$v \leqslant 140$	
Number of cars	236	b	295	300	

    i) Write down the values of a and b.

    ii) On graph paper, construct a cumulative frequency curve to represent this information. Use a scale of 1 cm for $10\ km\ h^{-1}$ on the horizontal axis and a scale of 1 cm for 20 cars on the vertical axis.

c) Use your graph to determine

   i) the percentage of cars travelling at a speed in excess of $105 \, \text{km h}^{-1}$;

   ii) the speed which is exceeded by 15% of the cars.

Part B

A company manufactures computer components. Three components A, B and C are needed to assemble a computer. The probabilities that each of these are faulty are 0.02, 0.1 and 0.08 respectively.

a) Find the probability that none of the components in a computer is faulty.

b) Find the probability that at least one component in a computer is faulty.

Ten computers are assembled. If any of the components in a computer are faulty the computer does not work.

c) Find the probability that exactly two of the computers do not work.

d) Find the probability that at most two of the computers do not work.

e) Find the expected number of computers that do not work.

3 a) Sketch the graph of $y = x \sin x$ for $0 \leqslant x \leqslant \pi$ and $0 \leqslant y \leqslant 2$.

   b) Find $\dfrac{dy}{dx}$.

> Note: This question requires the use of a graphic display calculator.

c) Show that $\dfrac{d^2 y}{dx^2} = -x \sin x + 2 \cos x$.

The graph has a point of inflexion between $x = 0$ and $x = \pi$.

d) Show that, at this point of inflexion $x = \dfrac{2}{\tan x}$.

e) Find the x-coordinate of this point of inflexion correct to 5 decimal places.

f) On the same diagram draw the line $y = \dfrac{x + 1}{2}$.

g) Find the x-coordinates of the points of intersection of the line and the curve correct to three decimal places.

The area between the line and the curve is rotated about the x-axis through 2π.

h) Write down an expression involving integrals for the volume generated.

i) Find this volume.

4 Part A

The matrix $\mathbf{A} = \begin{pmatrix} a & b & c \\ 2 & 0 & 1 \\ 0 & -1 & 2 \end{pmatrix}$.

a) Show that $\mathbf{A}$ has no inverse if $a - 2c = 4b$.

b) If $a = b = c = 1$, find $\mathbf{A}^{-1}$.

c) Hence solve the equation $\mathbf{AX} = \mathbf{B}$ where $\mathbf{B} = \begin{pmatrix} 1 \\ 2 \\ 3 \end{pmatrix}$.

Part B

The points $A(1, 2, 3)$ and $B(3, -1, 6)$ lie on the line L.

a) Find an equation for the line L in the form $\mathbf{r} = \mathbf{a} + t\mathbf{b}$.

b) Write down a vector in the direction of the line.

The point P has position vector $\overrightarrow{OP} = \begin{pmatrix} x \\ y \\ z \end{pmatrix}$ and OP is

perpendicular to the line L.

c) Show that $2x - 3y + 3z = 0$.

The point P lies on the line L.

d) Show that, at point P, $t = -\frac{5}{22}$.

5 A ball is dropped vertically from a great height. Its velocity v is given by

$$v = 50 - 50e^{-0.2t}, \ t \geqslant 0$$

where v is in metres per second and t is in seconds.

a) Find the value of v when

   i) $t = 0$;   ii) $t = 10$.

b) i) Find an expression for the acceleration, a, as a function of t.

ii) What is the value of a when $t = 0$?

c) i) As t becomes large, what value does v approach?

  ii) As t becomes large, what value does a approach?

  iii) Explain the relationship between the answers to parts i) and ii).

d) Let y metres be the distance fallen after t seconds.

  i) Show that $y = 50t + 250e^{-0.2t} + k$, where k is a constant.

  ii) Given that $y = 0$ when $t = 0$, find the value of k.

  iii) Find the time required to fall 250 m, giving your answer correct to **four** significant figures.

Answers

Answers are given only for questions with a numerical or algebraic answer.

Chapter 1

Exercise 1A

1 a) 6 b) 7 c) -2 d) 5 e) 2 f) $\frac{11}{2}$ **2** a) 3 b) -2 c) $\frac{13}{4}$ d) $-\frac{1}{10}$ e) -3 f) $-\frac{1}{3}$

3 a) 2 b) $-\frac{4}{15}$ c) 2 d) $\frac{29}{11}$ e) $-\frac{3}{4}$ f) $-\frac{27}{2}$ **4** a) 7 b) 3 c) $\frac{5}{13}$ d) $-\frac{11}{9}$

Exercise 1B

1 a) $-\frac{1}{2}$ b) $\frac{17}{2}$ c) $\frac{21}{2}$ d) $\frac{49}{6}$ **2** a) -4 b) $\frac{31}{11}$ c) $\frac{16}{13}$ d) $-\frac{17}{4}$

3 a) 4 b) $-\frac{7}{2}$ c) $\frac{16}{7}$ d) -2 e) 27 f) $-\frac{15}{13}$

4 $\frac{1}{3}$ **5** $-\frac{16}{29}$ **6** $-\frac{5}{8}$ **7** -2 **8** $\frac{18}{7}$ **9** $\frac{1}{2}$ **10** $\frac{28}{11}$ **11** -5 **12** $\frac{7}{2}$

Exercise 1C

1 a) $x > 4$ b) $x \geqslant -4$ c) $x \geqslant \frac{11}{2}$ d) $x > \frac{15}{4}$ e) $x < 5$ f) $x \geqslant \frac{2}{5}$

2 a) $x < 4$ b) $x < 2$ c) $x > 9$ d) $x \leqslant 0$ e) $x > \frac{19}{9}$ f) $x \geqslant -1$

3 a) $x < 10$ b) $x \leqslant 7$ c) $x \leqslant -19$ d) $x > -\frac{19}{9}$ e) $x > \frac{6}{5}$ f) $x \leqslant 8$ g) $x \leqslant -6$ h) $x > 2$

4 a) $1, 2, 3$ b) 1 c) $2, 3, 4, 5$ d) $-3, -2, -1$ e) $1, 2$ f) $2, 3, 4, 5, 6$ **6** 3

Exercise 1D

1 a) $3, (0, -2)$ b) $7, (0, 3)$ c) $\frac{5}{3}, (0, -\frac{2}{3})$ d) $\frac{1}{2}, (0, \frac{1}{6})$ e) $-\frac{1}{4}, (0, \frac{1}{2})$ f) $\frac{4}{3}, (0, \frac{5}{3})$ g) $-\frac{3}{2}, (0, \frac{5}{2})$ h) $-\frac{2}{3}, (0, \frac{4}{3})$

1 i) $\frac{3}{5}, (0, \frac{6}{5})$ j) $4, (0, -7)$ k) $-2, (0, \frac{4}{3})$ l) $\frac{8}{5}, (0, -\frac{12}{5})$

3 a) $2x - y + 4 = 0$ b) $x + y - 3 = 0$ c) $x - 2y + 4 = 0$ d) $3x + y - 6 = 0$ e) $2x - y + 4 = 0$

3 f) $3x - 4y - 12 = 0$ g) $3x + y - 6 = 0$ h) $4x + 5y + 20 = 0$ **4** $y = 3x - 1$

Exercise 1E

1 a) $(2, -1)$ b) $(1, 2)$ c) $(4, 1)$ d) $(-1, -2)$ e) $(3, -4)$ f) $(3, -1)$ g) $(2, -1)$ h) $(3, 2)$ i) $(-3, 2)$

1 j) $(1, -1)$ k) $(7, 6)$ l) $(2, -\frac{3}{2})$

2 a) $(4, 2)$ b) $(3, 4)$ c) $(2, 3)$ d) $(-2, -1)$ e) $(\frac{1}{2}, -\frac{3}{2})$ f) $(3, 2)$ g) $(\frac{1}{3}, -\frac{2}{3})$ h) $(\frac{1}{2}, \frac{1}{4})$

3 a) $(2, 5)$ b) $(6, 2)$ c) $(-1, 7)$ d) $(-2, 6)$ e) $(-1, -4)$ f) $(6, 2)$ g) $(-3, -2)$ h) $(0, 2)$

Exercise 1F

1 a) $2(x + 2y)$ b) $3(2a + 3b)$ c) $5(p - 2q)$ d) $4(2y - 3x)$ e) $2(m + 3n)$ f) $3(a - 3b)$ g) $2(a + 2b - 3c)$

1 h) $3(3x - y + 2z)$ **2** a) $x(5x + 1)$ b) $a(2a + 3)$ c) $p(6p - 5)$ d) $y(4y - 7)$ e) $2x(2x + 3y)$ f) $3a(a - 2b)$

2 g) $2b(c - 4a)$ h) $2a(2a - 3b + 4c)$ **3** a) $(a + b)(x + y)$ b) $(a + b)(x - y)$ c) $(p + 2q)(x + y)$

3 d) $(c - 2d)(x + 2y)$ e) $(a + b)(a + c)$ f) $(c - b)(c + d)$ g) $(p - 2x)((p + 5y)$ h) $(b - 3x)(b - 2y)$

4 a) $(x - 1)(x - 2)$ b) $(x + 3)(x + 1)$ c) $(x + 6)(x - 1)$ d) $(x + 3)^2$ e) $(x + 3)(x - 4)$ f) $(x + 2)(x - 6)$

4 g) $(x + 5)(x + 4)$ h) $(x + 3)(x - 2)$ i) $(x - 1)(x - 7)$ j) $(x + 2)(x - 5)$ k) $(x - 1)(x + 5)$ l) $(x + 5)(x + 2)$

5 a) $(2x - 3)(x + 2)$ b) $(2x + 1)(x + 1)$ c) $(3x + 1)(x - 2)$ d) $(3x - 1)(x - 3)$ e) $(5x - 1)(x - 2)$

5 f) $(5x + 3)(x + 4)$ g) $(2x - 5)(x - 6)$ h) $(3x + 5)(x + 7)$ i) $(7x + 2)(x - 3)$ j) $(2x + 3)(2x - 1)$

5 k) $(4x+3)(x-2)$ l) $(3x+5)(2x+1)$ **6** a) $(x+4)(x-4)$ b) $(x+3)(x-3)$ c) $(x+9)(x-9)$

6 d) $(x+12)(x-12)$ e) $(2x+3)(2x-3)$ f) $(5x+2)(5x-2)$ g) $(4-3x)(4+3x)$ h) $(7+8x)(7-8x)$

6 i) $(2x+9)(2x-9)$ j) $(5x+4y)(5x-4y)$ k) $(x+13y)(x-13y)$ l) $4(x+3y)(x-3y)$

Exercise 1G

1 a) $2, 3$ b) $-1, 4$ c) $2, 5$ d) $-3, -2$ e) $2, 4$ f) $-1, 6$ g) ± 3 h) $-4, 2$ i) $-3, 4$ j) $4, 5$ k) $0, 4$

1 l) $-1, 8$ **2** a) $-2, -\frac{1}{2}$ b) $\frac{1}{3}, 2$ c) $-1, \frac{5}{2}$ d) $-3, \frac{1}{5}$ e) $-1, -\frac{1}{4}$ f) $\frac{1}{3}, \frac{1}{2}$ g) $-\frac{2}{3}, 4$ h) $-3, \frac{5}{2}$

2 i) $\pm \frac{3}{4}$ j) $-\frac{5}{3}, 2$ k) $-3, \frac{2}{5}$ l) $\frac{1}{4}, \frac{3}{2}$ **3** a) $-5, 1$ b) $-4, 2$ c) $-\frac{1}{2}, 5$ d) $-12, 2$

4 a) $-6, 1$ b) $-4, 2$ c) $-2, \frac{3}{5}$ d) $-7, 3$ **5** $15\,\text{cm} \times 2\,\text{cm}$ **6** $\frac{2}{3}$ **7** $\frac{3}{4}$ **8** b) $x^2 - 6x + 9 = 0, 3$

Exercise 1H

1 a) $(x+2)^2 + 2$ b) $(x-3)^2 + 4$ c) $(x-5)^2 + 15$ d) $(x-\frac{1}{2})^2 - 5\frac{1}{4}$ e) $(x-2\frac{1}{2})^2 + 2\frac{3}{4}$ f) $(x-10)^2 - 97$

2 a) $2 \pm \sqrt{5}$ b) $-3 \pm \sqrt{7}$ c) $1 \pm \sqrt{2}$ d) $4 \pm \sqrt{19}$ e) $-\dfrac{1}{2} \pm \dfrac{\sqrt{5}}{2}$ f) $-\dfrac{3}{2} \pm \dfrac{\sqrt{5}}{2}$ g) $\dfrac{5}{2} \pm \dfrac{\sqrt{33}}{2}$ h) $\dfrac{1}{2} \pm \dfrac{\sqrt{13}}{2}$

2 i) $-\dfrac{5}{2} \pm \dfrac{\sqrt{21}}{2}$ j) $-6 \pm \sqrt{31}$ k) $\dfrac{9}{2} \pm \dfrac{\sqrt{41}}{2}$ l) $\dfrac{1}{4} \pm \dfrac{\sqrt{5}}{4}$

3 a) $2(x+2)^2 - 21$ b) $3(x-1)^2 - 1$ c) $9 - (x+2)^2$ d) $7 - 2(x-1)^2$ e) $2(x-1\frac{1}{2})^2 + \frac{1}{2}$ f) $28 - 5(x+1)^2$

4 a) $\dfrac{3}{4} \pm \dfrac{\sqrt{33}}{4}$ b) $1 \pm \sqrt{\dfrac{2}{3}}$ c) $-\dfrac{1}{2} \pm \sqrt{\dfrac{3}{2}}$ d) $-\dfrac{5}{6} \pm \dfrac{\sqrt{37}}{6}$ e) $-\dfrac{1}{10} \pm \dfrac{\sqrt{61}}{10}$ f) $\dfrac{3}{4} \pm \dfrac{\sqrt{17}}{4}$ g) $\dfrac{1}{4} \pm \dfrac{\sqrt{17}}{4}$

4 h) $-\dfrac{3}{8} \pm \dfrac{\sqrt{41}}{8}$ i) $1 \pm \dfrac{\sqrt{14}}{7}$ j) $\dfrac{1}{3} \pm \sqrt{\dfrac{11}{18}}$ k) $2 \pm \sqrt{\dfrac{3}{5}}$ l) $-\dfrac{9}{2} \pm \dfrac{\sqrt{39}}{2}$ **5** $(x+2)^2 + 3$ **6** $5(x-3)^2 + 2$

Exercise 1I

1 a) $-2.41, 0.414$ b) $-3.41, -0.586$ c) $-1.19, 4.19$ d) $0.628, 6.37$ e) $-4.19, 1.19$ f) $-9.10, 1.10$

2 a) $-1.62, 0.618$ b) $-1.36, 7.36$ c) $-4.30, -0.697$ d) $1.27, 4.73$ e) $1.84, 8.16$ f) $-13.5, 1.48$

3 a) $-2.35, 0.851$ b) $-1.18, 0.847$ c) $-2.09, 0.838$ d) $-2.78, -0.719$ e) $-0.690, 0.290$ f) $-1.24, 0.404$

4 a) $-2.27, 1.77$ b) $-0.729, 0.229$ c) $-1.77, -0.566$ d) $-1.39, 2.89$ e) $-1.23, -0.271$ f) $-1.27, 0.472$

5 a) $-1.65, 3.65$ b) $-9.22, 0.217$ c) $-13.3, -2.71$ d) $0.219, 2.28$ **6** a) none   b) two   c) one   d) two

6 e) one   f) two   **7** -71 **8** -16 **12** 4 or 16   **13** $-\frac{2}{9}$ or 2   **15** $a = -2$, roots $0, 2$; $a = 6$, roots $-4, -2$

Exercise 1J

1 a) $\pm 2, \pm 3$ b) $\pm \sqrt{3}$ c) $1, 3$ d) $-2, \sqrt[3]{3}$ e) $4, 9$ f) $1, 25$ **2** a) $\pm \sqrt{2}$ b) $1, -2$ c) $25, 49$ d) $\pm \sqrt{2}, \pm 2$

2 e) $4, 25$ f) $\pm \sqrt{3}$ **3** $1, -2$ **4** $-2, \frac{2}{3}$ **5** a) $2, 5$ b) $\pm 1, \pm 2$ **6** a) $-2, 7$ b) $-1, 2$ **7** $-2, -\frac{1}{2}, 1$

Exercise 1K

2 a) $y = 2$ at $x = -2$ b) $y = 4$ at $x = 3$ c) $y = 15$ at $x = 5$ d) $y = -6$ at $x = -1$ e) $y = 5\frac{3}{4}$ at $x = -1\frac{1}{2}$

2 f) $y = 2\frac{3}{4}$ at $x = 3\frac{1}{2}$ **4** a) $y = -17\frac{1}{2}$ at $x = -2\frac{1}{2}$ b) $y = 11$ at $x = -1$ c) $y = -7\frac{1}{16}$ at $x = -\frac{1}{8}$

4 d) $y = 8\frac{2}{3}$ at $x = \frac{1}{3}$ e) $y = 4\frac{23}{24}$ at $x = -\frac{1}{12}$ f) $y = 7\frac{4}{5}$ at $x = \frac{1}{5}$ **5** a) $y = 4$ at $x = -1$ b) $y = 9$ at $x = 2$

5 c) $y = 9$ at $x = 1$ d) $y = 3\frac{1}{2}$ at $x = \frac{1}{2}$ e) $y = 7\frac{9}{16}$ at $x = -\frac{3}{8}$ f) $y = 9\frac{1}{8}$ at $x = -\frac{5}{4}$

7 b) $100\,\text{m}^2$ **8** $x(40 - 2x)\,\text{m}^2$ **9** $133\frac{1}{3}\,\text{m}^2$ **10** $45\,\text{m}, 3\,\text{s}$ **12** b) $2x(1 - x)\,\text{m}^2$

Exercise 1L

1 a) $\dfrac{5x+7}{(x+3)(x-1)}$ b) $\dfrac{x-10}{(x+4)(x-3)}$ c) $\dfrac{2(x+13)}{(x+4)(x-5)}$ d) $\dfrac{x^2+6}{(x+2)(x-3)}$ e) $\dfrac{2x^2-5x-2}{(x+2)(x-2)}$

1 f) $\dfrac{4x+34}{(2x+5)(2x-3)}$ g) $\dfrac{5x^2+3x+4}{(x+4)(x-2)}$ h) $\dfrac{16x^2-25x+18}{(x^2+3)(2x-5)}$

2 $\frac{7}{5}$ **3** $\frac{5x-1}{(2x+1)(x-3)}, -\frac{9}{4}$ **4** $a=2, b=3$ **5** $A=1, B=1$

Revision exercise 1

1 $k=3$ **3** a) $2(x-2)^2-3$ b) -3 **4** a) $h=3, k=1$ **5** a) $p=-1$ b) $q=-6$

6 a) i) $(6,0)$ ii) $x=3$ b) $3, \frac{9}{2}$ **7** a) A$(-1,0)$ b) B$(2,0)$ c) $(0,2)$ **8** a) $p=-\frac{1}{2}, q=2$ b) $x=\frac{3}{4}$

9 a) negative b) negative c) zero d) positive

Chapter 2

Exercise 2A

1 $(12+4\sqrt{2}+4\sqrt{3})$ cm   **2** $(12+6\sqrt{2}+6\sqrt{6})$ cm   **3** 34 m   **4** 91.6 m   **5** a) 103° b) 180 km

6 a) 12 cm² b) 126 cm² c) 24 cm² d) 30 cm² e) 20.2 cm² f) 91.4 cm² g) 18.4 cm² h) 32.1 cm²

6 i) 13.6 cm² j) 89.3 cm²   **7** 26.7 cm²   **8** 13.8 m²   **9** a) 41.6° b) 17.9 cm²   **10** 42.7 cm²

12 b) 12.4 cm² c) 61.9 cm²   **13** 166 cm²   **14** 391 cm²   **15** a) 6.10 cm² b) 3.21 cm

16 16.8 cm², 5.26 cm   **17** 33.2 cm², 9.21 cm

Exercise 2B

1 a) $\frac{\pi}{6}$ rad b) $\frac{\pi}{2}$ rad c) $\frac{2\pi}{3}$ rad d) $\frac{\pi}{18}$ rad e) $\frac{4\pi}{9}$ rad f) $\frac{5\pi}{3}$ rad g) $\frac{\pi}{5}$ rad h) $\frac{4\pi}{3}$ rad i) $\frac{2\pi}{5}$ rad

1 j) 2π rad k) $\frac{19\pi}{10}$ rad l) $\frac{\pi}{180}$ rad   **2** a) 180° b) 45° c) 540° d) 30° e) 144° f) 15° g) 300°

2 h) 180° i) 75° j) 2° k) 270° l) 210°   **3** a) 229.2° b) 11.5° c) 246.4° d) 28.6° e) 40.1°

3 f) 171.9° g) 297.9° h) 120.3° i) 286.5° j) 2.3° k) 916.7° l) 57.3°   **4** a) $\frac{3\pi}{2}$ cm b) $\frac{15\pi}{4}$ cm²

5 a) 6π cm b) 27π cm²   **6** a) $\left(12+\frac{5\pi}{2}\right)$ cm b) $\frac{15\pi}{2}$ cm²   **7** a) $\left(14+\frac{5\pi}{2}\right)$ cm b) $\frac{35\pi}{4}$ cm²

8 a) $\frac{4\pi}{3}$ cm b) $\frac{16\pi}{3}$ cm²   **9** a) $(24+3\pi)$ cm b) 18π cm²   **10** 7.13 cm²   **11** $\frac{25}{3}(2\pi-3\sqrt{3})$ cm²

12 a) 1.70 cm² b) 12.5 cm   **13** a) 1.30 cm² b) 9.97 cm

14 a) $16\sqrt{3}$ cm² b) $\frac{32\pi}{3}$ cm² c) $(32\pi-48\sqrt{3})$ cm²

Exercise 2C

1 a) $x=8.96, y=8.65$ b) $x=6.60, \theta=39.9$ c) $x=10.6, y=9.13$ d) $x=8.01, \theta=27.3$

1 e) $x=7.67, \theta=60.1$ or $x=1.69, \theta=119.9$ f) $x=7.62, y=7.25$ g) $x=7.18, \theta=38.7$

1 h) $x=3.80, \theta=69.1$ or $x=1.23, \theta=110.9$ i) $x=3.80, y=10.0$ j) $x=13.8, y=11.7$

1 k) $x=5.50, \theta=102.8$ or $x=8.38, \theta=77.2$ l) $x=5.84, \theta=20.7$

Exercise 2D

1 a) 109.5° b) 13.8 c) 3.91 d) 52.2° e) 2.33 f) 8.94 g) 2.18 h) 7.43 i) 5.68 j) 53.6°

1 k) 163.2° l) 4.18

Exercise 2E

1 a) 4.70   b) 1.52   c) 8.18   d) 7.82   e) 1.23 or 1.91   f) 1.15   g) 0.631   h) 1.70   i) 5.69   j) 12.3

1 k) 5.35   l) 0.523 or 2.62   **2** a) 6.53 cm   b) 0.828 rad   **3** a) 1.31   b) 0.861 rad   **4** a) 146.8°   b) 3.33 cm

5 3.93 cm or 9.93 cm   **6** b) 48 km   c) 320.1°   **7** 1139 m at 345.8°   **8** b) 94.3 m at 328°   c) 154 m at 144.9°

9 a) 4.14 km   b) 2.93 km

Revision exercise 2

1 78.5 km   **2** a)38.2°   b) 17.3 cm^2 **3** a) 116°   b) 155 cm^2 **4** a) 111°   **5** a) 30 cm   b) 481.9 cm^2

6 a) 7.85 cm^2 b) 11.2 cm   **7** 12.3 cm^2 **8** 4.44 rad   **9** b) i) 180°   ii) 14.0°   iii) 1.69

10 ii) $\sqrt{x^2 + 100}$ c) 38.7°   d) 5.63   e) $\hat{\text{OPA}} = 0°$ ∴ O, A, P collinear   ii) $\frac{40}{3}$

Chapter 3

Exercise 3A

1 a) translation $\begin{pmatrix} 0 \\ 3 \end{pmatrix}$ b) translation $\begin{pmatrix} 1 \\ 0 \end{pmatrix}$ c) stretch $\times 3$ parallel to y-axis

1 d) stretch $\times 2$ parallel to the y-axis followed by a translation $\begin{pmatrix} 0 \\ 1 \end{pmatrix}$ **2** a) translation $\begin{pmatrix} 0 \\ 8 \end{pmatrix}$ b) translation $\begin{pmatrix} 3 \\ 0 \end{pmatrix}$

2 c) reflection in x-axis   d) reflection in x-axis followed by a translation $\begin{pmatrix} 1 \\ 0 \end{pmatrix}$ **3** a) translation $\begin{pmatrix} 0 \\ 2 \end{pmatrix}$

3 b) reflection in x-axis   c) translation $\begin{pmatrix} 3 \\ 0 \end{pmatrix}$ d) translation $\begin{pmatrix} -1 \\ 0 \end{pmatrix}$ **4** a) translation $\begin{pmatrix} 0 \\ 5 \end{pmatrix}$ b) translation $\begin{pmatrix} -1 \\ 0 \end{pmatrix}$

4 c) reflection in y-axis   d) stretch $\times 3$ parallel to y-axis   **5** a) translation $\begin{pmatrix} 0 \\ 1 \end{pmatrix}$ b) translation $\begin{pmatrix} -\frac{\pi}{2} \\ 0 \end{pmatrix}$

5 c) stretch $\times \frac{1}{2}$ parallel to x-axis   d) reflection in x-axis followed by a stretch $\times 2$ parallel to x-axis

14 stretch by a factor of 2 parallel to the y-axis, followed by a translation $\begin{pmatrix} 0 \\ 5 \end{pmatrix}$

15 a) $g(x) = (x + 1)^2 + 7$ b) reflection in x-axis, followed by a translation $\begin{pmatrix} -1 \\ 7 \end{pmatrix}$

16 b) $g(x) = -x^2, h(x) = 2 - x^2$ **17** b) $g(x) = \dfrac{3}{x + 4}, h(x) = -\dfrac{3}{x + 4}$ **18** $f(x - p) + q$

Exercise 3B

1 a) a) one-to-one   b) two-to-one   c) one-to-one   d) two-to-one   e) two-to-one   f) one-to-one

1 g) one-to-one   h) one-to-one   i) two-to-one   j) one-to-one   k) two-to-one   l) two-to-one

2 a) $f(x) \in \mathbb{R}, 4 < f(x) < 9$ b) $f(x) \in \mathbb{R}, f(x) \geq 7$ c) $f(x) \in \mathbb{R}, 1 < f(x) \leq 9$ d) $f(x) \in \mathbb{R}, \frac{1}{18} \leq f(x) \leq \frac{1}{3}$

2 e) $f(x) \in \mathbb{R}, f(x) \geq 9$ f) $f(x) \in \mathbb{R}, 4 < f(x) < 134$ g) $f(x) \in \mathbb{R}, -9 \leq f(x) \leq 0$ h) $f(x) \in \mathbb{R}, \frac{1}{10} < f(x) \leq \frac{1}{2}$

2 i) $f(x) \in \mathbb{R}, -4 < f(x) < \infty$ j) $f(x) \in \mathbb{R}, 2 \leq f(x) \leq 5$ k) $f(x) \in \mathbb{R}, 0 < f(x) \leq 20$ l) $f(x) \in \mathbb{R}, 0 < f(x) \leq \frac{1}{3}$

3 a) $4 \leq f(x) \leq 36$ b) $0 \leq g(x) \leq 9$ c) $0 \leq h(x) \leq 7$ d) $0 \leq f(x) \leq 4$ e) $1 \leq g(x) \leq 4$ f) $0 \leq h(x) \leq 16$

4 a) 2   b) $\frac{3}{4}$ **5** a) $-2, 8$ b) $-1, 5$ **6** a) $-\frac{1}{2}$ b) $-3.27, 0.766$ **7** a) $\frac{3}{2}$ b) $\frac{10}{3}$ c) $1, 4$

Exercise 3C

1 a) 7   b) 4   c) $\frac{1}{4}$ d) 19   e) 1   f) $\frac{1}{4}$ g) $\frac{1}{9}$ h) 23   i) 81   j) 12   k) $\frac{3}{2}$ l) $\frac{1}{33}$ **2** a) $3x^2 - 1$ b) $(3x - 1)^2$

2 c) $\dfrac{6}{x} - 1$ d) $\dfrac{2}{x^2}$ e) x^4 f) $9x - 4$ **3** a) i) $x^2 + 10x + 28$ ii) $x^2 + 8$ b) -2

4 a) $\dfrac{3}{x+5}, -2$ b) $\dfrac{3}{x}+5, 3$ **5** a) $-4, -1$ b) $\frac{3}{2}$ **6** $2x^2+5, 1 \leqslant x \leqslant 5, 7 \leqslant (g \circ f)(x) \leqslant 55$

7 $9x^4+6x^2-1, 0 \leqslant (q \circ p)(x) \leqslant 167$ **8** $\dfrac{1}{x^2+1}, 0 \leqslant (g \circ f)(x) \leqslant \frac{1}{17}$ **9** a) $\sqrt{x^2+1}, (f \circ g)(x) \in \mathbb{R}, (f \circ g)(x) > 1$

9 b) $x+1, (g \circ f)(x) \in \mathbb{R}, (g \circ f)(x) > 1$ **10** a) $9x+20, x$ b) $-\frac{5}{2}$ **11** $-2, 1$

Exercise 3D

1 a) $\dfrac{x-2}{3}$ b) $\dfrac{1+x}{5}$ c) $\dfrac{4-x}{3}$ d) $\dfrac{2}{x}, x \neq 0$ e) $\dfrac{3+x}{x}, x \neq 0$ f) $\dfrac{2x-5}{3x}, x \neq 0$ g) $\dfrac{2x}{1-x}, x \neq 1$ h) $\dfrac{5x}{2+x}, x \neq -2$

1 i) $\dfrac{x}{3-2x}, x \neq \dfrac{3}{2}$ **2** a) $\sqrt{x}, x \in \mathbb{R}, x > 4$ b) $\dfrac{1-2x}{x}, x \in \mathbb{R}, 0 < x < \frac{1}{2}$ c) $2+x^2, x \in \mathbb{R}, x > 1$

2 d) $\sqrt{\dfrac{x+1}{3}}, x \in \mathbb{R}, 2 < x < 47$ e) $\dfrac{x^2-3}{2}, x \in \mathbb{R}, x \geqslant 5$ f) $\dfrac{1}{x+3}, x \in \mathbb{R}, -\dfrac{11}{5} < x < -\dfrac{5}{2}$

2 g) $-2 + \sqrt{x-3}, x \in \mathbb{R}, x \geqslant 3$ h) $\sqrt{x-1}, x \in \mathbb{R}$ **3** a) $\dfrac{x+4}{3}, x \in \mathbb{R}$ c) 2

4 b) $\dfrac{10-x}{2}, x \in \mathbb{R}, x \leqslant 10$ c) $\dfrac{10}{3}$ **5** a) $\sqrt{x+6}, x \in \mathbb{R}, x > -6$ c) 3   **6** b) $2 + \sqrt{x}, x \in \mathbb{R}, x > 0$ c) 4

7 a) 2 b) $\dfrac{x+5}{2}, x \in \mathbb{R}; \dfrac{7-x}{4}, x \in \mathbb{R}$ c) -1, from a) $f(2) = g(2) = -1$ **8** a) $0 < h(x) < \frac{4}{3}$

8 b) $h^{-1}(x) = \dfrac{4}{x} - 3, x \in \mathbb{R}, x \neq 0$ c) 1   **9** a) $3x-5$ b) i) $\dfrac{x-1}{3}, x \in \mathbb{R}$ ii) $x+2, x \in \mathbb{R}$ iii) $\dfrac{x+5}{3}, x \in \mathbb{R}$

10 a) $g(x) \geqslant -3$ b) g^{-1} exists because g is one-to-one   **11** a) $\dfrac{x-3}{2}, x \in \mathbb{R}$ b) $\dfrac{1}{2(x+1)}, x \in \mathbb{R}, x \neq -1$ c) $\pm\sqrt{2}$

12 a) $3 + \dfrac{5}{x}, x \in \mathbb{R}, x \neq 0$ b) $\dfrac{5}{x^2+1}, x \in \mathbb{R}, x > 0$ c) $\frac{1}{3}, 3$

Revision exercise 3

1 a) 3 b) 4 c) 5   **2** a) $(x-1)^2, 4(x-1)^2, 4(x-1)^2 + 3$ **3** b) $(1, \frac{3}{2}), (2, 2)$ **4** a) -1, b) 16

5 $\dfrac{3-x^2}{2}, -11$ **6** a) i) $\frac{4}{3}$ ii) -4 b) $\frac{8}{3}\sqrt{x-4} + 3$ **7** a) $(x-3)^2 + 4$ c) $x \geqslant 4$ **8** $\dfrac{1 \pm \sqrt{17}}{2}$

9 a) -3 b) -3 d) 1   **10** a) $6, 4, 0, 2$

Chapter 4

Exercise 4A

1 $3x^2$ **2** $4x$ **3** $-2x$ **4** $3x^2 - 6$ **5** $-\dfrac{1}{x^2}$ **6** $\dfrac{1}{2\sqrt{x}}$

Exercise 4B

1 a) $4x^3$ b) $6x^5$ c) $12x$ d) $-15x^2$ e) 3 f) $12x^5$ g) $-14x$ h) 0 i) $2x^3$ j) $4x^5$ k) $-\frac{3}{4}x^2$ l) $\frac{2}{5}$

2 a) $-2x^{-3}$ b) $-4x^{-5}$ c) $-6x^{-4}$ d) $-4x^{-2}$ e) $-\dfrac{3}{x^4}$ f) $\dfrac{2}{x^3}$ g) $-\dfrac{9}{x^4}$ h) $\dfrac{2}{x^2}$ i) $-\dfrac{3}{x^3}$ j) $-\dfrac{27}{2x^4}$ k) $\dfrac{3}{x^5}$ l) $-\dfrac{2}{5x^2}$

3 a) $\frac{1}{2}x^{-\frac{1}{2}}$ b) $2x^{-\frac{2}{3}}$ c) $-\frac{2}{3}x^{-\frac{5}{3}}$ d) $2x^{-\frac{6}{5}}$ e) $\dfrac{7}{2\sqrt{x}}$ f) $\dfrac{1}{3\sqrt[3]{x^2}}$ g) $-\dfrac{2}{5\sqrt{x^3}}$ h) $\dfrac{2}{\sqrt[3]{x^4}}$ i) $-\dfrac{5}{4\sqrt{x^3}}$

Exercise 4C

1 a) $2x+2$ b) $6x-5$ c) $2x$ d) $-12x^2$ e) $2x+2$ f) $7x^6 + 12x^3$ g) $4x^3 - 6x$ h) $1 - \dfrac{1}{x^2}$ i) $10x + \dfrac{6}{x^4}$

2 a) $3 - 15x^2$ b) $\dfrac{3}{x^2}$ c) $-\dfrac{4}{x^2} - 6x^2$ d) $\dfrac{1}{\sqrt{x}}$ e) $\dfrac{1}{2\sqrt{x}} - \dfrac{1}{2\sqrt{x^3}}$ f) $-8x^{-3} - 3$ g) $x^{-\frac{2}{3}} + \frac{4}{3}x^{-\frac{1}{3}}$ h) $2x^{-\frac{1}{2}} + 2$

2 i) $4x^{-\frac{1}{3}} - 10x^{\frac{3}{2}}$ **3** a) 4 b) $\dfrac{1}{\sqrt{x}} - \dfrac{5}{2x^2}$ c) $8x + 7$ d) $2x^{-\frac{3}{5}} - 5x^{\frac{3}{2}}$ e) $\dfrac{3}{\sqrt{x}} + \dfrac{3}{x^3}$ f) $-14x^{-8} + 15x^{-4} + 1$

3 g) $-\dfrac{10}{x^3} + \dfrac{2}{x^2}$ h) $12x^{\frac{1}{5}}$ i) $-8x^{-5} + 8x^{-3}$ j) $-\frac{3}{2}x^{-\frac{3}{2}} - x^{-\frac{1}{2}}$ **4** a) $3x(x + 2)$ b) $2(1 - x)$ c) $12x^2 - 5x^4$

4 d) $\frac{1}{3}x^{-\frac{2}{3}}(8x - 5)$ $[= \frac{8}{3}x^{\frac{1}{3}} - \frac{5}{3}x^{-\frac{2}{3}}]$ e) $2x - 1$ f) $2(x + 4)$ g) $4x + 9$ h) $4(x - 3)$ i) $2(x + 3)$ j) $x^2(4x + 3)$

5 a) $6(x - 2)$ b) $3x^2(5x^2 - 1)$ c) $\dfrac{5x^2 + 3}{2\sqrt{x}}$ $[= \frac{5}{2}x^{\frac{3}{2}} + \frac{3}{2}x^{-\frac{1}{2}}]$ d) $\frac{1}{2}x^{-\frac{3}{2}}(x - 1)$ e) $\dfrac{x^2 - 7}{x^2}$ $[= 1 - 7x^{-2}]$

5 f) $-\dfrac{x + 10}{x^3}$ $[= -x^{-2} - 10x^{-3}]$ g) $\dfrac{3x^2 - 2}{x^2}$ $[= 3 - 2x^{-2}]$ h) $\dfrac{2(3x^3 + 7)}{x^3}$ $[= 6 + 14x^{-3}]$ i) $-\dfrac{3}{5x^2}$

6 a) $3x^2(4x - 1)$ b) $4x(2x - 1)(x - 1)$ $[-8x^3 \quad 12x^2 + 4x]$ c) $-\dfrac{1}{2\sqrt{x^3}}$ d) $\dfrac{3x^3 - 10}{x^3}$ $[= 3 - 10x^{-3}]$

6 e) $\dfrac{x^2 + 12}{x^2}$ $[= 1 + 12x^{-2}]$ f) $-\dfrac{2(x^2 - 13x + 30)}{x^4}$ $[= -2x^{-2} + 26x^{-3} - 60x^{-4}]$ g) $\dfrac{9x^2 - 1}{2x^2}$ $[= \frac{9}{2} - \frac{1}{2}x^{-2}]$

6 h) $\dfrac{5x - 3}{4\sqrt{x^3}}$ $[-\frac{5}{4}x^{-\frac{1}{2}} - \frac{3}{4}x^{-\frac{3}{2}}]$ i) $\dfrac{(3x - 1)(x + 1)}{2\sqrt{x^3}}$ $[= \frac{3}{2}x^{\frac{1}{2}} + x^{-\frac{1}{2}} - \frac{1}{2}x^{-\frac{3}{2}}]$ j) $\frac{1}{2}x^{\frac{1}{2}} - \frac{5}{6}x^{-\frac{3}{2}}$

Exercise 4D

1 a) $18x$ b) $2 - 120x^4$ c) $\dfrac{2}{x^3}$ d) $\dfrac{12}{x^4}$ e) $\dfrac{3}{4\sqrt{x^5}} + \dfrac{1}{4\sqrt{x^5}}$ f) $-\dfrac{2}{9\sqrt{x^5}} - 6x$ g) $10x(2x^7 + 3)$ h) $6(2x + 1)$ i) $-\dfrac{2}{x^3}$

1 j) $12x^{-3} - 30x^{-4}$ **2** a) $2ax + b, 2a$ b) $-\dfrac{a}{x^2} - \dfrac{2b}{x^3}, \dfrac{2a}{x^3} + \dfrac{6b}{x^4}$ c) $\dfrac{a}{2\sqrt{x}} - \dfrac{b}{2\sqrt{x^3}}, \dfrac{3b}{4\sqrt{x^5}} - \dfrac{a}{4\sqrt{x^3}}$

2   d) $2acx + ad + bc, 2ac$ e) $\dfrac{2ax + b}{c}, \dfrac{2a}{c}$

Exercise 4E

1 a) 6 b) 24 c) $\frac{1}{6}$ d) $-\frac{1}{9}$ e) 4 f) $\frac{1}{8}$ g) $\frac{2}{3}$ h) -1 i) 3 j) -5 **2** a) $(-2, -8), (2, 8)$ b) $(-1, 3)$

2 c) $(2, 17)$ d) $(-\frac{1}{2}, -8), (\frac{1}{2}, 8)$ e) $(-2, 4)$ f) $(-3, -34), (3, 32)$ g) $(0, 3), (\frac{2}{3}, \frac{77}{27})$ h) $(\frac{1}{4}, \frac{11}{2})$

2 i) $(-2, -3), (2, 1)$ j) $(-1, 2)$ **3** $a = \frac{3}{4}, b = -5$ **4** $A = \frac{17}{18}, B = -\frac{13}{3}$ **5** $a = -45, c = 24$

6 $a = 12, b = -18$ **7** $A = 2, B = 4$

Exercise 4F

1 a) $y = 4x - 1$ b) $y = 6x - 5$ c) $y = -x - 6$ d) $y = -2x + 16$ e) $y = -2x - 1$ f) $2y - 3x = 12$

1 g) $y = 5$ h) $y = -x - 3$ i) $y = 9x - 16$ j) $25y - 2x = 100$

2 a) $y = -x$ b) $3y + x = 8$ c) $2y - 3x + 5 = 0$ d) $y = -3x + 33$ e) $2y + x = 11$ f) $4y + x + 14 = 0$

3 $(0, -9)$ **4** a) $1, 3$ b) $y = 2, y = 6$ **5** $9y - x + 16 = 0, 9y - x = 464$ **7** a) $y = 6x - 4$ b) $(-2, -16)$

8 a) $y = x + 1$ b) $(1, 2), (3, 4)$ **9** $a = 3, b = 4$ **10** $c = -14, d = -20$

Exercise 4G

1 a) $2t + 3$ b) $7\,\mathrm{m\,s^{-1}}$ **2** a) $6t^2 + 5$ b) $59\,\mathrm{m\,s^{-1}}$ **3** a) $6t - 15t^2$ b) $6 - 30t$ **4** a) $1 + 3t^2$ b) $6t$

5 a) $6t - 2$ b) $34\,\mathrm{m\,s^{-2}}$ **6** a) $3t^2 - 6t + 6$ b) $6\,\mathrm{m\,s^{-2}}$ **7** a) $2t - 8$ b) 4 **8** a) $1, 3$ b) $9\,\mathrm{m}, 5\,\mathrm{m}$

9 $27\,\mathrm{m\,s^{-1}}$

Exercise 4H

1 a) $(2, -1)$ b) $(-3, -4)$ c) $(0, 6)$ d) $(\frac{5}{2}, -\frac{13}{4})$ e) $(\frac{3}{4}, -\frac{1}{8})$ f) $(-1, 4), (1, 0)$ g) $(-2, 40), (6, -216)$

1 h) $(1, 11), (-3, -21)$ i) $(0, 5), (2, 9)$ j) $(-1, 2), (0, 3), (1, 2)$ k) $(2, -45)$ l) $(-1, 8), (\frac{1}{2}, -\frac{7}{16}), (1, 0)$

2 a) $(1, 4)$, min b) $(-2, -2)$, min c) $(\frac{1}{2}, \frac{13}{4})$, max d) $(3, 21)$, min e) $(-\frac{5}{4}, -\frac{147}{8})$, min f) $(5, 0)$, min

2 g) $(2, -40)$, min; $(-6, 216)$, max h) $(\frac{1}{3}, \frac{40}{27})$, max; $(3, -8)$, min i) $(-5, -97)$, min; $(1, 11)$, max

2 j) $(-2, -13)$, min; $(0, 3)$, max; $(2, -13)$, min k) $(-3, -26)$, min; $(0, 1)$, saddle

2 l) $(-3, -127)$, min; $(1, 1)$, max; $(2, -2)$, min   **3** a) $(-1, 5)$, max; $(1, 1)$, min b) $(1, -4)$, min; $(3, 0)$, max

3 c) $(3, 8)$, saddle d) $(-2, -25)$, min; $(0, -9)$, max; $(2, -25)$, min e) $(0, 0)$, saddle; $(6, 432)$, max

3 f) $(-1, 0)$, saddle; $(\frac{5}{4}, -\frac{2187}{256})$, min   **4** a) $(-1, -2)$ max, $(1, 2)$ min b) $(2, 12)$ min c) $(6, \frac{1}{12})$ max

4 d) $(3, -\frac{1}{27})$ min e) $(-\frac{1}{2}, -16)$ min, $(\frac{1}{2}, 16)$ max f) $(2, -\frac{3}{8})$ min   **5** a) increase $x > 4$, decrease $x < 4$

5 b) increase $x < -1$, decrease $x > -1$ c) Increase $x < -1$ and $x > 3$, decrease $-1 < x < 3$

5 d) increase $-3 < x < 2$, decrease $x < -3$ and $x > 2$

5 e) increase $-3 < x < -2$ and $x > 2$, decrease $x < -3$ and $-2 < x < 2$

5 f) increase $x < -2$ and $x > 2$, decrease $-2 < x < 0$ and $0 < x < 2$

6 a) $(0, -2)$ b) $(-1, -3)$ c) $(0, -2)$ d) $(-2, 9)$ e) $(1, 4), (-1, -14)$ f) $(1, 9), (3, 25)$

Exercise 4I

1 20, \$10 000   **2** 4000, €20 800   **3** 55 km h^{-1} **4** 600 km   **5** 1.5 s, 11.25 m   **6** 320 m

7 a) $(10 - x)$ cm b) $(x^2 + (10 - x)^2)$ cm^2 c) 5   **8** a) $(9 - 4x)$ cm b) $(2x^2 - 12x + 27)$ cm^2 c) 3

9 a) $(25 - \frac{2}{3}x)$ cm b) $(50 - \frac{4}{3}x)$ cm c) $(1250 - \frac{200}{3}x + \frac{8}{9}x^2)$ cm^2 d) $\frac{300}{17}$ **10** a) $(500 - x)$ m b) 250, 62 500 m^2

11 a) $(40 - 2x)$ m b) 10, 200 m^2 **12** a) $\frac{8}{x^2}$ m b) $\left(2x^2 + \frac{32}{x}\right)$ m^2 c) 2   **13** a) $\frac{108}{x^2}$ m b) $\left(x^2 + \frac{432}{x}\right)$ m^2

13 c) 6   **14** a) $2x$ cm b) $\frac{288}{x^2}$ cm c) $\left(4x^2 + \frac{1728}{x}\right)$ cm^2 d) 6   **15** 2 m, 18 m^3 **16** 3 cm, 172.8 cm^3

17 a) $\frac{36}{x^2}$ cm b) $(40x^2)$ cents c) $\left(\frac{1080}{x}\right)$ cents d) $\left(\frac{4320}{x} + 80x^2\right)$ cents e) 3 f) \$21.60   **18** a) $\left(\frac{5}{6x^2}\right)$ m

18 b) $\$\left(\frac{80}{3x}\right)$ c) $\$(45x^2)$ d) $\frac{2}{3}$ e) \$60   **19** 10 cm $\times$ 10 cm $\times$ 7.5 cm   **20** $\frac{5}{3}$, $74\frac{2}{27}$ cm^3 **21** 1

Revision exercise 4

1 a) $9x^2 - 8x + 1$ b) $\frac{-4}{x^5} - \frac{25}{x^6}$ **2** a) $\frac{1}{2}x^{-\frac{1}{2}} - \frac{4}{3}x^{-\frac{2}{3}}$ b) $\frac{1}{3}x^{-\frac{4}{3}} + \frac{10}{3}x^{-\frac{1}{3}}$ **3** a) $4x + 7$ b) $-\frac{2}{x^3} - \frac{6}{x^4}$ c) $-\frac{1}{x^2}$

4 a) $4x - 12$ b) 8 c) $y - 9 = 8(x - 5)$ **5** a) $12x^2 - 48x + 45$ b) 9, $y = 9x - 7$ **6** $(3, 6)$

7 $x + 3y - 7 = 0$ **8** $a = 3, b = 4$ **9** $(\frac{2}{3}, 26\frac{23}{27})$ **10** a) -7.5 ms^{-1} b) 6 ms^{-2} c) 3 s d) 9 ms^{-2}

11 a) -10 b) $2\sqrt{5}$ s c) $-20\sqrt{5}$ ms^{-1} **12** a) $98 - 9.8t$ b) $t = 10$ (not 25) c) 490 m

Chapter 5

Exercise 5A

The constant of integration is omitted from the answers to questions **1** to **8**.

1 a) $\frac{1}{4}x^4$ b) $\frac{1}{5}x^5$ c) x^3 d) $2x^6$ e) $-2x^2$ f) $3x^5$ g) $\frac{1}{2}x^4$ h) $3x$ i) $\frac{1}{12}x^6$ j) $\frac{1}{6}x^4$ k) $-\frac{1}{9}x^3$ l) $\frac{2}{3}x$

2 a) $-x^{-1}$ b) $-\frac{1}{3}x^{-3}$ c) $-x^{-2}$ d) $2x^{-3}$ e) $-\frac{1}{2x^2}$ f) $\frac{1}{4x^4}$ g) $-\frac{3}{x}$ h) $\frac{1}{x^2}$ i) $-\frac{2}{3x^6}$ j) $-\frac{1}{2x^3}$ k) $\frac{5}{3x}$

2 l) $-\dfrac{2}{9x^3}$ **3** a) $\frac{3}{4}x^\frac{4}{3}$ b) $2x^\frac{3}{2}$ c) $3x^\frac{1}{3}$ d) $-5x^\frac{4}{5}$ e) $-2\sqrt{x^3}$ f) $\frac{4}{5}\sqrt[4]{x^5}$ g) $6\sqrt[3]{x^2}$ h) $-\frac{5}{2}\sqrt[5]{x^4}$ i) $\frac{6}{7}\sqrt{x}$ j) $\frac{9}{5}\sqrt[3]{x^2}$

3 k) $\frac{2}{5}\sqrt{x^5}$ l) $2\sqrt{x^3}$ **4** a) $\frac{1}{4}x^4 + x^2$ b) $x^3 - 2x^2$ c) $\frac{1}{4}x^4 - x$ d) $6x + \frac{1}{2}x^6$ e) $\frac{1}{3}x^3 - \frac{5}{2}x^2 + 3x$ f) $\frac{1}{9}x^9 + \frac{1}{3}x^6$

4 g) $\frac{1}{5}x^5 - \frac{3}{2}x^2 + 2x$ h) $\dfrac{1}{3}x^3 + \dfrac{1}{x}$ i) $x^5 + \dfrac{1}{x^2}$ j) $\dfrac{2}{7}x^7 - \dfrac{2}{x^4}$ k) $\dfrac{1}{3}x^3 + \dfrac{3}{x}$ l) $-\dfrac{5}{x} - 2x - \dfrac{1}{2}x^4$

5 a) $2\sqrt{x^3} - 4x$ b) $\frac{2}{3}\sqrt{x^3} + 2\sqrt{x}$ c) $\frac{9}{4}x^\frac{4}{3} - \frac{8}{5}x^\frac{5}{3}$ d) $10x^\frac{1}{2} + 3x^\frac{2}{3}$ e) $\dfrac{8}{3}\sqrt{x^3} + \dfrac{2}{3x}$ f) $\frac{3}{2}\sqrt[3]{x^4} - 12\sqrt{x}$

6 a) $\frac{3}{2}x^2 - \frac{1}{4}x^3$ b) $\frac{1}{4}x^4 + \frac{5}{3}x^3$ c) $\frac{1}{2}x^4 - \frac{1}{6}x^6$ d) $\frac{2}{5}\sqrt{x^5} + 2\sqrt{x^3}$ e) $\frac{6}{7}\sqrt{x^7} - \frac{6}{5}\sqrt{x^5} + 2\sqrt{x^3}$ f) $\frac{6}{7}x^\frac{7}{6} + \frac{9}{4}x^\frac{4}{3}$

6 g) $\frac{4}{3}x^\frac{9}{4} - \frac{24}{5}x^\frac{5}{4}$ h) $\frac{1}{3}x^3 + 4x^2 + 15x$ i) $\frac{1}{3}x^3 - 2x^2 + 4x$ j) $\frac{2}{3}x^3 + 10x^2 + 50x$ k) $\frac{1}{4}x^4 - \frac{2}{3}x^3 + \frac{1}{2}x^2$

6 l) $\frac{1}{2}x^2 + \frac{4}{3}\sqrt{x^3} - 15x$ **7** a) $\frac{5}{3}x^3 - 5x^2$ b) $x^6 - \frac{1}{4}x^4$ c) $\frac{2}{7}\sqrt{x^7} + \frac{2}{3}\sqrt{x^3}$ d) $\frac{6}{5}x^\frac{5}{3} + \frac{9}{2}x^\frac{2}{3}$ e) $x - \dfrac{5}{x}$ f) $-\dfrac{1}{x} + \dfrac{2}{x^2}$

7 g) $3x - \dfrac{5}{x}$ h) $\frac{4}{3}x^3 - \frac{5}{4}x^2$ i) $2\sqrt{x^5} - 8\sqrt{x}$ **8** a) $y = x^3 + x + 2$ b) $y = 2x^2 - 3x + 1$ c) $y = 2x^3 - 2x^2 - 12$

8 d) $y = 16 + 4x - 3x^2$ e) $y = 8 - \dfrac{2}{x} - x$ f) $y = \dfrac{5}{x^2} - 7$ g) $y = \frac{2}{3}\sqrt{x^3} - 5x + 9$ h) $y = \frac{1}{2}x^2 - \frac{1}{3}\sqrt{x^3} - \frac{14}{3}$

8 i) $y = x^3 - \frac{1}{4}x^4 + 4$ j) $y = \frac{4}{3}x^3 - 6x^2 + 9x + 4$ k) $y = x - 2\sqrt{x} + 6$ l) $y = \dfrac{1}{x^2} - \dfrac{1}{x} + \dfrac{11}{4}$ **9** $y = x^3 + 4x + 2$

10 $y = x^4 - 3x^2 + 4$ **11** $y = 4x^4 + x^2 + x + 2$ **12** $y = 9 - \dfrac{5}{x} - 4x$ **13** $f(x) = 2\sqrt{x^3} - 5x + 7$

14 $y = x^3 - 2x^2 + 5x + 1$ **15** $y = x^4 - 3x^2 + 3$ **16** $y = x^3 - 2x + \dfrac{2}{x}$ **17** $a = 3, b = -5, c = 4$

Exercise 5B

1 a) $15t^2 + 2$ b) $5t^3 + 2t$ **2** a) $16t^3 + 6t$ b) $4t^4 + 3t^2 + 1$ **3** a) $5t^4 + 20t - 13$ b) $t^5 + 10t^2 - 13t + 9$

4 a) $4\sqrt{t} + 5t^2$ b) $-57\frac{11}{16}$ **5** a) $s = 6 + 5t - t^2$ b) $2\frac{1}{2}$ c) $12\frac{1}{4}$ m   **6** a) $7 + 6t - t^2$ b) 7

7 a) $t^2 - 4t + 3$ b) $\frac{1}{3}t^3 - 2t^2 + 3t + 2$ c) $1, 3$ d) $3\frac{1}{3}, 2$ **8** a) $v = u + at, s = ut + \frac{1}{2}at^2$

Exercise 5C

1 a) $\frac{8}{3}$ b) 81 c) 45 d) 37 e) 21 f) -15 g) $\frac{3}{8}$ h) $\frac{3}{2}$ i) $12\frac{2}{3}$ j) 7 k) $\frac{3}{2}$ l) $11\frac{1}{4}$

2 a) 16 b) 70 c) -40 d) $-\frac{3}{4}$ e) $-\frac{9}{2}$ f) $-\frac{15}{2}$ g) 0 h) $-3\frac{63}{64}$ i) $11\frac{1}{5}$ j) -9 k) $\frac{4}{5}$ l) 24

3 a) $4\frac{2}{3}$ b) $16\frac{1}{2}$ c) 20 d) $20\frac{5}{6}$ e) $\frac{2}{5}$ f) $\frac{4}{3}$ g) 15 h) $4\frac{2}{3}$ i) 4 j) $53\frac{2}{5}$ k) $10\frac{2}{3}$ l) $6\frac{1}{4}$

4 a) $70\frac{1}{2}$ b) $19\frac{1}{2}$ c) $18\frac{1}{4}$ d) 24 e) 52 f) 14 **5** Area $= 50$ **6** $85\frac{1}{3}$ **7** Area $= \frac{1}{6}$

8 Area $= 2\frac{2}{3}$ **9** Area $= 6\frac{3}{4}$ **10** Area $= 9$ **11** $30\frac{3}{8}$ **12** b) $\frac{7}{12}$ c) $11\frac{1}{4}$ **13** b) $\frac{13}{60}$ c) $2\frac{14}{15}$

Exercise 5D

1 a) $P(1, 4), Q(2, 7)$ c) $\frac{1}{6}$ **2** a) $A(-2, 11), B(3, 6)$ c) $20\frac{5}{6}$ **3** a) $C(2, 20), D(4, 32)$ c) $2\frac{2}{3}$

4 a) $(-\frac{1}{2}, 2\frac{3}{4}), (3, -6)$ b) 2.655 **5** a) $(0, 0), (9, 3)$ b) $4\frac{1}{2}$ **6** 32 **7** $21\frac{1}{3}$ **8** 12 **9** 36

10 a) $P(2, 0), Q(4, 8)$ b) $10\frac{2}{3}$ **11** a) $A(3, 0), B(6, -18)$ b) 36 **12** a) $P(-1, 0), Q(1, 0), R(2, 3)$ b) $4\frac{1}{2}$

13 b) i) $-\frac{8}{3}$ ii) $\frac{5}{12}$ iii) $-2\frac{1}{4}$ iv) $3\frac{1}{12}$ v) $-5\frac{1}{3}$

Exercise 5E

1 a) 72π b) 625π c) 8π d) $\dfrac{7\pi}{24}$ e) 54π f) $\dfrac{158\pi}{3}$ g) 9π h) $\dfrac{348\pi}{5}$ i) $\dfrac{352\pi}{3}$ j) 42π

2 a) $P(-2, 4), Q(2, 4)$ b) $\dfrac{256\pi}{5}$ **3** a) $(-1, 2), (1, 2)$ b) $\dfrac{64\pi}{15}$ **4** $\dfrac{28\pi}{3}$ **5** 9π **6** a) $P(3, 9)$ b) $\dfrac{162\pi}{5}$

7 b) $\dfrac{\pi}{10}$ **8** a) $P(-2, 4), Q(2, 4)$ b) $\dfrac{256\pi}{3}$

Revision exercise 5

1 a) $\frac{2}{5}x^{\frac{5}{2}} + c$ b) $\frac{x^2}{2} + \frac{1}{2x^2} + c$ **2** $y = \frac{1}{4}x^4 + x^2 - x + 7$ **3** $y = -x^2 + 3x - 1$ **4** $\frac{48}{5}$ **5** a) 3

6 a) $-\frac{x^4}{4} + \frac{x^3}{3} + \frac{9x^2}{2} - 9x + c$ b) $a = 1, b = 3$ **8** $\frac{512}{9}\pi$ **9** a) $\frac{253}{12}$ b) $\frac{125}{12}$ c) areas are above and below x-axis

10 1 **11** a) $f(x) = \frac{1}{3}x^3 - x^2 - 3x + 2$ b) $2, 3\frac{2}{3}, 0$

Chapter 6

Exercise 6A

1 a) $3, 5, 7, 9$ b) $1, 4, 7, 10$ c) $3, 1, -1, -3$ d) $4, 7, 12, 19$ e) $1, \frac{1}{2}, \frac{1}{3}, \frac{1}{4}$ f) $\frac{1}{2}, \frac{2}{3}, \frac{3}{4}, \frac{4}{5}$ g) $\frac{1}{2}, \frac{1}{5}, \frac{1}{10}, \frac{1}{17}$

1 h) $3\frac{1}{2}, 3\frac{1}{6}, 3\frac{1}{12}, 3\frac{1}{20}$ i) $6, 24, 60, 120$ **2** a) $4n$ b) $2n + 3$ c) $5n - 1$ d) $3n + 5$ e) $\frac{1}{n+1}$ f) $\frac{1}{3n}$

2 g) $\frac{2}{3n+2}$ h) $\frac{n}{n+1}$ i) $\frac{n+1}{3n-2}$ **3** a) 2^n b) 5×2^n c) $5 \times 2^{n-1}$ d) $4 \times 3^{n-1}$ e) $2 \times (-3)^{n-1}$

3 f) $(-\frac{1}{2})^{n-1}$ g) n^2 h) $\frac{n}{(n+1)^2}$ i) $(-1)^n n(n+1)$ **4** a) $5, 7, 9, 11$ b) $3, 9, 15, 21$ c) $2, 1, 2, 1$ d) $3, 7, 15,$ 31

4 e) $5, 5, 5, 5$ f) $7, \frac{1}{7}, 7, \frac{1}{7}$ **5** a) $u_{n+1} = 3u_n - 3$ b) $u_{n+1} = 2u_n + 1$ c) $u_{n+1} = 5u_n - 4$ d) $u_{n+1} = 4u_n + 1$

5 e) $u_{n+1} = 3u_n - 4$ f) $u_{n+1} = 3u_n - 8$ **6** a) $1, 2, 3, 5, 8, 13, 21$ c) 610 **7** b) 171

Exercise 6B

1 a) $1^2 + 2^2 + 3^2 + 4^2 + 5^2$ b) $2 + 5 + 8 + 11 + 14 + 17$ c) $5 + 11 + 21 + 35$ d) $3^3 + 4^3 + 5^3 + 6^3$

1 e) $5 \times 2 + 6 \times 3 + 7 \times 4 + 8 \times 5 + 9 \times 6 + 10 \times 7$ f) $1^2 + 3^2 + 5^2 + 7^2 + 9^2 + 11^2$ g) $1 + \frac{1}{2} + \frac{1}{3} + \dots + \frac{1}{n}$

1 h) $3 + 3 + 3 + 3 + 3$ **2** a) $\sum_{r=1}^{5} r$ b) $\sum_{r=1}^{7} r^3$ c) $\sum_{r=1}^{7} (3r+4)$ d) $\sum_{r=3}^{20} \frac{1}{r}$ e) $\sum_{r=5}^{18} r(r+1)$ f) $\sum_{r=3}^{n} r^4$ g) $\sum_{r=5}^{n} \frac{r}{r^2-1}$

2 h) $\sum_{r=1}^{n} \frac{r}{(r+1)(r+2)}$ **3** 34 **4** 65 **5** 55

Exercise 6C

1 a) 3 b) -11 c) not d) not e) 0.1 f) not g) not h) -1 **2** a) 37 b) 65 c) -25 d) -61

2 e) $85 - 4n$ f) -7.1 g) $\frac{n}{6}$ h) $(2n-1)a$ **3** a) 55 b) 725 c) 837 d) 390 e) -580 f) $\frac{n}{2}(3n+11)$

3 g) -2775 h) $592\frac{1}{2}$ **4** a) 11 b) 21 c) 100 d) 44 e) 8 f) 19 g) 30 h) 28 **5** a) 5050 b) 234

5 c) 225 d) 650 e) -187 f) 35.4 g) 120 h) 96 **6** $7, 1590$ **7** $-14, 9, 265$ **8** a) $-9, 210$

9 $5, 1$ **10** $-10, 2, 570$ **11** 7 **12** 10 **13** 22nd **14** a) 36 months b) £2340

15 a) i) 2 ii) 42 b) 10

Exercise 6D

1 a) 3 b) -2 c) not d) not e) not f) $\frac{1}{2}$ g) -1 h) a **2** a) 1024 b) 729 c) 640 d) $51\frac{33}{128}$

2 e) $-4\frac{20}{27}$ f) $\frac{2}{625}$ g) $\frac{1}{2048}$ h) $-\frac{1}{243}$ **3** a) 3069 b) -1023 c) 2049 d) 1275 e) 1638 f) $1\,111\,111$

3 g) $1\frac{364}{729}$ h) $1 - (\frac{1}{2})^n$ **4** a) 5 b) 9 c) 7 d) 11 e) 6 f) 6 g) 7 h) 10 **5** a) 765 b) 2186

5 c) -728 d) 301 e) $53\frac{25}{27}$ f) $39\frac{11}{16}$ g) $\frac{364}{729}$ h) $2 - (\frac{1}{2})^n$ **6** $5, 3$ **7** $-4, 3$ **8** $\pm\frac{1}{2}, \pm384$

9 $\pm11, \pm77$ **10** 5115   **11** 8200   **12** $\frac{1}{7}$ **13** $\frac{1}{11}$ **14** $\frac{1}{29}, -\frac{5}{29}, \frac{25}{29}$ **15** 10   **16** 7

17 $r = -4: 8, -32, 128; r = 3: 8, 24, 72$ **18** $r = -5: 5, -25, 125; r = 4: 5, 20, 80$ **19** $5, \frac{2}{5}$ **20** $-3, -2$

21 $5 + 15 + 45 + 135 + 405$ **22** 2, 4   **23** £10 737 418.23   **24** $44 259.26   **25** £6375

Exercise 6E

1 a) 2 b) $\frac{3}{2}$ c) $\frac{1}{4}$ d) $\frac{4}{5}$ e) $\frac{1}{72}$ f) $\frac{1}{8}$ g) $\frac{9}{70}$ h) $\frac{49}{170}$ i) 6   **2** a) $\frac{5}{9}$ b) $\frac{8}{9}$ c) $\frac{8}{11}$ d) $\frac{34}{333}$ e) $\frac{22}{9}$ f) $\frac{443}{135}$

3 12   **4** $\frac{8}{3}$ **5** $\frac{9}{10}$ **6** $-\frac{1}{2}$ **7** $\frac{1}{2}$ **8** 18, 6, 2; 9, 6, 4   **9** $\frac{2}{3}$ **10** 3   **11** $\frac{1}{3}$ **12** $\frac{1}{2}$

Exercise 6F

1 a) 10 b) 15 c) 36 d) 6 e) 1 f) 66 g) 35 h) 100   **2** a) $1 + 4x + 6x^2 + 4x^3 + x^4$

2 b) $1 + 5x + 10x^2 + 10x^3 + 5x^4 + x^5$ c) $1 + 12x + 54x^2 + 108x^3 + 81x^4$ d) $1 - 3x + 3x^2 - x^3$

2 e) $1 - 8x + 24x^2 - 32x^3 + 16x^4$ f) $1 - 15x + 75x^2 - 125x^3$ g) $1 + 2x + \frac{3}{2}x^2 + \frac{1}{2}x^3 + \frac{1}{16}x^4$ h) $1 - \frac{2}{5}x + \frac{1}{25}x^2$

3 a) 35 b) 36 c) 80 d) 700 e) -540 f) -42 g) 96 h) 80 i) $\frac{3}{4}$

Exercise 6G

1 a) $8 + 12x + 6x^2 + x^3$ b) $81 + 108x + 54x^2 + 12x^3 + x^4$ c) $216 - 540x + 450x^2 - 125x^3$

1 d) $16 + 16x + 6x^2 + x^3 + \frac{1}{16}x^4$ e) $27x^3 + 54x^2y + 36xy^2 + 8y^3$ f) $32x^5 - 80x^4y + 80x^3y^2 - 40x^2y^3 + 10xy^4 - y^5$

1 g) $8x^3 + 60x^2y + 150xy^2 + 125y^3$ h) $81x^4 - 432x^3y + 864x^2y^2 - 768xy^3 + 256y^4$ **2** a) 1080 b) 44 800 c) 6048

2 d) 8960 e) -224 f) 20 000 g) $\frac{5}{2}$ h) $-\frac{8}{15}$ **3** a) $1 - 15x + 90x^2 - 270x^3$ b) $1 + 20x + 180x^2 + 960x^3$

3 c) $1 - 35x + 525x^2 - 4375x^3$ d) $32 - 240x + 720x^2 - 1080x^3$ e) $1024 - 1280x + 640x^2 - 160x^3$

3 f) $64 + 576x + 2160x^2 + 4320x^3$ g) $1 + 3x + 4x^2 + \frac{28}{9}x^3$ h) $4096 + 1536x + 240x^2 + 20x^3$

4 a) $1 + 4x^3 + 6x^6 + 4x^9 + x^{12}$ b) $1 + 9x^2 + 27x^4 + 27x^6$ c) $27 - 54x^3 + 36x^6 - 8x^9$ d) $1 + 2x + 3x^2 + 2x^3 + x^4$

4 e) $1 + 6x + 9x^2 - 4x^3 - 9x^4 + 6x^5 - x^6$ f) $4 + 12x + 5x^2 - 6x^3 + x^4$ g) $4 + 4x - 15x^2 - 8x^3 + 16x^4$

4 h) $9 + 12x + 10x^2 + 4x^3 + x^4$

5 a) $1 - 10x + 45x^2 - 120x^3 + 210x^4$ b) 0.904382   **6** a) $1 + 28x + 364x^2 + 2912x^3$ b) 1.319

7 a) $1 - 24x + 252x^2$ b) 0.97625   **8** a) $512 + 11 520x + 115 200x^2$ b) 523.64

9 $531 441 - 4 251 528x + 15 588 936x^2$, 527 205   **10** $3125 - 12 500x + 20 000x^2 - 16 000x^3$, 3002

11 $a = 5, n = 6$ **12** $b = -3, n = 5$ **13** $c = 5, n = 4$ **14** $a = -\frac{1}{2}, n = 16$

16 a) 178 b) 82 c) 7040 d) $60\sqrt{3}$

Revision exercise 6

1 a) 107 b) 13 332   **2** a) 6 b) 2001   **3** a) 330 b) 6 288 384   **4** a) 1023 b) 20   **5** a) $\frac{1000}{98}$ b) 4

6 a) 10 b) 2268   **7** a) 14 793 187 b) 13 271 941   **8** $216x^2y^2 + 96xy^3 + 16y^4$ **9** a) 14 b) 510

10 a) 1749 b) 14 c) 7.18%   **11** a) 390 b) i) $12 \times 1.1^2 = 14.52$ ii) 381.27 iii) 16th

12 a) $t_{21} = 64$ b) 3 is the constant difference between successive terms c) $\frac{n}{2}(3n + 5)$ d) 197

13 a) 1.5 b) 20 759 c) 1999 d) 61 958

13 e) In 2006 will sell 105 094 – more than one phone each man, woman and child!

Chapter 7

Exercise 7A

1 a) $6(2x - 1)^2$ b) $6(3x + 4)$ c) $20(5x - x)^3$ d) $-5(3 - x)^4$ e) $-18(4 - 3x)^5$ f) $8x(x^2 + 1)^3$ g) $6x^2(x^3 - 6)$

1 h) $-12x(1-2x^2)^2$ i) $-8x^3(4-x^4)$ j) $-90x^2(7-5x^3)^5$ k) $48x(6x^2-5)^3$ l) $-42x(9-7x^2)^2$

2 a) $-6(2x-5)^{-4}$ b) $-3(3x+2)^{-2}$ c) $-4x(x^2+3)^{-3}$ d) $6x^2(5-2x^3)^{-2}$ e) $-\dfrac{4}{(3+4x)^2}$ f) $\dfrac{2x}{(4-x^2)^2}$

2 g) $\dfrac{10}{(3-2x)^2}$ h) $-\dfrac{6}{(x+1)^3}$ i) $\dfrac{70x}{(2-x^2)^6}$ j) $\dfrac{6x}{(3x^2+8)^2}$ k) $-60x^2(5x^3-4)^{-5}$ l) $\dfrac{12x^3}{(5-3x^4)^3}$

3 a) $(2x-1)^{-\frac{1}{2}}$ b) $-\frac{1}{3}(6-x)^{-\frac{2}{3}}$ c) $2x^2(x^3-2)^{-\frac{1}{3}}$ d) $x^4(4-x^5)^{-\frac{6}{5}}$ e) $\dfrac{2}{\sqrt{4x-5}}$ f) $\dfrac{2x}{3\sqrt[3]{(x^2+3)^2}}$

3 g) $\dfrac{1}{\sqrt{(5-2x)^3}}$ h) $-\dfrac{4x}{\sqrt[3]{(x^2+5)^4}}$

4 a) $4(2x+1)(x^2+x-1)^3$ b) $\dfrac{3x^2-6}{2\sqrt{x^3-6x}}$ c) $\dfrac{3-2x}{(x^2-3x+5)^2}$ d) $-\dfrac{2}{\sqrt{x^3}}\left(\dfrac{1}{\sqrt{x}}-1\right)^3$

4 e) $20x(x^2-1)(x^4-2x^2+3)^4$ f) $-\dfrac{6}{x^2}\left(1+\dfrac{3}{x}\right)$ g) $4\left(\dfrac{1}{\sqrt{x}}-1\right)(2\sqrt{x}-x)^3$ h) $\frac{3}{4}(2-x^2)(6x-x^3)^{-\frac{3}{4}}$

Exercise 7B

The constant of integration is omitted from the answers.

1 a) $\frac{1}{10}(2x-3)^5$ b) $\frac{1}{15}(5x+8)^3$ c) $\frac{1}{18}(3x-4)^6$ d) $(x-7)^3$ e) $-\frac{1}{6}(4-x)^6$ f) $\frac{1}{28}(6-7x)^4$ g) $-\frac{1}{6}(3x-4)^{-2}$

1 h) $\frac{2}{3}(5-9x)^{-1}$ i) $-\dfrac{1}{12(2x-1)^6}$ j) $\dfrac{3}{1-x}$ k) $\frac{1}{3}\sqrt{(2x-3)^3}$ l) $18\sqrt[3]{(x-4)^2}$

2 a) $\frac{1}{12}(2x-7)^6$ b) $\sqrt{2x-1}$ c) $\frac{1}{8}(x^2+2)^4$ d) $-\frac{1}{18}(4-3x^2)^6$ e) $\frac{1}{9}(x^3-4)^3$ f) $\dfrac{2}{3-x^2}$ g) $\frac{1}{6}\sqrt{(x^4-1)^3}$

2 h) $-\frac{1}{2}\sqrt[3]{(2-3x^2)^4}$ i) $\frac{1}{4}(x^4-2)^3$ j) $\dfrac{5}{3-x^5}$ k) $\frac{1}{6}(x^2+1)^3$ l) $\frac{1}{5}x^5+\frac{2}{3}x^3+x$

Exercise 7C

1 a) $2x-1$ b) $4x-9$ c) $12x+7$ d) $-(1+2x)$ e) $3x^2+8x-2$ f) $12x^2-2x-20$ g) $3x^2+6x-15$

1 h) $9x^2-22x+31$ i) $x(5x^3+3x-10)$ **2** a) $6x(x+1)(x+3)^3$ b) $x^2(2+x)(6+5x)$ c) $x^3(21x-4)(3x-1)^2$

2 d) $6x(2x+5)(4x+5)$ e) $3x^2(12x^2-1)(4x^2-1)^2$ f) $20x(2-x^3)(1-2x^3)$ g) $6x(25x^2+1)(5x^2+1)^3$

2 h) $x^6(14-95x^3)(2-5x^3)^3$ i) $x(8x^2+5x-2)(x^2+x-1)^2$ **3** a) $(5x-4)(x+2)(x-5)^2$ b) $2(5x+11)(2x-1)^2(x+4)$

3 c) $4(35x-9)(5x+2)^3(4x-3)^2$ d) $-2(7+10x)(2-x)^5(5+2x)^3$ e) $-(107+315x)(3+5x)(4-7x)^6$

3 f) $4(4x^2-3x+2)(x^2+1)(2x-3)^3$ g) $3(15x^2+18x-10)(5x+9)^2(x^2-2)^2$

3 h) $4(32x^2-35x-18)(2x^2-3)^4(4x-7)^5$ i) $x(52x^3+45x-16)(x^3-1)^2(4x^2+5)$

4 a) $\dfrac{3x+2}{2\sqrt{x+1}}$ b) $\dfrac{3(2-x)}{\sqrt{3-x}}$ c) $\dfrac{3(3x+5)}{\sqrt{5+2x}}$ d) $\dfrac{x(5x+12)}{2\sqrt{x+3}}$ e) $\dfrac{2x(3-5x)}{\sqrt{3-4x}}$ f) $\dfrac{30x+7}{2(5x+3)^{\frac{1}{2}}}$ g) $-\dfrac{9x+14}{\sqrt{2x+5}}$

4 h) $\dfrac{(35x-4)(5x-4)^2}{2\sqrt{x}}$ i) $\dfrac{(15x-19)(3x+5)}{2\sqrt{x-2}}$ **5** a) $-\dfrac{2}{(x-2)^2}$ b) $-\dfrac{4}{(x-1)^2}$ c) $-\dfrac{7}{(4+x)^2}$ d) $\dfrac{13}{(x+4)^2}$

5 e) $\dfrac{10}{(x+2)^2}$ f) $\dfrac{11}{(2-5x)^2}$ g) $\dfrac{x(x+6)}{(x+3)^2}$ h) $\dfrac{x(8-x)}{(4-x)^2}$ i) $\dfrac{x^2(4x-9)}{(2x-3)^2}$

6 a) $\dfrac{(9x+2)(3x-2)}{2\sqrt{x^3}}$ b) $\dfrac{(25x-1)(5x+1)^2}{2\sqrt{x^3}}$ c) $\dfrac{(19x^2+4)(x^2-4)^4}{2\sqrt{x^3}}$ d) $-\dfrac{2x+1}{2\sqrt{x}(2x-1)^2}$ e) $\dfrac{3x-12\sqrt{x}-2}{2\sqrt{x}(2+x)^3}$

6 f) $\dfrac{5(4x+12\sqrt{x}+1)}{\sqrt{x}(5-4x)^4}$ g) $\dfrac{3}{2\sqrt{x}-2\sqrt{(x+1)^3}}$ h) $\dfrac{11}{2\sqrt{x}-3\sqrt{(2x+5)^3}}$

7 a) $x^2(9-5x)(3-x)$ b) $-\dfrac{1}{(2x-1)^2}$ c) $\dfrac{(25x-1)(5x-1)}{2\sqrt{x}}$ d) $\dfrac{\sqrt{x}+2}{(\sqrt{x}+1)^2}$ e) $(25x-24)(5x+3)^2(x-2)$

7 f) $-\dfrac{7}{2\sqrt{3x}-2\sqrt{(x-3)^3}}$ g) $\dfrac{7(3-x)x^2}{\sqrt{7-2x}}$ h) $\dfrac{x(4-x)}{(2-x)^2}$ i) $(7-9x)(3-x)^3(2+x)^4$

8 a) $\dfrac{3(x^2-4x+1)(x-1)^2}{(x-3)^2}$ b) $\dfrac{30-12x^2+x^3}{\sqrt{(5-x^2)(6-x)^2}}$ c) $\dfrac{x^2(7\sqrt{x^5}-40x^2-20\sqrt{x}+120)}{2\sqrt{(4-x^2)(5-\sqrt{x})^2}}$

Exercise 7D

1 $y+4x=16,\,4y-x=30$ **2** $2y+x=9,\,y=2x-3$ **3** $(-\tfrac{3}{4},-\tfrac{3}{4})$ **4** $9\tfrac{11}{13}$ **5** $(-4,\tfrac{4}{3}),(2,\tfrac{2}{3})$ **6** $(2,2)$

7 $(0,0),(4,-8)$ **8** $(-1,-1)$ **9** $(-3,0)$, min; $(\tfrac{1}{3},\tfrac{500}{27})$, max **10** $(-2,\tfrac{1}{4})$, max **11** $(10,2\sqrt{5})$, min

12 $(-2,0)$ max; $(-\tfrac{4}{5},-\tfrac{148\,176}{3125})$, min; $(1,27)$, max; $(2,0)$, min **14** $100,£15\,000$ **15** $25\,\text{cm}^2$

16 b) $a=2,\,b=-3,\,y''(3)=\tfrac{2}{3}>0$ **17** b) $\dfrac{32\sqrt{3}}{9}$

Revision exercise 7

1 a) $8x(1+x^2)^3$ b) $\dfrac{-2}{\sqrt{3-4x}}$ c) $\dfrac{-3}{\sqrt{x}(1+\sqrt{x})^4}$ d) $x^2(5x+2)(25x+6)$ e) $\dfrac{2-3x}{(3-2x)^{\frac{1}{2}}}$ f) $\dfrac{-1-x^2}{(x^2-1)^2}$

1 g) $\dfrac{5x(1+3x^{\frac{1}{2}})}{2(1+x^{\frac{1}{2}})^2}$ **2** a) $\dfrac{5+4x-x^2}{(x^2+5)^2}$ **3** a) $4x(15-x)(20-x)$ b) $12x^2-280x+1200$ c) $3032\,\text{cm}^3$

4 a) $a=4-\dfrac{x}{2}$ b) $\dfrac{9x^2}{4\pi}+2(4-\tfrac{1}{2}x)^2$ **5** a) $x^2(x-2)(5x-6)$ $4x(5x^2-12x+6)$

6 a) $\dfrac{x(2-x)}{(1-x)^2},\dfrac{2}{(1-x)^3}$ b) 0 or 2 c) 0 **7** a) i) $x=-\tfrac{5}{2}$ ii) $y=\tfrac{3}{2}$ b) $\dfrac{19}{(2x+5)^2}$

8 a) i) $a=-1,\,b=3$ ii) $y=1$ c) $(-1,2)$

Chapter 8

Exercise 8A

1 a) i) $1,-1$ ii) $\dfrac{\pi}{2},\dfrac{3\pi}{2}$ iii) 2π b) i) $1,-1$ ii) $\dfrac{\pi}{4},\dfrac{3\pi}{4},\dfrac{5\pi}{4},\dfrac{7\pi}{4}$ iii) π c) i) $1,-1$ ii) $0,\dfrac{\pi}{4},\dfrac{\pi}{2},\dfrac{3\pi}{4},\pi,$

1 c) ii) $\dfrac{5\pi}{4},\dfrac{3\pi}{2},\dfrac{7\pi}{4},2\pi$ iii) $\dfrac{\pi}{2}$ d) i) infinite ii) $0,\dfrac{\pi}{2},\pi,\dfrac{3\pi}{2},2\pi$ iii) $\dfrac{\pi}{2}$ e) i) $3,-3$ ii) $\dfrac{\pi}{2},\dfrac{3\pi}{2}$ iii) 2π

1 f) i) $1,-1$ ii) $0,\pi,2\pi$ iii) 2π g) i) $1,-1$ ii) $\dfrac{3\pi}{4},\dfrac{7\pi}{4}$ iii) 2π h) i) infinity ii) $\dfrac{\pi}{4},\dfrac{5\pi}{4}$ iii) π

2 a) $1,90°;-1,270°$ b) $4,0°;2,180°$ c) $8,270°;2,90°$ d) $1,70°;-1,250°$ e) $7,220°;-1,40°$

2 f) $10,330°;-4,150°$ g) $\tfrac{1}{2},270°;\tfrac{1}{4},90°$ h) $6,0°;2,90°$

3 a) $-\sin 20°$ b) $-\cos 60°$ c) $-\tan 20°$ d) $\cos 50°$ e) $\tan 40°$ f) $-\cos 50°$ g) $-\sin 20°$ h) $\cos 80°$

4 a) $\dfrac{1}{\sqrt{2}}$ b) $-\tfrac{1}{2}$ c) $\dfrac{1}{\sqrt{3}}$ d) $-\dfrac{\sqrt{3}}{2}$ e) -1 f) $\dfrac{\sqrt{3}}{2}$ g) $-\dfrac{1}{\sqrt{2}}$ h) $\sqrt{3}$

Exercise 8B

1 a) $17.5°,162.5°$ b) $45.6°,314.4°$ c) $63.4°,243.4°$ d) $120°,240°$ e) $200.5°,339.5°$ f) $98.1°,278.1°$

1 g) $66.4°,293.6°$ h) $270°$ **2** a) $0.93,2.21$ b) $1.37,4.91$ c) $1.25,4.39$ d) $2.21,4.07$ e) $3.99,5.44$

2 f) $1.74,4.88$ g) $0.45,5.83$ h) $0.30,2.84$ **3** a) $93.1°,166.9°$ b) $58.5°,261.5°$ c) $126.0°,306.0°$

3 d) 254.4°, 345.6° e) 74.1°, 254.1° f) 29.5°, 238.5° g) 229.2°, 280.8° h) 97.8°, 277.8°

4 b) 3, 5 c) 9, 19   **5** b) 0.5, 2.5; 5 c) 1   **6** b) 0.5 m c) 7, 20.5 d) 14 s

Exercise 8C

1 a) 1.05, 5.24, 2.09, 4.19 b) 1.25, 4.39, 1.89, 4.19 c) 0, 3.14, 6.28 d) 1.57, 4.71, 1.23, 5.05 e) 0, 3.14, 6.28, 2.90, 6.04

1 f) 0.52, 2.62, 0.34, 2.80 g) 1.05, 5.24, 1.32, 4.97 h) 1.11, 4.25, 1.82, 4.96   **2** a) 8.7°, 81.3° b) 4.7°, 64.7°, 124.7°

2 c) 7.5°, 37.5°, 97.5°, 127.5° d) 56.8°, 123.2° e) 110.9°, 159.1° f) 13.3°, 58.7°, 85.3°, 130.7°, 157.3°

2 g) 24.3°, 65.7°, 114.3°, 155.7° h) 18°, 90°, 162°   **3** a) $\dfrac{\pi}{3}, \dfrac{2\pi}{3}$ b) $-\dfrac{\pi}{6}, -\dfrac{5\pi}{6}$ c) $\pm\dfrac{\pi}{3}, \pm\dfrac{2\pi}{3}$ d) $\pm\dfrac{\pi}{3}, \pm\dfrac{2\pi}{3}$

3 e) $-\dfrac{5\pi}{8}, -\dfrac{\pi}{8}, \dfrac{3\pi}{8}, \dfrac{7\pi}{8}$ f) $\pm\dfrac{\pi}{6}, \pm\dfrac{\pi}{3}, \pm\dfrac{2\pi}{3}, \pm\dfrac{5\pi}{6}$ g) $-\dfrac{5\pi}{6}, \dfrac{\pi}{6}$ h) $-\dfrac{5\pi}{8}, \dfrac{3\pi}{8}$

Exercise 8D

1 a) $\frac{4}{5}$ b) $\frac{3}{4}$ **2** a) $\frac{5}{13}$ b) $\frac{5}{12}$ **3** a) $\dfrac{3\sqrt{10}}{10}$ b) $\dfrac{\sqrt{10}}{10}$ **4** a) $\dfrac{\sqrt{5}}{3}$ b) $\dfrac{2\sqrt{5}}{5}$ **5** a) $\dfrac{\sqrt{15}}{4}$ b) $\sqrt{15}$

6 a) $\frac{7}{25}$ b) $\frac{24}{25}$ **7** a) 71.6°, 251.6° b) 59.0°, 239.0° c) 135°, 315° d) 33.7°, 213.7° e) 63.4°, 243.4°

7 f) 149.0°, 329.0° g) 78.7°, 258.7° h) 23.2°, 203.2°   **8** a) −2.62, −0.52, 0.34, 2.80 b) −1.05, 0, 1.05

8 c) −2.09, −1.91, 1.91, 2.09 d) −0.52, −2.61 e) 0.52, 1.57, 2.62 f) −1.23, 1.23 g) −1.05, −0.72, 0.72, 1.05

8 h) 0.73, 2.41

Exercise 8E

1 a) $\frac{24}{25}$ b) $-\frac{7}{25}$ **2** a) $-\frac{119}{169}$ b) $\frac{120}{169}$ **3** a) $\frac{4}{5}$ b) $-\frac{3}{5}$ **4** a) $\dfrac{\sqrt{3}}{2}$ b) $\frac{1}{2}$ **5** a) $-\frac{527}{625}$ b) $\frac{336}{625}$

6 a) $\frac{12}{13}$ b) $\frac{-5}{13}$ **7** a) 0°, 180°, 360°, 80.4°, 279.6° b) 90°, 270°, 41.8°, 138.2°   c) 90°, 270°, 210°, 330°

7 d) 60°, 300°, 109.5°, 250.5° e) 48.6°, 131.4° f) 90°, 194.5°, 345.5° g) 104.5°, 255.5°

7 h) 0°, 90°, 180°, 270°, 360°, 45°, 135°, 225°, 315°   **8** a) $0, \pi, 2\pi, \dfrac{\pi}{3}, \dfrac{5\pi}{3}$ b) $\dfrac{\pi}{2}, \dfrac{3\pi}{2}, \dfrac{\pi}{3}, \dfrac{2\pi}{3}$ c) $0, \pi, 2\pi, \dfrac{2\pi}{3}, \dfrac{4\pi}{3}$

8 d) $\dfrac{\pi}{2}, \dfrac{7\pi}{6}, \dfrac{11\pi}{6}$ e) $\dfrac{2\pi}{3}, \dfrac{4\pi}{3}$ f) $0, 2\pi, \dfrac{\pi}{3}, \dfrac{5\pi}{3}$ g) $0, 2\pi, \dfrac{2\pi}{3}, \dfrac{4\pi}{3}$ h) $\dfrac{\pi}{2}, \dfrac{3\pi}{2}, \dfrac{\pi}{4}, \dfrac{3\pi}{4}, \dfrac{5\pi}{4}, \dfrac{7\pi}{4}$

Revision exercise 8

1 a) 1 b) −1   **2** a) $\tan(x^2)$ b) 3   **3** a) $\dfrac{2\pi}{3}, \dfrac{4\pi}{3}$ **4** a) $-3\cos^2 x + 4\cos x + 3$ b) 0°, 70.5°

5 a) $2\cos^2 x + \cos x - 1 = 0$ b) $(2\cos x - 1)(\cos x + 1 = 0)$ **6** a) $(3\sin x - 2)(\sin x - 3)$

6 b) i) $\sin x = \frac{2}{3}$ or 3 ii) 41.8°, 138.2°   **7** a) i) −1 ii) 4π b) 4 (0, 1.257, 3.770, 4.189)

8 a) 12.3 b) 2.0 c) 12   **9** a) 30° b) $\dfrac{\sqrt{3}}{3}$ **10** $\frac{-120}{169}$ **11** $a = -3, b = 2$ **12** $m = 10, n = 6$

13 a) I b) III c) IV

Chapter 9

Exercise 9A

The constant of integration is omitted from the answers to questions **3** and **6**.

1 a) $3\cos 3x$ b) $-2\sin 2x$ c) $5\cos 5x$ d) $-6\cos 6x$ e) $-14\sin 7x$ f) $30\sin 5x$ g) $4\cos\frac{1}{2}x$ h) $-\sin(x + 3)$

1 i) $\cos(x-4)$ **2** a) $2x\cos(x^2)$ b) $-3x^2\sin(x^3)$ c) $-4x\sin(x^2-1)$ d) $18x^2\cos(2x^3+3)$ e) $8x\cos(1-x^2)$

2 f) $72x^3\sin(4-3x^4)$ g) $2(x-1)\sin(x^2-2x)$ h) $3(x^2-2x)\cos(x^3-3x^2)$ i) $\dfrac{3\cos\sqrt{x}}{\sqrt{x}}$

3 a) $\sin 2x$ b) $-\tfrac{1}{4}\cos 4x$ c) $-\tfrac{3}{5}\cos 5x$ d) $\tfrac{1}{2}\sin(2x-1)$ e) $2\cos(3x+2)$ f) $-\dfrac{4}{5}\cos\left(\dfrac{5x-\pi}{4}\right)$ g) $\tfrac{1}{2}\sin(x^2)$

3 h) $-2\cos(x^4)$ i) $\tfrac{3}{2}\sin(x^2-7)$ **4** a) $2\sin x\cos x$ b) $-3\sin x\cos^2 x$ c) $-\dfrac{\sin x}{2\sqrt{\cos x}}$ d) $\dfrac{2\sin x}{\cos^3 x}$

4 e) $14\cos x\sin^6 x$ f) $18\sin x\cos^5 x$ g) $20\cos 5x\sin^3 5x$ h) $-3\sin\tfrac{1}{2}x\cos^5\tfrac{1}{2}x$ i) $-\dfrac{4\sin 4x}{\sqrt{\cos 4x}}$

5 a) $2\cos x(1+\sin x)$ b) $4\sin x(3-\cos x)^3$ c) $-18\sin x(5+3\cos x)^5$ d) $3(\cos x-2\sin 2x)(\sin x+\cos 2x)^2$

5 e) $\dfrac{\sin x}{(1+\cos x)^2}$ f) $-\dfrac{3\cos x}{\sqrt{1-6\sin x}}$ g) $-\dfrac{9\sin 3x}{(1+\cos 3x)^2}$ h) $\dfrac{12\cos 6x}{\sqrt{(1-\sin 6x)^3}}$ i) $3\sin 2x(1+\sin^2 x)^2$

6 a) $\sin^4 x$ b) $-\tfrac{1}{3}\cos^3 x$ c) $\tfrac{1}{6}(4-\cos x)^6$ d) $\tfrac{1}{2}(3+\sin x)^4$ e) $\dfrac{1}{1+\cos x}$ f) $2\sqrt{4-\sin x}$ g) $\tfrac{1}{3}\sin^6 3x$

6 h) $\tfrac{1}{3}(x-\sin x)^3$ i) $\tfrac{1}{8}\sin^2 2x$

7 a) $\sin x+x\cos x$ b) $x(2\cos x-x\sin x)$ c) $\cos 3x-3x\sin 3x$ d) $3x^2(\sin 6x+2x\cos 6x)$

7 e) $\sin^4 x(\sin x+5x\cos x)$ f) $6x\cos^3 2x(\cos 2x-4x\sin 2x)$ g) $\dfrac{\sin x-x\cos x}{\sin^2 x}$ h) $-\dfrac{2(x+1)\sin 2x+\cos 2x}{(x+1)^2}$

7 i) $\dfrac{\cos x}{(1+\sin x)^2}$ j) $\dfrac{2(1+\sin 2x)}{\cos^2 2x}$ k) $\dfrac{1+2x\sin x\cos x+\cos^2 x}{(1+\cos^2 x)^2}$ l) $\dfrac{1+\sin x+\cos x}{(1+\cos x)^2}$

Exercise 9B

The constant of integration is omitted from the answers to question **3**.

1 a) $2\sec^2 2x$ b) $5\sec^2 5x$ c) $3\sec^2(3x-2)$ d) $4\sec^2(4x+1)$ e) $3x^2\sec^2(x^3)$ f) $5x^4\sec^2(x^5)$ g) $4x^3\sec^2(x^4-1)$

1 h) $-6x\sec^2(2-3x^2)$ **2** a) $3\sec^2 x\tan^2 x$ b) $7\sec^2 x\tan^6 x$ c) $6\sec^2 x\tan^2 x-3\sec^2 x$ d) $2\sec^2 x(1+\tan x)$

2 e) $\dfrac{\sec^2 x}{\sqrt{1+\tan x}}$ f) $\dfrac{15\sec^2 x}{(4-3\tan x)^2}$ **3** a) $\tfrac{1}{3}\tan 3x$ b) $\tfrac{1}{6}\tan 6x$ c) $\tfrac{1}{5}\tan(5x-2)$ d) $\tfrac{1}{2}\tan(x^2)$

4 a) $\tan x+x\sec^2 x$ b) $3x^2\tan x+x^3\sec^2 x$ c) $5x^4\tan 3x+3x^5\sec^2 3x$ d) $2x\tan 5x+5x^2\sec^2 5x$

4 e) $\dfrac{\tan x-x\sec^2 x}{\tan^2 x}$ f) $\dfrac{2x\sec^2 2x-2\tan 2x}{x^3}$ g) $\dfrac{4\sec^2 4x\tan 6x-6\sec^2 6x\tan 4x}{\tan^2 6x}$ h) $\dfrac{2\sec^2 x}{(1-\tan x)^2}$

Exercise 9C

1 $6y-9x=3\sqrt{3}-\pi$ **2** $y=-x,\ y=x-2\pi$ **3** $\left(\dfrac{\pi}{2},\dfrac{\pi}{8}\right);\dfrac{\pi^2}{32}$ **4** $24y+6\sqrt{3}x=12+\sqrt{3}\pi;\ 6\sqrt{3}y-24x=3\sqrt{3}-4\pi$

5 $\left(-\dfrac{\pi}{3},-\sqrt{3}\right),\left(\dfrac{\pi}{3},\sqrt{3}\right)$ **6** $\left(\dfrac{\pi}{3},\dfrac{\sqrt{3}}{3}\right),\left(\dfrac{5\pi}{3},-\dfrac{\sqrt{3}}{3}\right)$ **7** $\left(\dfrac{2\pi}{3},\dfrac{5\sqrt{3}}{3}\right),\left(\dfrac{4\pi}{3},\dfrac{5\sqrt{3}}{3}\right)$ **8** $(0,0),\left(\dfrac{2\pi}{3},\dfrac{3\sqrt{3}}{4}\right)$

9 $\left(\dfrac{\pi}{4},2\sqrt{2}\right)$ **10** $(0,0),(2\pi,\pi)$ **11** a) $\left(\dfrac{\pi}{6},\dfrac{\pi}{6}+\sqrt{3}\right)$, max; $\left(\dfrac{5\pi}{6},\dfrac{5\pi}{6}-\sqrt{3}\right)$, min

11 b) $(0,1)$, min; $\left(\dfrac{\pi}{3},\dfrac{3}{2}\right)$, max; $(\pi,-3)$, min; $\left(\dfrac{5\pi}{3},\dfrac{3}{2}\right)$, max; $(2\pi,1)$, min   c) $\left(\dfrac{\pi}{2},1\right)$, max; $\left(\dfrac{3\pi}{2},-\dfrac{1}{3}\right)$, min

11 d) $\left(\dfrac{\pi}{6},\dfrac{3\sqrt{3}}{16}\right)$, max; $\left(\dfrac{\pi}{2},0\right)$, saddle; $\left(\dfrac{5\pi}{6},-\dfrac{3\sqrt{3}}{16}\right)$, min; $\left(\dfrac{7\pi}{6},\dfrac{3\sqrt{3}}{16}\right)$, max; $\left(\dfrac{3\pi}{2},0\right)$, saddle; $\left(\dfrac{11\pi}{6},-\dfrac{3\sqrt{3}}{16}\right)$, min

12 a) $\dfrac{\pi}{2} - 1$ b) $\frac{1}{3}$ c) $\frac{3}{16}$ d) -2 **13** 2   **14** $\dfrac{5(\sqrt{3} - 1)}{2}$ **15** a) $A\left(\dfrac{\pi}{6}, \dfrac{1}{2}\right), B\left(\dfrac{5\pi}{6}, \dfrac{1}{2}\right)$ b) $\sqrt{3} - \dfrac{\pi}{3}$

16 a) $P\left(\dfrac{\pi}{3}, \dfrac{\sqrt{3}}{2}\right)$ b) $\frac{1}{4}$ **17** $\dfrac{\pi}{2\sqrt{3}}$ **18** b) $\dfrac{75\sqrt{3}}{4}\,\mathrm{m}^2$ **19** b) $100\,\mathrm{cm}^2$ **20** b) $\dfrac{2\sqrt{3}\pi a^3}{9}$

Revision exercise 9

1 a) $-2\sin 2x$ b) $2\sin x \cos x$ c) $3\sec^2(x + 2)$ **2** 10   **3** a) $-\frac{1}{3}\cos(3x + 7) + c$ b) $4\sin\left(x + \dfrac{\pi}{3}\right) + c$

4 a) $4\sin x \cos x \cos 3x - 6\sin^2 x \sin 3x$ b) $2\cos(1 - 3x) + c$ **5** $\sin x - 3$ **6** $a = 3, b = 2$

7 a) $3\cos(3x - 1)$ b) 3 c) 3 in total; at $x = \frac{1}{3}$ (given) $x = \dfrac{2\pi + 1}{3}$ and $x = \dfrac{4\pi + 1}{3}$ **8** $-4.95\,\mathrm{ms}^{-1}$

9 b) $\displaystyle\int_0^1 \pi \sin^2 2x \, dx$ c) 2.22   **10** a) $a = \pi$ b) i) $\sin x + x \cos x$

11 a) $15(1 + \cos x + \sin x)$ b) $15(\cos x - \sin x)$ c) $\dfrac{\pi}{4}$

Chapter 10

Exercise 10A

1 a) x^9 b) p^2 c) $9k^6$ d) $y^{\frac{5}{6}}$ e) c^4 f) $\dfrac{3h^6}{2}$ g) $2d$ h) $16p^{-4}$ **2** a) 2 b) 3 c) 27 d) 32 e) 25

2 f) 343 g) $\frac{1}{5}$ h) $\frac{4}{9}$ **3** a) $\frac{1}{7}$ b) $\frac{1}{9}$ c) $\frac{1}{2}$ d) $\frac{1}{125}$ e) $\frac{3}{2}$ f) $\frac{1}{9}$ g) $\frac{2}{3}$ h) $\frac{2}{5}$ **4** a) 5 b) $\pm\frac{1}{7}$ c) -7

4 d) $\frac{9}{5}$ e) $\frac{1}{2}$ f) $\pm\frac{1}{5}$ g) $\pm\frac{1}{2}$ h) $\pm\frac{5}{3}$ **5** a) 9 b) 32 c) $\frac{4}{49}$ d) $\frac{1}{256}$ e) $\pm\frac{1}{2}$ f) $\frac{1}{5}$ g) $\frac{1}{7}$ h) $\pm\frac{2}{3}$

6 a) ± 27 b) 16 c) 16 d) $\pm\frac{1}{729}$ e) $\frac{1}{25}$ f) 81 g) $-\frac{1}{216}$ h) -32 **7** a) $-1, 8$ b) $\frac{1}{16}, 256$

Exercise 10B

1 a) $2\sqrt{3}$ b) $5\sqrt{2}$ c) $4\sqrt{7}$ d) $11\sqrt{3}$ e) $16\sqrt{5}$ f) $2\sqrt{2}$ g) $2\sqrt{2}$ h) $9\sqrt{5}$ **2** a) $\dfrac{3\sqrt{2}}{2}$ b) $\dfrac{5\sqrt{3}}{3}$ c) $\dfrac{\sqrt{6}}{3}$

2 d) $\dfrac{\sqrt{14}}{2}$ e) $2\sqrt{35}$ f) $\dfrac{\sqrt{30}}{4}$ g) $\dfrac{\sqrt{6}}{3}$ h) $\dfrac{3\sqrt{10}}{5}$ **3** a) $2 + \sqrt{3}$ b) $\dfrac{3 - \sqrt{5}}{4}$ c) $\dfrac{5 + \sqrt{7}}{9}$ d) $\dfrac{6 - \sqrt{3}}{11}$

3 e) $3 + 2\sqrt{2}$ f) $\dfrac{13 + 2\sqrt{2}}{23}$ g) $-(17 + 8\sqrt{5})$ h) $\dfrac{6 - \sqrt{6}}{2}$ **4** $\sqrt{2}$

Exercise 10C

1 a) $\log_2 32 = 5$ b) $\log_3 81 = 4$ c) $\log_4(\frac{1}{16}) = -2$ d) $\log_9 729 = 3$ e) $\log_6 36 = 2$ f) $\log_7(\frac{1}{343}) = -3$

1 g) $\log_{12} 1 = 0$ h) $\log_{10} 1\,000\,000 = 6$ i) $\log_2(\frac{1}{512}) = -9$ j) $\log_{16} 4 = \frac{1}{2}$ k) $\log_{1000} 10 = \frac{1}{3}$ l) $\log_{\frac{1}{2}}(\frac{1}{8}) = 3$

2 a) 3 b) 5 c) 2 d) 3 e) 1 f) 2 g) -2 h) -4 i) -4 j) 0 k) 10 l) -5 **3** a) $\log a + \log b$

3 b) $\log a - \log b$ c) $2\log a + \log b$ d) $\frac{1}{2}\log a$ e) $-2\log a$ f) $\log a + \frac{1}{2}\log b$ g) $3\log a - \log b$

3 h) $2\log a - 3\log b$ i) $\frac{1}{2}\log a - \frac{1}{2}\log b$ j) $-\log a - 4\log b$ k) $-\frac{1}{2}\log a - \frac{1}{2}\log b$ l) $\frac{1}{3}\log a + \frac{1}{6}\log b$

4 a) $\log 12$ b) $\log 14$ c) $\log 5$ d) $\log 6$ e) $\log 30$ f) $\log 2$ g) $\log 3$ h) $\log 10$ i) $\log 4$ j) $\log 5$

4 k) $\log\left(\dfrac{a^2}{bc}\right)$ l) $\log\left(\dfrac{a\sqrt{b}}{c^3}\right)$ **5** $x = 10\,000, y = 1000$

Exercise 10D

1 a) 2.32 b) 1.77 c) 1.52 d) 0.65 e) 3.86 f) 2.57 g) 0.71 h) 0.24

2 a) 0.79 b) 0.70 c) 0.67 d) 2.91 e) 2.77 f) -1.02 g) 2.58 h) 0.22   **3** a) 1.22 b) -0.83

Exercise 12B

1 a) $\begin{pmatrix} 4 & -6 \\ 11 & -15 \end{pmatrix}$ b) $\begin{pmatrix} 8 & 7 \\ -10 & 38 \end{pmatrix}$ c) $\begin{pmatrix} 22 & 5 \\ -20 & 5 \end{pmatrix}$ d) $\begin{pmatrix} 1 & 11 \\ 14 & 61 \end{pmatrix}$ e) $\begin{pmatrix} 2 & 13 \\ 3 & -38 \end{pmatrix}$ f) $\begin{pmatrix} -5 & -30 \\ 15 & 10 \end{pmatrix}$ g) $\begin{pmatrix} -5 & 2 \\ -35 & 59 \end{pmatrix}$

1 h) $\begin{pmatrix} 9 & -40 \\ 17 & -10 \end{pmatrix}$ i) $\begin{pmatrix} 7 & 2 & 3 \\ 5 & 4 & 1 \\ -8 & 2 & -3 \end{pmatrix}$ j) $\begin{pmatrix} 17 & 5 & 1 \\ 2 & -1 & -2 \\ 12 & 3 & 6 \end{pmatrix}$ k) $\begin{pmatrix} 4 & 3 & 0 \\ -8 & -4 & -4 \\ 7 & 3 & -3 \end{pmatrix}$ l) $\begin{pmatrix} -1 & -1 & 8 \\ 0 & -7 & 7 \\ -6 & 3 & 9 \end{pmatrix}$

2 a) $\begin{pmatrix} 2 & -2 & 10 \\ 4 & -10 & 8 \end{pmatrix}$ b) $\begin{pmatrix} 8 & 14 \\ 4 & 8 \end{pmatrix}$ c) $\begin{pmatrix} 5 & -8 & 19 \\ 0 & -4 & -8 \\ 1 & 4 & 15 \end{pmatrix}$ d) $\begin{pmatrix} 8 & 1 \\ 4 & 5 \end{pmatrix}$ e) impossible f) $\begin{pmatrix} 13 \\ 4 \end{pmatrix}$ g) impossible h) $\begin{pmatrix} 7 \\ -11 \\ 4 \end{pmatrix}$

2 i) impossible j) $\begin{pmatrix} 1 & 4 & 10 \\ -6 & 1 & -6 \\ -6 & 10 & 16 \end{pmatrix}$ k) $\begin{pmatrix} 7 & 4 & 14 \\ -6 & 10 & 16 \end{pmatrix}$ l) $\begin{pmatrix} 14 & 4 \\ -8 & 2 \\ 14 & -2 \end{pmatrix}$ **3** a) $\begin{pmatrix} 18 & 16 \\ 42 & -45 \end{pmatrix}$ b) $\begin{pmatrix} 18 & 27 & 60 \\ -3 & 36 & 41 \end{pmatrix}$

3 c) $\begin{pmatrix} 44 & -11 \\ -22 & -33 \\ 22 & 0 \end{pmatrix}$ d) $\begin{pmatrix} 22 & 24 & -20 & 7 \\ 16 & 27 & -5 & -14 \end{pmatrix}$ e) $\begin{pmatrix} 8 & 34 & 56 \\ 28 & 12 & 43 \\ 15 & 47 & 60 \end{pmatrix}$ f) $\begin{pmatrix} -13 & 9 & -21 \\ 73 & -3 & -30 \\ -22 & -1 & 22 \end{pmatrix}$ **5** a) $\begin{pmatrix} -11 & 10 & -2 \\ 3 & 3 & 2 \\ 3 & 4 & -13 \end{pmatrix}$

5 b) $\begin{pmatrix} 1 & 4 & 10 \\ 3 & 12 & -1 \\ -6 & 7 & -2 \end{pmatrix}$ c) $\begin{pmatrix} 13 & 0 & 0 \\ 0 & 13 & 0 \\ 0 & 0 & 13 \end{pmatrix}$, $k = 31$ **6** a) $A^2 = \begin{pmatrix} 7 & 7 & 2 \\ 17 & 28 & 16 \\ -10 & -2 & 9 \end{pmatrix}$, $A^3 = \begin{pmatrix} 31 & 42 & 20 \\ 86 & 157 & 104 \\ -44 & -20 & 23 \end{pmatrix}$ b) $k = 17$

Exercise 12C

1 a) 13 b) 11 c) -17 d) 8 e) -26 f) -6 g) 15 h) -41 **2** a) $3x - 10$ b) 4 **3** a) $2a + 15$

3 b) -7 **4** $8\frac{1}{2}$ **5** $-3\frac{2}{3}, 4$ **6** $-1, 3$ **7** a) $\frac{1}{2}\begin{pmatrix} 1 & -1 \\ -3 & 5 \end{pmatrix}$ b) $\frac{1}{7}\begin{pmatrix} 4 & -1 \\ -5 & 3 \end{pmatrix}$ c) $\frac{1}{4}\begin{pmatrix} 2 & -3 \\ 2 & -1 \end{pmatrix}$ d) $\frac{1}{2}\begin{pmatrix} 0 & -2 \\ 1 & 6 \end{pmatrix}$

7 e) $\frac{1}{8}\begin{pmatrix} 3 & 2 \\ 5 & 6 \end{pmatrix}$ f) $\frac{1}{13}\begin{pmatrix} -2 & 1 \\ -3 & 8 \end{pmatrix}$ g) $\frac{1}{8}\begin{pmatrix} 7 & 4 \\ -2 & 0 \end{pmatrix}$ h) $\frac{1}{17}\begin{pmatrix} 3 & 2 \\ -5 & 9 \end{pmatrix}$ **8** a) $\begin{pmatrix} 1 & 1 \\ -1.5 & 2 \end{pmatrix}$ b) $\begin{pmatrix} 0.75 & 1 \\ -0.25 & 0 \end{pmatrix}$ c) $\begin{pmatrix} 0.4 & 0.1 \\ 0.2 & 0.3 \end{pmatrix}$

8 d) $\begin{pmatrix} 0.4 & -0.2 \\ -1.4 & 1.2 \end{pmatrix}$ e) $\begin{pmatrix} 0.6 & -0.5 \\ 0.2 & 0 \end{pmatrix}$ f) $\begin{pmatrix} -1.5 & -1 \\ -2.5 & -2 \end{pmatrix}$ g) $\begin{pmatrix} 0.7 & 0.8 \\ 0.4 & 0.6 \end{pmatrix}$ h) $\begin{pmatrix} 0.4 & -0.6 \\ 0.2 & 0.2 \end{pmatrix}$ **9** a) $\begin{pmatrix} 1 & 0 \\ 0 & 1 \end{pmatrix}$ b) $\begin{pmatrix} 14 & 25 \\ 5 & 9 \end{pmatrix}$

10 a) $\begin{pmatrix} 1 & 0 \\ 0 & 1 \end{pmatrix}$ b) $\begin{pmatrix} -3 & 2 \\ -14 & 9 \end{pmatrix}$ **11** a) $\begin{pmatrix} 3 & -5 \\ -1 & 2 \end{pmatrix}$ b) $\begin{pmatrix} 31 & 20 \\ 13 & 7 \end{pmatrix}$ **12** a) $\frac{1}{3}\begin{pmatrix} 3 & -2 \\ -6 & 5 \end{pmatrix}$ b) $\begin{pmatrix} 4\frac{2}{3} & -6\frac{1}{3} \\ -8\frac{2}{3} & 14\frac{1}{3} \end{pmatrix}$

13 a) $\begin{pmatrix} 1.5 \\ 0 \end{pmatrix}$ b) $\begin{pmatrix} 2.2 \\ -0.2 \end{pmatrix}$ c) $\begin{pmatrix} -43 \\ 11 \end{pmatrix}$ d) $\begin{pmatrix} 12 \\ 1 \end{pmatrix}$ e) $\begin{pmatrix} 2.9 \\ 0.7 \end{pmatrix}$ f) $\begin{pmatrix} 10 \\ -17 \end{pmatrix}$ g) $\begin{pmatrix} -8.5 \\ -5 \end{pmatrix}$ h) $\begin{pmatrix} -13\frac{2}{3} \\ 25 \end{pmatrix}$

14 a) $\begin{pmatrix} 1 & -4 \\ 0.5 & -2.5 \end{pmatrix}$ b) $\begin{pmatrix} 13 & -63 \\ 7.5 & -36.5 \end{pmatrix}$ **15** $\begin{pmatrix} 1.6 & 1.8 \\ 2.9 & 3.2 \end{pmatrix}$

Exercise 12D

1 a) -6 b) -18 c) -12 d) -12 e) 0 f) 26 g) -30 h) 54 **2** a) $16 - 2x$ b) 5

3 a) $10 - 5p$ b) -4 **4** a) $16x - 27$ b) $x = 2$, $C = \begin{pmatrix} 5 & 2 & -1 \\ 4 & 2 & 7 \\ 2 & 1 & 6 \end{pmatrix}$ **5** a) $16 + 8p - 6p^2$ b) $2, -\frac{2}{3}$

6 a) $10x^2 - 20x - 8$ b) $3, -1$ **7** a) $\frac{1}{2}\begin{pmatrix} 1 & 1 & -1 \\ 1 & -1 & 1 \\ -2 & 2 & 0 \end{pmatrix}$ b) $\begin{pmatrix} -20 & 4 & -9 \\ -5 & 1 & -2 \\ 11 & -2 & 5 \end{pmatrix}$ c) $\frac{1}{2}\begin{pmatrix} 1 & 1 & -1 \\ -10 & 4 & 2 \\ 7 & -3 & -1 \end{pmatrix}$

7 d) $\frac{1}{4}\begin{pmatrix} 4 & 3 & -10 \\ -8 & -3 & 18 \\ 0 & -1 & 2 \end{pmatrix}$ e) $\begin{pmatrix} -1 & 1 & -1 \\ 0 & -1 & 2 \\ 1 & -1 & 2 \end{pmatrix}$ f) $\frac{1}{3}\begin{pmatrix} -2 & -1 & 7 \\ 0 & 3 & -6 \\ 1 & -1 & 1 \end{pmatrix}$ g) $\frac{1}{7}\begin{pmatrix} 3 & 4 & -4 \\ 0 & -7 & 14 \\ -1 & 1 & -1 \end{pmatrix}$ h) $\frac{1}{33}\begin{pmatrix} 7 & 4 & 1 \\ -10 & -1 & 8 \\ -11 & -11 & 22 \end{pmatrix}$

8 b) $\begin{pmatrix} 12 & 20 & 23 \\ 2 & -9 & -10 \\ 1 & -16 & -18 \end{pmatrix}$ **9** b) $\begin{pmatrix} 12 & 1 & 2 \\ 2 & 1 & -2 \\ -15 & 0 & 19 \end{pmatrix}$ **10** a) $\begin{pmatrix} 0 & -1 & 2 \\ 1 & -1 & -3 \\ -2 & 3 & 5 \end{pmatrix}$ b) $\begin{pmatrix} 7 \\ -11 \\ 19 \end{pmatrix}$

11 a) $\begin{pmatrix} 2 \\ 1 \\ -1 \end{pmatrix}$ b) $\begin{pmatrix} 8 \\ 5 \\ -2 \end{pmatrix}$ c) $\begin{pmatrix} 6 \\ 2 \\ 1 \end{pmatrix}$ d) $\begin{pmatrix} 5 \\ 2 \\ -2 \end{pmatrix}$

Revision exercise 12

1 a) 1 b) $\begin{pmatrix} 0 \\ 1 \end{pmatrix}$ **2** a) $\begin{pmatrix} 1 & 0 \\ 0 & 1 \end{pmatrix}$ b) $\begin{pmatrix} a & \sqrt{1-a^2} \\ \sqrt{1-a^2} & -a \end{pmatrix}$

3 a) $\begin{pmatrix} -1 & 0 \\ 0 & -1 \end{pmatrix}$, $\begin{pmatrix} 1 & 0 & 0 \\ 0 & 1 & 0 \\ 0 & 0 & 1 \end{pmatrix}$ b) $\begin{pmatrix} 0 & 3 & 2 \\ -1 & 1 & 0 \end{pmatrix}$ **4** $-7, 1$ **5** $\begin{pmatrix} 1 & 3 \\ 4 & 12 \end{pmatrix}$ **6** a) $5a - 6, a = \frac{6}{5}$

6 b) $\begin{pmatrix} 4-3a & a^2-a \\ 3-3a & a^2-9a+9 \end{pmatrix}, a = 1$ **7** b) -1 c) i) $\begin{pmatrix} 0 \\ 0 \end{pmatrix}$ **8** a) $1, \begin{pmatrix} 2 & -5 \\ -1 & 3 \end{pmatrix}$ b) $2, 1$

9 a) $2a = 42$ b) -21 c) $0.238, 0.810, 0.262$ **10** b) $-2, 6, 5$

Chapter 13

Exercise 13A

1 a) 5 b) $\sqrt{74}$ c) 3 d) $\sqrt{61}$ e) 13 f) $2\sqrt{5}$ g) $\sqrt{130}$ h) $\sqrt{83}$ **2** ± 7 **3** ± 4 **4** ± 3

5 $\frac{4}{5}\mathbf{i} - \frac{3}{5}\mathbf{j}$ **6** $\frac{5}{\sqrt{89}}\mathbf{i} - \frac{8}{\sqrt{89}}\mathbf{j}$ **7** $\begin{pmatrix} -\frac{7}{\sqrt{130}} \\ \frac{9}{\sqrt{130}} \end{pmatrix}$ **8** $\frac{3}{\sqrt{38}}\mathbf{i} - \frac{2}{\sqrt{38}}\mathbf{j} + \frac{5}{\sqrt{38}}\mathbf{k}$ **9** $\frac{1}{\sqrt{14}}\mathbf{i} - \frac{3}{\sqrt{14}}\mathbf{j} + \frac{2}{\sqrt{14}}\mathbf{k}$

10 $\begin{pmatrix} -\frac{3}{13} \\ \frac{12}{13} \\ -\frac{4}{13} \end{pmatrix}$ **11** $12\mathbf{i} - 6\mathbf{j} + 4\mathbf{k}$ **12** $\mathbf{i} - 2\mathbf{k}$ **13** $\begin{pmatrix} \sqrt{5} \\ -\frac{3\sqrt{5}}{5} \\ \frac{\sqrt{5}}{5} \end{pmatrix}$ **14** $5\mathbf{i} + 2\mathbf{j} + 3\mathbf{k}$ **15** $3\mathbf{i} + 3\mathbf{j} - 13\mathbf{k}$

16 $\begin{pmatrix} 8 \\ 1 \\ -8 \end{pmatrix}$ **17** $-3\mathbf{i} + 10\mathbf{j}$ **18** $10\mathbf{i} - 11\mathbf{j} - \mathbf{k}$ **19** $\begin{pmatrix} -7 \\ -6 \\ 1 \end{pmatrix}$ **20** $-7, -3, 1$

Exercise 13B

1 a) i) $\frac{1}{2}\mathbf{a}$ ii) $\mathbf{b} - \mathbf{a}$ iii) $\frac{1}{2}(\mathbf{b} - \mathbf{a})$ iv) $\frac{1}{2}\mathbf{b}$ b) $\overrightarrow{OB}$ is parallel to $\overrightarrow{PQ}$, and $OB = 2PQ$

2 a) $\mathbf{b} - \mathbf{a}$ b) $\mathbf{b} - \frac{1}{3}\mathbf{a}$ c) $\frac{1}{2}\mathbf{a} - \frac{1}{2}\mathbf{b}$ d) $\frac{1}{6}\mathbf{a} + \frac{1}{2}\mathbf{b}$ **3** a) i) $\mathbf{c} - \mathbf{a}$ ii) $\frac{1}{4}\mathbf{c}$ iii) $-\frac{1}{4}\mathbf{a}$ iv) $\frac{1}{4}(\mathbf{c} - \mathbf{a})$

3 b) $\overrightarrow{ST}$ and $\overrightarrow{AC}$ are parallel since both are multiples of $\mathbf{c} - \mathbf{a}$, and $ST : AC = 1 : 4$ **4** a) $\frac{1}{2}\mathbf{a}$ b) $\mathbf{a} - \mathbf{b}$ c) $\frac{3}{2}\mathbf{a} - \mathbf{b}$

4 d) $\frac{1}{2}\mathbf{a} - \mathbf{b}$ **5** a) i) $\mathbf{b} - \mathbf{c}$ ii) $\frac{4}{5}\mathbf{b} - \frac{1}{5}\mathbf{c}$ iii) $\mathbf{b} - \frac{3}{5}\mathbf{c}$ iv) $\frac{1}{5}\mathbf{b} - \frac{4}{5}\mathbf{c}$ b) $\frac{11}{10}\mathbf{b} - \mathbf{c}$ **6** a) i) $\mathbf{b} - \mathbf{a}$

6 a) ii) $3\mathbf{b} - 3\mathbf{a}$ iii) $4\mathbf{b} - 3\mathbf{a}$ iv) $\frac{3}{4}\mathbf{a}$ **7** a) i) $\frac{3}{4}\mathbf{a}$ ii) $\frac{1}{3}\mathbf{a}$ iii) $\frac{5}{12}\mathbf{a}$ **8** a) $\begin{pmatrix} 2 \\ 1 \\ 3 \end{pmatrix}, \begin{pmatrix} 4 \\ 2 \\ 6 \end{pmatrix}$ b) $1 : 2$ **11** 8, 9

Exercise 13C

1 a) -9 b) -9 c) 26 d) -13 e) 25 f) 13 **2** a) -3 b) -3 c) 12 d) 21 e) 9 f) 14

3 a) 1 b) 3 c) 3 d) 2 e) -13 f) 17 **4** a) -45 b) -43 c) 24 d) 2 e) -19 f) 91

5 a) perpendicular b) parallel c) parallel d) neither e) parallel f) perpendicular g) parallel h) neither

6 a) $75.7°$ b) $45°$ c) $119.2°$ d) $66.4°$ e) $53.1°$ f) $115.3°$ g) $95.3°$ h) $69.4°$ **7** a) 10 b) $1\frac{1}{4}$

8 a) -12 b) 12 **9** $2\frac{4}{7}$ **10** 18 **11** 2 or -5 **12** 1 or 3 **13** $\dfrac{1}{9\sqrt{5}}(7\mathbf{i} + 10\mathbf{j} + 16\mathbf{k})$

Exercise 13D

1 $\mathbf{r} = (4 + t)\mathbf{i} + (3 - 2t)\mathbf{j}$ **2** $\mathbf{r} = (5 + 4t)\mathbf{i} - (2 + 3t)\mathbf{j} + (3 + t)\mathbf{k}$ **3** $\mathbf{r} = (5 + t)\mathbf{i} - (1 + t)\mathbf{j}$

4 $\mathbf{r} = (2 + 3t)\mathbf{i} + (6 - 8t)\mathbf{j}$ **5** $\mathbf{r} = (7t - 1)\mathbf{i} + (2 + t)\mathbf{j} + 3(t - 1)\mathbf{k}$ **6** $\mathbf{r} = (4 + 3t)\mathbf{i} + (1 - t)\mathbf{j}$

7 $\mathbf{r} = (3 + 2t)\mathbf{i} + (5 - t)\mathbf{j}$ **8** $\mathbf{r} = (4 + 2t)\mathbf{i} - (1 + t)\mathbf{j}$ **9** $\mathbf{r} = (5 + t)\mathbf{i} - 2(1 + t)\mathbf{j}$ **10** $10\mathbf{i} + 4\mathbf{j}$

11 $2\mathbf{i} - 3\mathbf{j}$ **12** a) $2\mathbf{i} + 5\mathbf{j} + 9\mathbf{k}$ b) $15.6°$ **13** a) $\mathbf{i} + 3\mathbf{j} + 5\mathbf{k}$ b) $58.5°$ **14** a) $\mathbf{r}_t = 9t\mathbf{i} + (5 + 3t)\mathbf{j}$

14 b) $\mathbf{r}_p = (4 - 3s)\mathbf{i} + (3 + 9s)\mathbf{j}$ c) $(3, 6)$ **15** a) $\mathbf{r}_t = (t - 1)\mathbf{i} + 2(t - 1)\mathbf{j}$ b) $\mathbf{r}_p = -2s\mathbf{i} + (5 + s)\mathbf{j}$ c) $(2, 4)$

16 a) $\dfrac{\sqrt{3}}{3}$ **17** a) i) $\dfrac{\sqrt{2}}{3}$

Exercise 13E

1 a) $(35, 43)$ b) $5\,\text{ms}^{-1}$ c) $4x - 3y - 11 = 0$ e) $(77, 99)$ **2** a) $13\,\text{ms}^{-1}$ b) $(50, 37, 149)$

3 a) $(60, 72)$ b) $(80, -8)$ c) $\begin{pmatrix} -10 \\ 1 \end{pmatrix}$

Revision exercise 13

1 a) $\begin{pmatrix} 10 \\ 0 \\ 0 \end{pmatrix}$ $\begin{pmatrix} 0 \\ 15 \\ 0 \end{pmatrix}$ $\begin{pmatrix} 0 \\ 0 \\ 22 \end{pmatrix}$ $\begin{pmatrix} 10 \\ 15 \\ 0 \end{pmatrix}$ $\begin{pmatrix} 10 \\ 0 \\ 22 \end{pmatrix}$ $\begin{pmatrix} 0 \\ 15 \\ 22 \end{pmatrix}$ $\begin{pmatrix} 10 \\ 15 \\ 22 \end{pmatrix}$ b) 28.4 **2** b) $\begin{pmatrix} 2 \\ \frac{13}{4} \end{pmatrix}$

3 a) $2x^2 + 7x - 15 = 0$ b) $-5, \frac{3}{2}$ **4** a) $5\mathbf{i} + 12\mathbf{j}$ b) $-\dfrac{26\sqrt{5}}{5}\mathbf{i} + \dfrac{52\sqrt{5}}{5}\mathbf{j}$ **5** a) $\sqrt{78}$ b) $a = 1, b = 2$

6 $\begin{pmatrix} x \\ y \\ z \end{pmatrix} = \begin{pmatrix} 1 \\ 2 \\ 3 \end{pmatrix} + t\begin{pmatrix} -2 \\ -2 \\ 1 \end{pmatrix}$ **7** a) -800 b) $104.3°$ **8** a) $47°$ b) $(1, 0, -1)$ c) $\mathbf{r} = \begin{pmatrix} 1 \\ 5 \\ 0 \end{pmatrix} + q\begin{pmatrix} 1 \\ 2 \\ 3 \end{pmatrix}$

8 e) $P(1, 5, 0), 47°$ **9** a) $3.54\,\text{m}$ b) $1.22\,\text{ms}^{-1}$ c) $10x - 7y = 20$ d) $\left(\dfrac{170}{29}, \dfrac{160}{29} \right)$ e) $1.24\,\text{ms}^{-1}$

10 a) $9\mathbf{i} + 9\mathbf{j}, (-4, 6)$ b) $\mathbf{r} = \begin{pmatrix} -4 \\ 6 \end{pmatrix} + s\begin{pmatrix} 9 \\ 9 \end{pmatrix}$ c) $\lambda = \dfrac{-4}{9}$ d) i) 5 ii) $157°$

11 ii) $\begin{pmatrix} 288 \\ 84 \end{pmatrix}$ iii) 50 min. b) $20.6°$ c) i) $\begin{pmatrix} 99 \\ 158 \end{pmatrix}$ iii $75\,\text{km}$ d) $180\,\text{km}$

Chapter 14

Exercise 14A

1 a) $\frac{1}{2}$ b) $\frac{4}{13}$ c) $\frac{2}{13}$ d) $\frac{17}{26}$ **2** a) $\frac{1}{2}$ b) $\frac{1}{5}$ c) $\frac{1}{10}$ d) $\frac{3}{5}$ **3** a) $\frac{11}{36}$ b) $\frac{1}{9}$ c) $\frac{1}{18}$ d) $\frac{13}{36}$ **4** $\frac{7}{12}$

5 $\frac{23}{30}$ **6** 0.16 **7** $\frac{7}{20}$ **8** $\frac{1}{15}$ **9** $\frac{19}{21}$ **11** $\frac{1}{8}$ **12** $\frac{5}{12}$ **14** $\frac{2}{9}$ **15** a) $\frac{19}{37}$ b) $\frac{12}{37}$ c) $\frac{6}{37}$ d) $\frac{12}{37}$

16 11% **17** $\frac{1}{12}$ **18** 0.6 **19** $\frac{3}{32}$ **20** 0.45

Exercise 14B

1 $\frac{2}{5}$ **2** $\frac{1}{3}$ **3** $\frac{2}{11}$ **4** $\frac{1}{3}$ **5** $\frac{3}{25}$ **6** $\frac{1}{2}$ **7** a) HHH, HHT, HTH, THH, HTT, THT, TTH, TTT b) $\frac{1}{4}$

8 a) HHHH, HHHT, HHTH, HTHH, THHH, HHTT, HTHT, HTTH, THHT, THTH, TTHH, HTTT, THTT, TTHT, TTTH, TTTT b) $\frac{2}{5}$

9 a)

	1	2	3	4	5
1		×	×	×	×
2	×		×	×	×
3	×	×		×	×
4	×	×	×		×
5	×	×	×	×	

b) $\frac{1}{2}$ **10** a)

	A	B	C	D	E
A	×	×	×	×	×
B	×	×	×	×	×
C	×	×	×	×	×
D	×	×	×	×	×
E	×	×	×	×	

b) $\frac{1}{4}$

11 $\frac{5}{9}$ **12** $\frac{1}{10}$ **13** $\frac{31}{36}$ **14** $\frac{3}{7}$ **15** $\frac{7}{9}$ **16** $\frac{2}{7}$ **18** $\frac{5}{18}$ **19** b) $\frac{8}{15}$

20 b) $\frac{21}{50}$ **21** b) $\frac{1}{2}$ **22** b) $\frac{23}{48}$ c) $\frac{8}{23}$ **23** a) $\frac{37}{60}$ b) $\frac{12}{37}$ **24** a) $\frac{103}{180}$ b) $\frac{30}{103}$ **25** $\frac{3}{5}$ **26** $\frac{15}{23}$ **27** $\frac{5}{11}$

Revision exercise 14

1 a) $\frac{19}{50}$ b) $\frac{13}{46}$ **2** b) $\frac{11}{36}$ **3** b) i) 2 ii) $\frac{1}{18}$ c) $n(A \cap B) \neq 0$ or $P(A \cap B) \neq 0$ **4** $\frac{44}{65}$ **5** a) $\frac{22}{23}$ b) $\frac{693}{2300}$

6 a) $\frac{3}{4}$ b) $\frac{3}{4}$ **7** a) 0.58 b) 0.3 **8** a) 0.36 b) 92% **9** a) $a = 21, b = 11, c = 17$

9 b) i) $\frac{1}{8}$ ii) $\frac{21}{32}$ c) i) $\frac{315}{1247}$ ii) $\frac{932}{1247}$ **10** b) i) 0.108 ii) 0.376 iii) 0.769

Chapter 15

Exercise 15A

1 a)

Points	74	75	76	77	78	79	80	81	82	83	84	85	86
Frequency	1	1	0	3	1	4	5	6	4	2	2	0	1

2 a)

Size	2	3	4	5	6	7	8	9	10	11	12
Frequency	2	2	4	8	4	8	6	2	2	0	2

3 a) 95 b) 20% **4** a) 750 b) 2300

5

tens	units
4	5, 5, 8
5	6, 7, 8
6	0, 2, 5, 7, 7
7	2, 3, 3, 3, 4, 6, 7
8	1, 1, 2, 2, 4, 5, 8, 8
9	1, 1, 3, 4

6

tens	units
0	0, 0, 8
1	2, 5, 7
2	3, 4, 6, 7, 9
3	2, 3, 4, 6, 7
4	1, 6, 8
5	
6	3

10 a) 9 b) 40

Exercise 15B

1 2 **2** Sunday **3** orange **4** 50 **5** 67 kg **6** 82.5 kg **7** 23.9 s **8** 9 yrs **9** 178 cm

10 164 g **11** £10 656.25 **12** 2.856 kg **13** 120.4 **14** a) 50 b) 64.2 km h^{-1}

Exercise 15C

1 a) 7, 2.38 b) 8, 2.39 c) 4, 3.32 d) 2, 3.03 **2** a) 19.5, 9.73 b) 20.4, 6.34 c) 24.52, 2.35

3 a) £4980, £3499.94 b) £3650, £2377.50

3 c) Cars in the second garage are on average cheaper, and their is less variety in price.

4 a) 20   b) 191 cm   c) 10.7 cm   **5** a) 3, 2   b) i) 13, 2   ii) 47, 2   iii) 30, 200   iv) 49, 18

Exercise 15D

1 b) 64 g   c) 67.5 g, 58 g, 10.5 g   **2** b) 188 cm   c) 195 cm, 173 cm, 22 cm   **3** b) 82%   c) 51°

4 b) 922 h   c) 1032 h, 833 h, 199 h   **5** b) 1.2 min   c) 17.2 min   **6** b) 1.8 yrs   c) 8.5 yrs

7 b) English: median 76   IQR 11
Maths: median 56   IQR 26

Pupils tended to score lower marks on average in the Maths exam, and in the Maths exam there was also a greater spread of marks.

Exercise 15E

1 a)

IQ	90–99	100–109	110–119	120–129	130–139	140–149	150–159
Frequency	2	3	2	5	6	4	2

2 a)

Length (mm)	80–99	100–119	120–139	140–159	160–179	180–199
Frequency	2	2	3	4	5	4

3 a)

Time (s)	55–59	60–64	65–69	70–74	75–79	80–84
Frequency	2	3	4	3	3	1

4 a)

Mass (kg)	3.00–3.49	3.50–3.99	4.00–4.49	4.50–4.99	5.00–5.49
Frequency	2	10	9	6	3

5 b) 67.5   **6** b) 178.8   **7** b) 20 km h^{-1}

Exercise 15F

1

x	0	1	2	3
$P(X=x)$	$\frac{1}{8}$	$\frac{3}{8}$	$\frac{3}{8}$	$\frac{1}{8}$

2

x	0	1	2
$P(X=x)$	$\frac{5}{18}$	$\frac{10}{18}$	$\frac{3}{18}$

3

x	0	1	2	3
$P(X=x)$	$\frac{703}{1700}$	$\frac{741}{1700}$	$\frac{234}{1700}$	$\frac{22}{1700}$

4

x	2	3	4	5	6	7	8	9	10	11	12
$P(X=x)$	$\frac{1}{36}$	$\frac{2}{36}$	$\frac{3}{36}$	$\frac{4}{36}$	$\frac{5}{36}$	$\frac{6}{36}$	$\frac{5}{36}$	$\frac{4}{36}$	$\frac{3}{36}$	$\frac{2}{36}$	$\frac{1}{36}$

5

x	15	20	25	30	35
$P(X=x)$	$\frac{2}{12}$	$\frac{2}{12}$	$\frac{4}{12}$	$\frac{2}{12}$	$\frac{2}{12}$

6

x	1	4	9	16	25	36
$P(X=x)$	$\frac{1}{6}$	$\frac{1}{6}$	$\frac{1}{6}$	$\frac{1}{6}$	$\frac{1}{6}$	$\frac{1}{6}$

7

x	0	1	8	27
$P(X=x)$	$\frac{1}{8}$	$\frac{3}{8}$	$\frac{3}{8}$	$\frac{1}{8}$

8 a) 0.5, 2.3   b) $\frac{1}{5}, 3\frac{4}{5}$ c) 0.16, 2.38   d) $\frac{4}{9}, \frac{2}{3}$ e) 0.19, 2.12   f) 0.28, 7.72   **9** a) $\frac{1}{15}$ b) $3\frac{2}{3}$

10 a) $\frac{1}{91}$ b) $4\frac{11}{13}$ **11** a) $\frac{12}{25}$ b) $\frac{48}{125}$ **12** a) $\frac{1}{40}$ b) $2\frac{1}{4}$ **13** a) $\frac{1}{42}$ b) 1   **14** $a = 0.4, b = 0.1$

Exercise 15G

1 $\frac{25}{216}$ **2** $\frac{20}{243}$ **3** $\frac{256}{625}$ **4** 0.171   **5** 0.194   **6** 0.258   **7** 0.227   **8** 0.211   **9** 0.146   **10** 0.0386

11 a) i) 0.282   ii) 0.377   iii) 0.230   iv) 0.889   b) 1.2

Exercise 15H

1 a) 0.9922   b) 0.8925   c) 0.9922   d) 0.7939   e) 0.8749   f) 0.9319   g) 0.1312   h) 0.1955   i) 0.9014

1 j) 0.1949   k) 0.0267   l) 0.8000   **2** a) 1.405   b) 1.126   c) 1.175   d) 1.751   e) -0.6128 f) -0.385

2 g) -1.555 h) -2.326 i) 1.881   j) 1.227   k) 1.341   l) 0.739   **3** a) 0.9032   b) 0.9854   c) 0.0630

3 d) 0.0287   e) 0.2119   f) 0.9495   g) 0.9884   h) 0.3399   i) 0.0865   j) 0.2733   k) 0.8798   l) 0.7814

4 a) 0.1056   b) 0.5987   c) 0.3085   d) 0.9599   **5** a) 0.8413   b) 0.0082   c) 0.9192   d) 0.0359

6 a) 0.3794   b) 0.2778   c) 0.3674   d) 0.3759   **7** a) 0.7333   b) 0.3612   c) 0.9848   d) 0.6826

8 a) 0.1974   b) 0.2206   c) 0.2873   d) 0.3994   **9** 0.7642   **10** 20.71%   **11** 0.7796   **12** 0.7683

13 a) 9.34%   b) 65.83%   c) 24.83%   **14** a) 25.46%   b) 45.08%   c) 29.46%   **15** 184.6 cm

16 $80 \, \text{km h}^{-1}$ **17** 207   **18** -5.36 **19** 0.4   **20** 27.3   **21** $\mu = 2845.9 \, \text{g}, \sigma = 227.7 \, \text{g}$

Revision exercise 15

1 a) 4   b) 3.8   **2** $a = 3$ **3** 675   **4** a) \$59   b) 62, 102, 120, 130, 134   c) i) 33 min   ii) 52

5 a) i) 24   ii) 154   b) 40 km   c) 7%   **6** a) i) 10   ii) 24   b) 62.6 cm, 20.3 cm   c) negative skew   d) 65

7 b) 68, 35   **8** a) 0.279   b) 0.113   c) 0.649   d) 0.351   **9** b) 2.2   c) 0.931

10 a) i) 0.3085   ii) 0.3085   iii) 0.3829   b) 0.4425   **11** b) 0.1481   **12** a) 0.1   b) 10   d) 0.7392

Practice Paper One

1 a) i) $u_n = a(\frac{1}{3})^{n-1}$ ii) $S_{10} = \frac{3a}{2}\left(1 - \frac{1}{3^{10}}\right)$ b) $a = 20$ **2** a) $1 - \frac{6}{x^4}$ b) $\frac{x^2}{2} - \frac{1}{x^2} + c$

3 a) $f(x) = (x-4)^2 - 1$ b) $(4, -1)$ c) $x = 4$ d) $(3, 0), (5, 0)$ **4** a) translation $\begin{pmatrix} 0 \\ 3 \end{pmatrix}$ b) translation $\begin{pmatrix} -3 \\ 0 \end{pmatrix}$

4 c) reflection in $y = x$ **5** a) $A = 4\,500\,000, k = 0.015$ b) 54th year   **6** a) $v = -2t^2 + 40$ b) $-40 \, \text{ms}^{-1}$

6 c) $s = -\frac{2}{3}t^3 + 40t$ **7** a) $\overrightarrow{OV} = \begin{pmatrix} 0 \\ 0 \\ 4 \end{pmatrix} \overrightarrow{OB} = \begin{pmatrix} 5 \\ -5 \\ 0 \end{pmatrix}$ b) $\sqrt{66} \approx 8.12$ cm   **8** $(-3, 0), (3, 0)$ **9** $200 \, \text{cm}^3$

10 a) 4   b) 256°   **11** a) $x^4 - 12x^2 + 54 - \frac{108}{x^2} + \frac{81}{x^4}$ b) 54

13 a)

x	1	2	3	4	5	6
$P(X = x)$	$\frac{1}{9}$	$\frac{2}{9}$	$\frac{2}{9}$	$\frac{1}{9}$	$\frac{2}{9}$	$\frac{1}{9}$

b) 600   **14** 2.5   **15** c) iii) only   d) 0.704

Practice Paper Two

1 c) $f^{-1}(x) = \frac{x-2}{3}$ d) $3x^2 - 1$ **2 Part A** a) $98.2 \, \text{km h}^{-1}$ b) i) $a = 165, b = 275$ c) i) 33   ii) $114 \, \text{km h}^{-1}$

2 Part B a) 0.811   b) 0.189   c) 0.301   d) 0.711   e) 1.89

3 b) $x \cos x + \sin x$ e) 1.07687   g) 1.179, 2.348   h) $\int_{1.179}^{2.348} \pi (x \sin x)^2 \, dx - \int_{1.179}^{2.348} \pi \left(\frac{x+1}{2}\right)^2 \, dx$ i) 2.61

4 Part A b) $\begin{pmatrix} -0.2 & 0.6 & -0.2 \\ 0.8 & -0.4 & -0.2 \\ 0.4 & -0.2 & 0.4 \end{pmatrix}$ c) $\begin{pmatrix} 0.4 \\ -0.6 \\ 1.2 \end{pmatrix}$ **4 Part B** a) $r = \begin{pmatrix} 1 \\ 2 \\ 3 \end{pmatrix} + t \begin{pmatrix} 2 \\ -3 \\ 3 \end{pmatrix}$ b) $\begin{pmatrix} 2 \\ -3 \\ 3 \end{pmatrix}$

5 a) i) $0 \, \text{km h}^{-1}$ ii) $43.2 \, \text{km h}^{-1}$ b) i) $10e^{-0.2t}$ ii) $10 \, \text{km h}^{-2}$ c) i) $50 \, \text{km h}^{-1}$ ii) $0 \, \text{km h}^{-2}$

5 c) iii) when acceleration is zero velocity is $50 \, \text{km h}^{-1}$ d) ii) -250 iii) 9.207 sec

Mathematical notation

N	the set of positive integers and zero, $\{0, 1, 2, 3, ...\}$
Z	the set of integers, $\{0, \pm1, \pm2, \pm3, ...\}$
Z^+	the set of positive integers, $\{1, 2, 3, ...\}$
Q	the set of rational numbers
Q^+	the set of positive rational numbers, $\{x \mid x \in Q, x > 0\}$
R	the set of real numbers
R^+	the set of positive real numbers, $\{x \mid x \in R, x > 0\}$
$\{x_1, x_2, ...\}$	the set with elements, $x_1, x_2, ...$
$n(A)$	the number of elements in the finite set A
$\{x \mid \ \}$	the set of all x such that
$\in$	is an element of
$\notin$	is not an element of
$\varnothing$	the empty (null) set of
U	the universal set
$\cup$	union
$\cap$	intersection
$\subset$	is a proper subset of
$\subseteq$	is a subset of
A'	the complement of the set A
$a \mid b$	a divides b
$a^{\frac{1}{n}}, \sqrt[n]{a}$	a to the power of $\frac{1}{n}$, nth root of a (if $a \geq 0$ then $\sqrt[n]{a} \geq 0$)
$a^{\frac{1}{2}}, \sqrt{a}$	a to the power of $\frac{1}{2}$, square root of a (if $a \geq 0$ then $\sqrt{a} \geq 0$)
$\|x\|$	the modulus or absolute value of x, ie x for $x \geq 0, x \in R, \quad -x$ for $x < 0, x \in R$
$\approx$	is approximately equal to
$>$	is greater than
$\geq$	is greater than or equal to
$<$	is less than
$\leq$	is less than or equal to
$\ngtr$	is not greater than
$\nless$	is not less than
u_n	the nth term of a sequence or series
d	the common difference of an arithmetic sequence
r	the common ratio of a geometric sequence
S_n	the sum of the first n terms of a sequence, $u_1 + u_2 + ... + u_n$
S_∞	the sum to infinity of a sequence, $u_1 + u_2 + ...$
$\sum_{i=1}^{n} u_i$	$u_1 + u_2 + ... + u_n$
$\prod_{i=1}^{n} u_i$	$u_1 \times u_2 \times ... \times u_n$
$\binom{n}{r}$	the rth binomial coefficient, $r = 0, 1, 2, ...$, in the expansion of $(a + b)^n$
$f: A \to B$	f is a function under which each element of set A has an image in set B
$f: x \mapsto y$	f is a function under which x is mapped to y
$f(x)$	the image of x under the function f
f^{-1}	the inverse function of the function f
$f \circ g$	the composite function of f and g
$\lim_{x \to a} f(x)$	the limit of $f(x)$ as x tends to a
$\dfrac{dy}{dx}$	the derivative of y with respect to x
$f'(x)$	the derivative of $f(x)$ with respect to x

$\dfrac{d^2y}{dx^2}$	the second derivative of y with respect to x
$f''(x)$	the second derivative of $f(x)$ with respect to x
$\int y\,dx$	the indefinite integral of y with respect to x
$\int_a^b y\,dx$	the definite integral of y with respect to x between the limits $x = a$ and $x = b$
e^x	exponential function of x
$\log_a x$	logarithm to the base a of x
$\ln x$	the natural logarithm of x, $\log_e x$
sin, cos, tan	the circular functions
A(x, y)	the point A in the plane with Cartesian coordinates x and y
[AB]	the line segment with end points A and B
AB	the length of [AB]
(AB)	the line containing points A and B
$\hat{A}$	the angle at A
$\hat{CAB}$	the angle between [CA] and [AB]
$\triangle$ABC	the triangle whose vertices are A, B and C
v	the vector v
$\overrightarrow{AB}$	the vector represented in magnitude and direction by the directed line segment from A to B
a	the position vector $\overrightarrow{OA}$
i, j, k	unit vectors in the directions of the Cartesian coordinate axes
$|\mathbf{a}|$	the magnitude of **a**
$|\overrightarrow{AB}|$	the magnitude of $\overrightarrow{AB}$
v · w	the scalar product of **v** and **w**
$\boldsymbol{A}^{-1}$	the inverse of the non-singular matrix $\boldsymbol{A}$
$\boldsymbol{A}^{T}$	the transpose of the matrix $\boldsymbol{A}$
det $\boldsymbol{A}$	the determinant of the square matrix $\boldsymbol{A}$
$\boldsymbol{I}$	the identity matrix
P(A)	probability of event A
P(A')	probability of the event 'not A'
P$(A|B)$	probability of the event A given B
$x_1, x_2, \ldots$	observations
$f_1, f_2, \ldots$	frequencies with which the observations $x_1, x_2, \ldots$ occur
B(n, p)	binomial distribution with parameters n and p
N(μ, σ^2)	normal distribution with mean μ and variance σ^2
$X \sim$ B(n, p)	the random variable X has a binomial distribution with parameters n and p
$X \sim$ N(μ, σ^2)	the random variable X has a normal distribution with mean μ and variance σ^2
μ	population mean
σ^2	population variance, $\sigma^2 = \dfrac{\sum_{i=1}^{k} f_i(x_i - \mu)^2}{n}$, where $n = \sum_{i=1}^{k} f_i$
σ	population standard deviation
$\bar{x}$	sample mean
s_n^2	sample variance, $s_n^2 = \dfrac{\sum_{i=1}^{k} f_i(x_i - \bar{x})^2}{n}$, where $n = \sum_{i=1}^{k} f_i$
s_n	standard deviation of the sample
Φ	cumulative distribution function of the standardised normal variable with distribution N$(0, 1)$

Glossary of command terms

Write down	Obtain the answer(s), usually by extracting information. Little or no calculation is required. Working does not need to be shown.
Calculate	Obtain the answer(s) showing all relevant working. 'Find' and 'determine' can also be used.
Find	Obtain the answer(s) showing all relevant working. 'Calculate' and 'determine' can also be used.
Determine	Obtain the answer(s) showing all relevant working. 'Find' and 'calculate' can also be used.
Differentiate	Obtain the derivative of a function.
Integrate	Obtain the integral of a function.
Solve	Obtain the solution(s) or root(s) of an equation.
Draw	Represent by means of a labelled, **accurate** diagram or graph, using a pencil. A ruler (straight edge) should be used for straight lines. Diagrams should be drawn to scale. Graphs should have points correctly plotted (if appropriate) and joined in a straight line or smooth curve.
Sketch	Represent by means of a diagram or graph, labelled if required. A sketch should give a general idea of the required shape of the diagram or graph. A sketch of a graph should include relevant features such as intercepts, maxima, minima, points of inflexion and asymptotes.
Plot	Mark the position of points on a diagram.
Compare	Describe the similarities and differences between two or more items.
Deduce	Show a result using known information.
Justify	Give a valid reason for an answer or conclusion.
Show that	Obtain the required result (possibly using information given) without the formality of proof. 'Show that' questions should not generally be 'analysed' using a calculator.
Hence	Use the preceding work to obtain the required result.
Hence or otherwise	It is suggested that the preceding work is used, but other methods could also receive credit.

Area under the standard normal curve

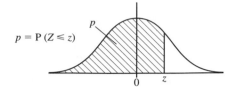

$p = P\,(Z \le z)$

z	0	0.01	0.02	0.03	0.04	0.05	0.06	0.07	0.08	0.09
0.0	0.5000	0.5040	0.5080	0.5120	0.5160	0.5199	0.5239	0.5279	0.5319	0.5359
0.1	0.5398	0.5438	0.5478	0.5517	0.5557	0.5596	0.5636	0.5675	0.5714	0.5753
0.2	0.5793	0.5832	0.5871	0.5910	0.5948	0.5987	0.6026	0.6064	0.6103	0.6141
0.3	0.6179	0.6217	0.6255	0.6293	0.6331	0.6368	0.6406	0.6443	0.6480	0.6517
0.4	0.6554	0.6591	0.6628	0.6664	0.6700	0.6736	0.6772	0.6808	0.6844	0.6879
0.5	0.6915	0.6950	0.6985	0.7019	0.7054	0.7088	0.7123	0.7157	0.7190	0.7224
0.6	0.7257	0.7291	0.7324	0.7357	0.7389	0.7422	0.7454	0.7486	0.7517	0.7549
0.7	0.7580	0.7611	0.7642	0.7673	0.7704	0.7734	0.7764	0.7794	0.7823	0.7852
0.8	0.7881	0.7910	0.7939	0.7967	0.7995	0.8023	0.8051	0.8079	0.8106	0.8133
0.9	0.8159	0.8186	0.8212	0.8238	0.8264	0.8289	0.8315	0.8340	0.8365	0.8389
1.0	0.8413	0.8438	0.8461	0.8485	0.8508	0.8531	0.8554	0.8577	0.8599	0.8621
1.1	0.8643	0.8665	0.8686	0.8708	0.8729	0.8749	0.8770	0.8790	0.8810	0.8830
1.2	0.8849	0.8869	0.8888	0.8907	0.8925	0.8944	0.8962	0.8980	0.8997	0.9015
1.3	0.9032	0.9049	0.9066	0.9082	0.9099	0.9115	0.9131	0.9147	0.9162	0.9177
1.4	0.9192	0.9207	0.9222	0.9236	0.9251	0.9265	0.9279	0.9292	0.9306	0.9319
1.5	0.9332	0.9345	0.9357	0.9370	0.9382	0.9394	0.9406	0.9418	0.9429	0.9441
1.6	0.9452	0.9463	0.9474	0.9484	0.9495	0.9505	0.9515	0.9525	0.9535	0.9545
1.7	0.9554	0.9564	0.9573	0.9582	0.9591	0.9599	0.9608	0.9616	0.9625	0.9633
1.8	0.9641	0.9649	0.9656	0.9664	0.9671	0.9678	0.9686	0.9693	0.9699	0.9706
1.9	0.9713	0.9719	0.9726	0.9732	0.9738	0.9744	0.9750	0.9756	0.9761	0.9767
2.0	0.9773	0.9778	0.9783	0.9788	0.9793	0.9798	0.9803	0.9808	0.9812	0.9817
2.1	0.9821	0.9826	0.9830	0.9834	0.9838	0.9842	0.9846	0.9850	0.9854	0.9857
2.2	0.9861	0.9864	0.9868	0.9871	0.9875	0.9878	0.9881	0.9884	0.9887	0.9890
2.3	0.9892	0.9896	0.9898	0.9901	0.9904	0.9906	0.9909	0.9911	0.9913	0.9916
2.4	0.9918	0.9920	0.9922	0.9925	0.9927	0.9929	0.9931	0.9932	0.9934	0.9936
2.5	0.9938	0.9940	0.9941	0.9943	0.9945	0.9946	0.9948	0.9949	0.9951	0.9952
2.6	0.9953	0.9955	0.9956	0.9957	0.9959	0.9960	0.9961	0.9962	0.9963	0.9964
2.7	0.9965	0.9966	0.9967	0.9968	0.9969	0.9970	0.9971	0.9972	0.9973	0.9974
2.8	0.9974	0.9975	0.9976	0.9977	0.9977	0.9978	0.9979	0.9979	0.9980	0.9981
2.9	0.9981	0.9982	0.9983	0.9983	0.9984	0.9984	0.9985	0.9985	0.9986	0.9986
3.0	0.9987	0.9987	0.9988	0.9988	0.9988	0.9989	0.9989	0.9989	0.9990	0.9990
3.1	0.9990	0.9991	0.9991	0.9991	0.9992	0.9992	0.9992	0.9992	0.9993	0.9993
3.2	0.9993	0.9993	0.9994	0.9994	0.9994	0.9994	0.9994	0.9995	0.9995	0.9995
3.3	0.9995	0.9995	0.9996	0.9996	0.9996	0.9996	0.9996	0.9996	0.9996	0.9997
3.4	0.9997	0.9997	0.9997	0.9997	0.9997	0.9997	0.9997	0.9997	0.9997	0.9998
3.5	0.9998	0.9998	0.9998	0.9998	0.9998	0.9998	0.9998	0.9998	0.9998	0.9998

Index